AF677322

Theoretical Nuclear Physics in Italy

Proceedings of the 10th Conference on
Problems in Theoretical Nuclear Physics

THEORETICAL NUCLEAR PHYSICS IN ITALY

Cortona, Italy 6–9 October 2004

edited by

S. Boffi, A. Covello, M. Di Toro
A. Fabrocini, G. Pisent & S. Rosati

World Scientific

NEW JERSEY · LONDON · SINGAPORE · BEIJING · SHANGHAI · HONG KONG · TAIPEI · CHENNAI

Published by

World Scientific Publishing Co. Pte. Ltd.
5 Toh Tuck Link, Singapore 596224
USA office: 27 Warren Street, Suite 401-402, Hackensack, NJ 07601
UK office: 57 Shelton Street, Covent Garden, London WC2H 9HE

British Library Cataloguing-in-Publication Data
A catalogue record for this book is available from the British Library.

THEORETICAL NUCLEAR PHYSICS IN ITALY
Proceedings of the 10th Conference on Problems in Theoretical Nuclear Physics

Copyright © 2005 by World Scientific Publishing Co. Pte. Ltd.

All rights reserved. This book, or parts thereof, may not be reproduced in any form or by any means, electronic or mechanical, including photocopying, recording or any information storage and retrieval system now known or to be invented, without written permission from the Publisher.

For photocopying of material in this volume, please pay a copying fee through the Copyright Clearance Center, Inc., 222 Rosewood Drive, Danvers, MA 01923, USA. In this case permission to photocopy is not required from the publisher.

ISBN 981-256-208-7

Printed by FuIsland Offset Printing (S) Pte Ltd, Singapore

PREFACE

These Proceedings contain the invited and contributed papers presented at the 10th Conference on Problems in Theoretical Nuclear Physics held in Cortona, Italy, from October 6th to October 9th, 2004, in the *Villa Passerini*, also called *il Palazzone*, a prestigious palace built by G. B. Caporali about 1515 and presently a summer center of the Scuola Normale Superiore, Pisa. The traditional goal of this biennial Conference is to offer the Italian theorists working on nuclear physics an opportunity for reviewing their activity and to strengthen the collaboration between different groups. In addition, in recent times it was also promoted to a mid-term review of the Research Project of National Interest (PRIN) entitled "*Theoretical Physics of the Nucleus and the Many-Body Systems*" and financially supported through the years 2003–2005 by the Italian Ministry of Education, University and Research (MIUR).

The Conference was attended by about 80 scientists, most of them coming from the 17 Italian Universities (Bologna, Cagliari, Catania, Firenze, Genova, Lecce, Milano, Napoli, Padova, Pavia, Perugia, Pisa, Roma Tor Vergata, Torino, Torino Politecnico, Trento, Trieste) taking part of the above project.

The atomic nucleus accounts for over 99% of the mass of the atom and of all visible mass in the universe. It represents a unique laboratory for studying different fundamental physics phenomena. Its basic constituents, the nucleons, are protons and neutrons obeying Fermi statistics and occurring in a finite number in a spacially confined region. As such the nucleus is a very peculiar many-body system exhibiting microscopic and mesoscopic features like few- and many-body quantum phenomena governed by the interplay of the electromagnetic, weak and strong interactions. The nuclear dynamics is driven by an effective force between nucleons depending on distance, temperature, density and angular momentum with its origin in the quantum chromodynamics (QCD) of their constituent quarks and gluons.

However, in the nuclear regime QCD cannot be solved perturbatively, and the relatively small number of nucleons does not allow the use of statistical methods available in other fields. Although the ultimate goal of nuclear physics is to furnish a unified theory of nuclear matter and finite nuclei, the complexity of the nuclear problem requires a variety of aspects to be simultaneously attacked and special techniques and methods to be devised for modelling the nuclear behaviour in specific situations.

The investigation field must then take advantage from a choral effort of the different research groups in close contact with one another, each one differently characterized, while addressing the common goal. Advanced computing facilities and strict contacts with experimental groups working in the main international laboratories are an essential support for such a research that, therefore, unavoidably has to be performed within an international framework. For a long time almost all theoretical groups in Italy working on nuclear physics, in particular those groups inside the present PRIN, are already working along these lines.

The programme of the Conference, under the supervision of the Organizing Committee (S. Boffi, A. Covello, A. Fabrocini, M. Di Toro, G. Pisent and S. Rosati) focussed on the following topics: nuclear dynamics and structure, fundamental interactions and nuclear physics, nuclear astrophysics, non-nuclear complex systems that can be studied with methods developed in theoretical nuclear physics.

These various subjects have been reviewed during the Conference by general talks given by G. Colò (Nuclear Structure), G.Pollarolo (Nuclear Dynamics), A. Kievsky (Few-Nucleon Systems), A. Bonasera (Highlights on Heavy Ion Reactions around the Fermi Energy), G.P. Co' (Nuclear Physics with Electroweak Probes), F. Becattini (Quark Gluon Plasma and Relativistic Heavy Ion Collisions), A. Drago (Nuclear Astrophysics). In addition, other 29 contributions from the PRIN collaboration have been presented, most of them by young participants.

One session was devoted to European prospects by illustrating the programmes of two Integrated Infrastructure Initiatives on hadronic and nuclear-structure physics. The two talks by C. Guaraldo and C. Scheidenberger were followed by a presentation by P. Gianotti on the physics made possible by the PANDA experiment and by a review of a possible theoretical initiative on hadronic physics by P. Mulders.

All talks are included in these Proceedings in order of presentation.

I would like to express my gratitude to the Authors of the general reports for accepting the demanding work of summarizing the achievements in the different fields in a concise and critical way, and to the Authors of the session on the European prospects for giving us the flavor of their future activity. I would also like to acknowledge the hospitality of the *Palazzone* where in a calm atmosphere lively discussions were possible among the participants.

On behalf of the Organizing Committee
Sigfrido Boffi
December, 2004

CONTENTS

Theoretical Nuclear Physics In Italy

NUCLEAR STRUCTURE

G. COLÒ

Dipartimento di Fisica, Università degli Studi and INFN, Sezione di Milano, via Celoria 16, 20133 Milano (Italy)
E-mail: colo@mi.infn.it

The intense activity of the Italian community devoted to nuclear structure, carried out during the last two years within the framework of the research project "Theoretical Physics of the Nucleus and of the Many-Body Systems", is reviewed. The important role of the many international collaborations, and of the fruitful exchanges with the experimental groups, is emphasized.

1. Introduction

The aim of this report is to review the activity of the Italian physicists during the last two years, in the field of nuclear structure. In the Conferences on "Problems in Theoretical Nuclear Physics" held in Cortona every two years, it has become customary to have overview talks concerning the main subjects of investigation of the nuclear physics community. The present contribution is the logical continuation of that kind of effort [1,2,3].

The research in nuclear structure is one of the most traditional and extensively pursued among those of the Italian PRIN ("Progetto di Ricerca di Interesse Nazionale", that is, Research Project of National Interest) named "Fisica Teorica del Nucleo e dei Sistemi a Molti Corpi" ("Theoretical Physics of the Nucleus and of the Many-Body Systems"). Nuclear structure physics deals with basic, yet still not completely solved problems like the nature of the nucleon-nucleon (NN) interaction and the hierarchy of many-body correlations in the nuclear medium. Under these two short definitions, many longstanding questions are included which have been debated during several decades; although the questions are not new, quite relevant developments have taken place in nuclear structure during the last years, so that we can, in a way, describe this realm of physics as a traditional field which has started displaying a new landscape.

The connection with the experimental progress is of course of paramount importance. The possibility to produce unstable nuclei has given a real new impetus to nuclear structure. Whereas only the stable nuclei could be studied until a few years ago, nowadays it is possible to observe isotopes which have finite lifetimes, and, still with limitations, perform measurements of their specific properties. In this sense, we can say that one of the most important present experimental efforts is that devoted to finding the actual limits of nuclear existence. These limits, the so-called neutron and proton drip lines, are in fact currently known only up to $Z \sim 8$. "Halo" systems have been observed, whose size is larger than the size predicted by standard rules like $R = r_0 A^{1/3}$. Also the shell structure of the nuclei close to the drip lines has shown surprises. To this kind of new discoveries we have alluded, when we have mentioned the new landscape of nuclear structure physics.

But there are other aspects which have contributed to change the character of the nuclear structure research in recent years. For instance, the increasingly tight connections with hadron physics - both phenomenological hadrodynamics and more "fundamental" QCD. More generally, also in non-relativistic studies the paradigm of Density Functional Theory (DFT) is becoming, more than in the past, a reference point. These aspects will be discussed below in the present contribution.

While the above statements could appear very general and only applicable to the broad international nuclear physics community, it is interesting to notice that on the smaller Italian scale many - if not all - of these aspects are pursued, at least to some extent. In fact, the connections of Italian scientists with many different international collaborations are working quite efficiently. It can be certainly stated that within the PRIN, the simultaneous presence and complementarity of different physical issues, motivations and theoretical tools, provides a relevant added value. At the same time, it is true that the research in nuclear structure is exposed to some risk of fragmentation.

The main groups in Italy which do research in nuclear structure are in Bologna, Catania, Lecce, Milano, Napoli, Padova, Pisa, Torino, Trento. The size of the groups can vary but even the smallest ones are recognized in the international context, as it will be evident from our discussion below, and from the references which will be quoted (all from relevant international journals).

We will try, in the following Sections, to describe the different researches carried out by the teams we have mentioned, within the framework of a unitary discussion. We have found that this unitarity emerges in a natural way. We start from the basic problem of the nature of the NN interaction and the properties of nuclear matter and of the ground state of magic nuclei, and we evolve then towards models for complex nuclei and/or excited states.

2. The nucleon-nucleon interaction and the nuclear models for the ground and low-lying states

Speaking of the NN interaction, many years ago H. Bethe remarked that "more man-hours have been devoted to this problem than to any other scientific question" [4]. In recent years, the quality of the bare interactions which have become available, has improved quite significantly. Their accuracy in reproducing the NN scattering observables has come down to a $\chi^2/datum \sim 1$ [5].

There is a traditional picture in nuclear physics which is the counterpart of many other "reductionist" pictures. It is often said that the bare NN force should be derived from the underlying quark dynamics inside the isolated nucleons (and, to some extent, this problem has started to be attacked by chiral theories), and that the effective interaction in the heavy nuclei, in turn, has to be derived from the bare NN force. Like other reductionist pictures, this appears too simplistic.

2.1. *Linking the nuclear models to QCD*

Among the works which testify to the importance of establishing a bridge between nuclear matter and the quark dynamics at the many-body-level (without any reductionist attitude in the above sense), we quote the recent papers by P. Finelli and collaborators [6]. A specific contribution is included in this volume [7]. The basic motivation of these works is to link directly the parameters of the relativistic mean-field (RMF) models to QCD observables.

The RMF models have overcome technical difficulties in recent years and have become increasingly popular and successful [8]. In these models, the nucleons are described as Dirac particles which interact through the exchange of effective mesons. The nucleon-meson couplings describe successfully the nuclear saturation, as the result of the balance between the attraction and repulsion associated respectively with the scalar and vector couplings. However, all the parameters entering the Lagrangian of the model, $L_{\rm RMF}$, are purely phenomenological. In particular, pions are completely absent; or to say it better, they are hidden in the parameters defining the fictitious σ-meson. In the works by Finelli *et al.*, the various coupling constants g of the relativistic Lagrangian are written as a sum of two terms, $g = g^{(0)} + g^{(\pi)}$, one derived from the changes of the quark condensates at finite density through the so-called QCD sum rules, and the other associated with the pionic fluctuations that are calculated using Chiral Perturbation Theory (CPT). The quality of the results for nuclear matter and finite nuclei is good, in the sense that, e.g., the error in reproducing the ground-state energies and radii of nuclei is of the order of 0.5%. This is quite remarkable. However, some problems shoud still be solved, including conceptual questions about the approximations made. In particular, some parameters are still treated in a phenomenological way.

An example of a calculation of the Bologna group, employing a standard RMF Lagrangian, can be found in Ref. [9]. The Catania group is instead active in investigating the role of the so-called δ-meson in the RMF framework [10]. The standard Lagrangians include both scalar and vector mesons in the isoscalar channel, but only a vector meson in the isovector channel, namely the ρ-meson. The introduction of a isovector-scalar particle may be demanded in analogy with the isoscalar channel, and has consequences in the behaviour of the symmetry energy, especially at densities above $\sim 1.5\varrho_0$ (ϱ_0 being the saturation density): at $2\varrho_0$ the inclusion of the δ-meson increases the symmetry energy by about 25%. The knowledge of the symmetry energy as function of the density is necessary to make predictions for exotic nuclei or even more exotic systems like the neutron stars.

2.2. *From the bare to the effective forces within the non-relativistic framework*

Within the non-relativistic description of the nucleus the saturation comes from the balance of a short-range repulsion and a longer-range attraction. Realistic bare forces are characterized by a very strong repulsive core. Since the old times of nuclear physics, this has prevented the use of the realistic forces in Hartree-Fock (HF) calculations. The effective forces used in these calculations cannot be quantitatively related to the bare interaction. In this respect, it has been recognized as very important by the international community the discovery that, if one is interested in low-energy observables, a low-momentum interaction V_{low-k} can be used, which does not have a repulsive core. V_{low-k} is derived from the bare NN forces using renormalization group techniques, and it has a number of attractive features:

- it is "universal", in the sense that starting from different bare interactions essentially a unique V_{low-k} can be derived;
- it allows HF calculations and further corrections (in the sense of

the Goldstone diagrams expansion for the ground-state energy);

- it can also be used to build an effective force for shell-model calculations.

The Napoli group has been very active, both in the construction of V_{low-k} [11] and in calculations which employ and test it [12]. They have made for the first time HF calculations with the V_{low-k} interaction, and shown that they are feasible and constitute a basis for further corrections. They have studied the convergence properties of the Goldstone expansion and they have arrived at the noticeable results that the experimental ground-state energies and radii of ^{4}He, ^{16}O and ^{40}Ca can be described with an accuracy of the order of 1%, and that the theoretical findings depend only slightly from the bare NN potential used as a starting point [13].

At the same time, for heavy nuclei the Napoli group has a longstanding experience in shell-model calculations. In this case, there are standard procedures to derive, from a bare iteraction, an effective interaction suited for the model space which has been chosen. These procedures are based on the so-called folded diagram method and are explained, for instance, in some of the references quoted in [12]. In their recent works the group has started from the V_{low-k} interaction.

Several shell-model calculations have been performed, for instance around the regions of ^{100}Sn and ^{132}Sn. These respectively proton-rich and neutron-rich nuclei are unstable and, as mentioned in the Introduction, understanding the shell structure far from stability is one of the new frontiers of nuclear physics. In many cases, the results of shell-model calculations are able to reproduce the experimental data for the low-lying levels (sometimes with striking exceptions, like in the case of the famous "cluster" states). This demonstrates the soundness of the theory and its predictive power [14]. This predictive power allows close collaborations with the experimental groups. In the stable Sn isotopes, a new generation of high resolution (p,t) experiments are presently being performed. In ^{114}Sn there is a one-to-one correspondence between the experimental levels and the outcome of the shell-model calculations [15]. In this sense, it can be said that these calculations appear to be instrumental for the interpretation of the experiments. Other examples in this respect concern recent results coming from γ-spectroscopy [16].

In some cases, when the quality of the results is not fully satisfactory, this may point to the fact that the assumed core of the shell-model calculation is actually not closed. It should also be reminded that in the shell model calculations the consistency between single-particle levels and two-body matrix elements is lost, and that high-lying states are not yet numerically treatable. For the former problem, the HF calculations mentioned above may play a crucial role. As far as the numerical treatments are concerned, a substantial reduction of the model space could be envisaged by improving the algorithms used to diagonalize the Hamiltonian (cf., e.g., Ref. [17]).

Recently, the Napoli group has also performed calculations for odd-odd nuclei, close to ^{100}Sn, ^{132}Sn and ^{208}Pb. These calculations, and the comparison with available experimental data, provide information about the particle-particle and particle-hole matrix elements around closed cores. These matrix elements are sensitive to core-polarization processes. In fact, the detailed analysis of Ref. [18] shows that some particle-particle multiplets are markedly affected by the coupling with the low-lying states of the core. This shows a possible, yet not exploited so far, link with the calculations based on the particle-vibration coupling which are discussed in the next Sections 3 and 4.

2.3. *Correlations in nuclear matter*

In this subsection, we come back to the problem of linking the bare and the effective interactions. The traditional solution of this problem lies in the Brückner theory. According to this theory, in the presence of a bare interaction characterized by a strong repulsive core, as in the nuclear medium, the scattering amplitude should be used as an effective potential. This can be derived for instance by performing the sum of the "ladder" diagrams and obtaining the so-called G-matrix. The G-matrix can be used in HF calculations: this is the well-known Brückner-Hartree-Fock (BHF) method. In the BHF scheme, the total energy is written in terms of diagrams which include only two hole lines. Further corrections are evaluated by means of an expansion in terms of the number of hole lines. The convergence is obtained at the level of three holes in nuclear and neutron matter. As it is well known, in all the Brückner calculations it is impossible to reproduce correctly the equation of state (EoS) of uniform matter, in particular the empirical saturation point of symmetric nuclear matter, without the contribution of the three-body forces.

One of the problems of the Brückner-type calculations is the choice of the auxilary potential that is used to evaluate the single-particle wavefunctions. Although in principle this choice should not affect the result of the calculations, in practice this is not always the case. The work of the Catania group has shown that the so-called "continuous" choice results in a better convergence than the "gap" choice [19]. The calculations are extended up to rather high densities, of the order of 6 times the saturation density. Another contribution has been the full calculation of the hole spectral function in nuclear matter at the level of 1 particle-2 holes (1p-2h) in the self-energy [20]. This work has shown the importance of keeping the full non-locality of the G-matrix. Spin- and isospin-polarized nuclear matter has been calculated in Ref. [21].

As already mentioned, one of the key problems for nuclear physics nowadays, is how to put constraints on the EoS of neutron-rich, or pure neutron matter. In Ref. [22], neutron matter is calculated using a Brückner calculation which includes up to three hole lines diagrams. A specific issue is addressed, namely the contribution to the total energy associated to the spin-orbit force. This can be as large as 20%-25% and its importance increases with density.

An interesting aspect of this work is that it includes comparisons with the EoS of neutron matter obtained using other sophisticated techniques, like variational methods or Green's function Montecarlo (GFMC) approaches - which have been quite successful recently in explaining also finite nuclei [23] up to $A \sim 12$ and are discussed in the contribution included in the present volume and devoted to few-body systems [24]. The differences in the total energies predicted by the various methods, which may become of the order of 15%-20% at large densities (i.e., $\sim 3\varrho_0$), gives an idea of the present overall accuracy of theory.

Among the methods based on Montecarlo techniques, the so-called auxiliary field diffusion Montecarlo (AFDMC) [25] has been applied to neutron matter and small neutron droplets in Ref. [26]. The results are quite promising, since the essence of the method consists in simplifying the the spin-isospin part of the propagator, and this allows extensions to larger systems than those currently accessible by Montecarlo techniques.

We conclude by mentioning the last attempt aimed to applying to medium-heavy nuclei another theory initially developed for the study of nuclear systems starting directly from the bare NN interaction: the correlated basis function (CBF) theory. The calculations have been carried out by colleagues from Lecce and Pisa [27]. It is quite interesting that the CBF theory, with a realistic Argonne potential, has been used to calculate nuclei as heavy as ^{208}Pb; however, there are still strong approximations (for instance, the three-body force is not yet included) which prevent from considering the results obtained as fully realistic.

2.4. *Deriving an energy functional*

From all the above examples, it appears quite clearly that the calculations of nuclear and neutron matter using realistic forces have reached a satisfactory level so that one can hope they contain the proper amount of short-range correlations. In order to transfer the scheme to finite nuclei, DFT within the local-density approximation (LDA) should provide a quite natural framework, as it has been known and exploited for many years within the realm of condensed matter physics.

In this spirit, the approach of Ref. [28] is rather interesting. The authors have performed Brückner calculations for infinite matter at different values of the density (in the low-density regime, that is, well below ϱ_0) and of the neutron-proton asymmetry. The values of the total energy that are obtained, do not depend significantly from the starting bare NN interaction, which is indeed a benchmark of the satisfactory level reached by the Brückner calculations. These values of E/A can be compared with those predicted by the effective functionals which are usually employed, like Skyrme and Gogny. The correlation energy which is derived from the calculation can be used to write an energy functional which in turn permits to write down the Kohn-Sham equations. However, the effective potential does not simply include the functional derivative $\frac{\delta}{\delta\varrho}$ of E/A, but also a component depending on $\nabla\varrho$ which is purely phenomenological and is written in terms of a parameter which must be adjusted. From the calculations, a global accuracy of the order of the order of 1% is found for the ground-state energies and radii.

We should also add a general comment at this point. As already repeated, calculations based on the Brückner theory include only a specific class of correlations, which are believed to be the most important to produce the nuclear saturation but are not certainly the only ones. We do miss, generally speaking, a theoretical scheme which is able to accomodate on the same footing other classes of diagrams like, e.g., the "ring" diagrams. The sum of the ring diagrams produces the so-called Random Phase Approximation (RPA), or polarization, propagator and the associated effective interaction.

In the next Section we are going to show that the problem of the relative contributions of bare interactions and polarization terms, has been attacked and solved in the pairing channel. In the mean-field channel, concerning the ground-state energies of nuclei, a starting point for a similar analysis can be found in Ref. [29]. The problem of the RPA correlation energies has been studied also by other groups very recently [30].

3. The nucleon-nucleon interaction and the pairing problem

$^1S_0$ pairing, and the associated superfluidity, is an important property of both infinite matter and finite nuclei. Altough it has been discussed for several decades now, it still captures the interest of theorists because of many fundamental reasons.

There are effective forces, like the Gogny force, which are quite successful in giving an economic description of pairing also in finite nuclei within the Hartree-Fock-Bogoliubov (HFB) framework. The Gogny force is believed to be similar, to some extent, to a bare force in the $^1S_0$ channel [31], but this is not a rigorous and accurate statement. In particular, it comes only from calculations in infinite matter. Studies of, e.g, a finite slab have also been performed [32] and show different surface properties of the bare interactions and v_{Gogny}. Other effective interactions, like zero-range density dependent ones, have been employed to describe pairing in nuclei within the HFB scheme. It should be added that standard HF plus Bardeen-Cooper-Schrieffer (BCS) calculations can replace the more complicated HFB in many cases. A final general remark is that there are basic problems in trying to describe pairing in relativistic theories [33], so that RMF Lagrangians are usually complemented by Gogny pairing. Having in mind these general considerations, we swicth to the discussion of pairing in terms not of effective forces, but of bare forces plus polarization contributions.

Pairing gaps Δ in infinite matter and finite nuclei can be calculated using only a bare interaction. We do not consider here the calculations done for infinite matter since they are connected with the superfluid properties of neutron stars, and they are discussed in the contribution to the present volume which is devoted to nuclear astrophysics [34]. In Ref. [35] it has been shown that the solution of so-called generalized BCS equations (which include the couplings between pairs of particles in time-reversal states having different radial quantum numbers), with only the v_{14} Argonne NN force, accounts for half of the observed pairing gap in the paradigmatic superfluid ^{120}Sn nucleus. The experimentally observed value of Δ (1.4 MeV) can only be reproduced if the polarization contributions are included. Among them, the most important one is the induced interaction, that is, the exchange of low-lying collective (mainly surface) nuclear phonons between pairs of nucleons [36]. In Ref. [35] the Dyson-Gor'kov formalism is employed, which allows treating on the same footing the induced interaction and the other diagrams associated with the couplings of the nucleons with the collective nuclear vibrations, that is, self-energies and vertex corrections. In fact, a complete treatment of the particle-phonon couplings, leads to changes in the single-particle energies and renormalizes in an important way the occupation factors around the Fermi energy. These effects are quite important to be taken into account since the pairing gap is sensitive to their influence.

Another contribution, along the same line, of the Milano group, is that of Ref. [37] which concerns a light neutron-rich system: ^{12}Be. Like the even more exotic ^{11}Li, which had been studied in Ref. [38], this nucleus possesses weakly bound neutrons which display delocalized wavefunctions extending faraway from the nuclear core and forming a two-neutron halo. The problem of the two neutrons outside the ^{10}Be core can be formulated on the basis of the available states: $s^2_{1/2}$, $p^2_{1/2}$ and $d^2_{5/2}$. If a bare interaction is diagonalized in this space, the two-neutron separation energy S_{2n} turns out to be -6.24 MeV, in disagreement with experiment (-3.67 MeV). Only when the particle-vibration coupling is included the result (-3.58 MeV) can reproduce the experimental finding. Also the spectroscopic factors are nicely accounted for.

The general conclusion is that models based on bare forces plus the particle-vibration coupling can reproduce the observed pairing properties in very different situations, ranging to stable isotopes to weakly bound systems. The message is qualitatively important and clear, despite the fact that the models still contain phenomenological ingredients, i.e., are not entirely self-consistent.

There are completely different approaches to the pairing problem in nuclei. Nuclear superfluidity, viewed as the emergence of a condensate of $L = 0$ bosons from a fermionic system, can be seen as a paradigmatic example of bosonization. There are models, like the Interacting Boson Model (IBM), in which the entire nuclear dynamics is described in terms of effective bosons having different angular momenta. In the IBM, the link of the parameters of the model with the underlying NN interaction is usually lost - although there have been, in the past, attempts to show that models based on particle-vibration coupling, supplemented by some *ad-hoc* ansatz, can give as a result an Hamiltonian which resembles the IBM one [39]. There is a present line of research aimed to trying to recover the link between the fermionic Hamiltonian and the IBM one, using very general methods, and exploiting the pairing channel as a guideline and benchmark.

The Torino group has re-analyzed the behaviour of two pairs of particles governed by the simple Hamiltonan $H_{mean\ field} + H_{pairing}$, with the aim of deriving new and simple analytical expressions [40]. In Ref. [41], a method to derive the parameters of $H_{pairing}$ from the fermionic NN interaction, based on the path integral formalism, is introduced. The Goldstone nature of the pairs emerges naturally and the results of BCS are reproduced. This results paves the way for a generalization, which is suited not only for the general IBM Hamiltonian but also for any problem of bosonization of a fermionic Hamiltonian both in the non-relativistic framework and in that of a relativistic quantum field theory [42].

4. Collective models

4.1. *Algebraic models*

The algebraic models mentioned in the last part of the previous Section, have been traditionally used since the early days of nuclear physics when dealing of excited states of collective character. In fact, in the case of very complex level structures, group theory can help to classify the patterns in terms of dynamical symmetries of the Hamiltonian - in the nucleus as well as in other systems like molecules, clusters or solids. The IBM, mentioned above, and its further improvements, describes the nucleus in terms of effective bosons, built with the N valence nucleons of a series of nuclei, and whose interactions are defined by a number of parameters which must be fitted to the experimental data. In the simplest case, the bosons carry either 0 or 2 angular momentum. All the linear combinations of these boson operators obey the $U(6)$ commutation rules. There are situations in which the eigenvalues of the Hamiltonian can be classified according to specific subroups chains of $U(6)$. The relation between these classifications and the nuclear shapes is still a subject of interest, as a complete understanding is not yet reached.

Within this framework, we quote a few recent papers of the Padova group. In Ref. [43] it is shown that, in the large-N limit of the IBM, the transition point between two subgroups chains of $U(6)$, namely those including $U(5)$ (spherical case) and $O(6)$ (γ-unstable case), is associated to eigenlevels which correspond to the solution of the Bohr differential equation with a β^4 potential. Other correspondences between new analytic solutions of the Bohr Hamiltonian and algebraic structures have been found [44], with applications to the spectrum of ^{234}U [45].

This latter nucleus is clearly a system for which fully microscopic models are quite hard to apply. For odd-odd nuclei, extensions of the IBM have been proposed, like the Interacting Boson Fermion-Fermion Model (IBFFM) in which two fermions outside an even-even core (described in terms of the IBM) can be treated. Identical rotational bands in ^{134}Pr are analyzed in this way in the work of Ref. [46]. However for odd and odd-odd systems also microscopic theories can be applied. An example is the microscopic mean field plus BCS analysis of odd deformed nuclei, aimed to predicting the anisotropies of their α-decay and reported in [47].

4.2. *Microscopic models*

After the digression devoted to algebraic models, we come back to the main route of the present contribution by dealing with the issue of how microscopic models based on effective interactions are used to describe collective nuclear modes.

We have explained in the previous pages that, both in the non-relativistic and in the relativistic framework, each effective interaction identifies an energy functional $E[\varrho]$ which is built in finite nuclei by taking the expectation value of the effective Hamiltonian over a proper combination of independent particle wavefunctions (within the HF formalism in the case of non-relativistic Skyrme or Gogny two-body forces, and within Hartree in the case of the effective RMF Lagrangians). The nuclear ground-state is defined as the equilibrium point of this functional, that is, is associated with the density which minimizes $E[\varrho]$. Small oscillations around this equilibrium point correspond to the vibrational nuclear states. They are usually described within the harmonic picture, that is, using linear response theory. In nuclear physics, this is the so-called RPA which has been already mentioned in our discussion. In fact, the nuclear vibrational modes (or phonons) have been considered, while treating the polarization contributions to nuclear masses or pairing in the previous Section, as mediators of the particle-vibration coupling. Within RPA, the nuclear phonons are described as coherent superpositions of 1p-1h states. Ref. [48] provides a general review about the mean field description of the ground and excited states.

There is a rich phenomenology of collective modes identified by different quantum numbers. At energies above the particle emission threshold, between 10 and 30 MeV, in most of the cases highly collective modes show up, the so-called giant resonances. The isovector giant resonance (IVGDR), characterized by $\Delta L = 1$, $\Delta S = 0$ and $\Delta T = 1$ (but $\Delta T_z = 0$), is the oldest and more studied example, thanks to the fact that very selective experimental probes are available to excite this mode.

Isoscalar ($\Delta T = 0$) quadrupole and octupole spectra, having respectively $\Delta L = 2$ and $\Delta L = 3$, are characterized both by a giant resonance and by low-lying collective states. Whereas the giant resonances are associated with high-lying transitions, and have a smooth A-dependence, the low-lying states arise from single-particle transitions around the Fermi energy. Accordingly, the latter are much more sensitive to the details of the shell structure (for instance, they are markedly affected by the pairing interaction in superfluid nuclei whereas pairing is, as a rule, unimportant for the giant resonances). All this may explain why in the semiclassical approach of Ref. [49], which is based on the solution of the linearized Vlasov equations with moving surface boundary conditions, the result for the energy location of the giant resonances is in agreement with the empirical findings, at odds with the result for the low-lying strength.

In superfluid nuclei, the vibrational states have to be described as superpositions of two-quasiparticle configurations. The corresponding theory, that is, the extension of RPA to the superfluid case, is called quasiparticle RPA (QRPA). The theory is well known from textbooks, but fully self-consistent QRPA calculations are not easy to develop due to technical difficulties. In the case of the Gogny interaction, whose performance in the description of nuclear pairing has been discussed above, the first self-consistent QRPA calculations based on full HFB, are reported in Ref. [50] and critically discussed, pointing, e.g., to difficulties in the description of the detailed properties of the low-lying states.

4.3. *Calculations beyond mean field*

The low-lying quadrupole states display strong anharmonicities. In Ref. [35], already discussed above, the Nuclear Field Theory (NFT), namely the model based on the systematic treatment of the coupling between single-particle and collective degrees of freedom [51], has been successfully applied to the description of the first excited quadrupole state 2_1^+ in ^{120}Sn. In Ref. [52] a different model, based however on a similar philosophy, the Quasiparticle Phonon Model (QPM) [53] has been appled to ^{92}Zr. An effective Hamiltonian H based on a Woods-Saxon potential describing the single-particle motion plus a separable multipole-multipole interaction, is diagonalized in a model space which includes one-, two- and three-phonon states. The interesting result of this calculation is that the nucleus under study seems to be at variance with simple expectations. In fact, in the simplest vibrational model, the first quadrupole state is a one-phonon state, and at about twice its energy a triplet of two-phonon states with quantum numbers 0^+, 2^+ and 4^+ should appear. The result for ^{92}Zr is that both states 2_1^+ and 2_2^+ are one-phonon states, with admixture of collective and non-collective components. Moreover, the proton-neutron symmetry is seriously broken (at variance again with a simple picture, that of the isoscalar character of the low-lying states). The general message emerging from NFT and QPM calculations is that the quest for anharmonicities in the low-energy part of the nuclear spectrum, should be pursued.

Another contribution of the Milano group concerns the extensions of self-consistent RPA and QRPA aimed to including the contribution of 2 particle-2 hole (2p-2h) or four quasiparticles configurations. In fact, the mean field models like RPA and QRPA do not take into account the spreading effects, whereas giant resonances are known empirically to possess a sizeable spreading width $\Gamma^{\downarrow}$ associated with the coupling to complex configurations, mainly those consisiting of p-h pairs plus a collective low-lying state (the giant resonances have also an escape width $\Gamma^{\uparrow}$ associated to particle emission, which is more important for light nuclei than for heavy ones) [54]. Although this picture is well established for stable nuclei, only scarce information is available about spreading effects in unstable isotopes. In Ref. [55], a model based on QRPA plus the coupling with four quasiparticle-type states (made up with two quasiparticles and a collective low-lying phonon) has been implemented. The model, called QRPA-PC (QRPA with phonon coupling), utilizes Skyrme effective forces and can be applied on the same footing to magic as well open-shell isotopes; it reproduces quite well the dipole spectrum of ^{120}Sn and it has been therefore used for a prediction of the dipole strength in the unstable ^{132}Sn system. The corresponding experiment has been performed at the GSI laboratory and the results are expected to appear soon.

4.4. *Giant resonances based on excited states*

All the works quoted so far in this Section, concern the response of spherical nuclei excited starting from their ground state. The response of deformed, fast rotating nuclei is studied in Ref. [56] as function of the rotational frequency ω, within a RPA approach based on the cranked Nilsson model. In the paper, predictions for the evolution of the different multipole strengths are illustrated. The most significant result concerns the magnetic dipole

(M1) strength. The magnetic dipole operator is known to include an orbital contribution as well as a spin contribution. It is found that, in correspondence with the superfluid-normal phase transition at large ω and with the associated "backbending" of the extracted moment of inertia, there is a large enhancement (up to a factor ~ 4) of the orbital M1 strength.

The Catania group has been interested for quite a time in the properties of the two-phonon states, namely the double giant resonances. These states have been experimentally observed in the last two decades, and serious questions about their deviations from the linear and harmonic picture have been raised. In Ref. [57], the authors have found that large anharmonicities arise from the coupling of two- to three-phonon states, mainly when these states include giant quadrupole and monopole resonances.

The work of Ref. [57] is based on a microscopic theory employing an effective Skyrme interaction but relies at the same time on boson-mapping techniques. There is some interest in exploring whether these techniques allow extensions of the usual RPA. In particular, RPA is based on the so-called quasi-boson approximation and more exact theories should be envisaged. These kinds of problems are dealt with in Ref. [58], where extensions of RPA built using the boson-mapping, are tested within the solvable Lipkin model.

4.5. *From the compressional modes to the nuclear incompressibility*

The giant monopole resonance (GMR), which is an isotropic ($\Delta L = 0$) compression of the whole nucleus and is therefore often called the "breathing mode", is a systematic properties of nuclei, being rather fragmented in the light systems and concentrated in a single peak in the heavier ones. Its importance stems from the fact that its study gives unique insight into a basic properties of the EoS of nuclear matter, namely the nuclear incompressibility K_∞ which is related to the curvature of the E/A around its minimum.

Theories based on an energy functional $E[\varrho]$, either RMF or non-relativistic Skyrme and Gogny models, allow extracting a value of K_∞ from the measured energy of the GMR. In fact, it is possible to build, within each class of functionals, a set of different parametrizations which differ only in the value of K_∞. After performing calculations of the GMR in, e.g., ^{208}Pb (where experiments are more accurate), one can select the functional which best reproduces the GMR energy - and choose the associated value of K_∞ as the best value. Until 2003, the biggest problem in this procedure has been its model dependence: Skyrme calculations predicted 210-220 MeV, while Gogny calculations pointed to 230 MeV and the RMF models to values between 250 and 270 MeV. The contribution of the Milano group has been to discover that the discrepancy between Skyrme and Gogny does not exist and that the correct value extracted from both models is 230 MeV, and to point out that the still existing discrepancy between RMF and non-relativistic models depend on our poor knowledge of the surface and symmetry properties of nuclei [59].

In principle, also the isoscalar giant dipole resonance (ISGDR), which is a non-isotropic ($\Delta L = 1$) compression mode, provides an alternative way to extract information on K_∞. There are several complications, compared to the case of the GMR. Experiments are more difficult, in the sense that the disentanglement of the ISGDR strength from the other multipoles is far from being trivial. There is low-lying ISGDR strength, whose nature is under debate. The work of Ref. [60] has tried to elucidate the nature of this low-lying strength, without finding a simple association with a macroscopic picture (and questioning therefore the experimental analysis which is based on the use of macroscopic form factors). Also, the dependence of the ISGDR properties on the inputs of the starting Hamiltonian has been studied; a marked sensitivity to the single-particle spectrum has been observed.

5. Conclusions

Our review of the various activities of the Italian nuclear structure community has been very wide. Nonetheless, we have tried to show the existence of several connections between the studies carried out by different groups. With some exceptions, this has allowed a relatively unitary presentation.

We conclude by mentioning a few, relevant challenges for the future (based on a personal choice). Of course, the ultimate goal of nuclear structure is fixing the parameters defining the nuclear Hamiltonian and finding many-body techniques which are as accurate and as efficient as possible to deduce the properties of nuclei, including the exotic ones. Progress has been made along different lines. We know, more than in the past, the properties of the relativistic models and their link with QCD. We dispose of complementary approaches, like the shell-model and the mean field based on effective functionals, in similar mass regions. We manage to calculate in rather sophisticated ways the processes which go beyond mean field and which are hard to introduce in the shell-model description, like the polarization phenomena due to the coupling of particles and collective modes. But we still miss a unified, DFT-based, model which includes both the short-range correlations associated with a realistic NN force and the longer-range correlations associated to polarization processes, consistently calculated.

The studies of the nuclear theorists are focusing on the properties of isospin-asymmetric systems. This is quite natural, in view of the experimental progress devoted to approaching the nuclear drip lines, and because of the links between nuclear structure and nuclear astrophysics. However, the symmetry energy, in particular its behaviour around, and well above the saturation density, is too poorly known. Putting constraints on this basic observable is one of the most important challenges for the next years.

The close collaboration with the experimental groups should force the nuclear theorists to treat in a more unfied way the structure and reaction models. We have mentioned in this contribution the outcome of transfer reactions, and we have touched upon the problem of the ISGDR. Still, in nuclear physics, part of the experimental analysis consists in applying model hypotheses, which are sometimes inconsistent or less precise than the theories which have been illustrated in this contribution or which are available in the literature.

Nuclear structure is a complex field with many facets. The variety of nuclear systems, with their shapes and excitation modes, prevents from drawing here really conclusive statements. However, the long list of references which follows should communicate an optimistic view about the ongoing activities.

Acknowledgments

Writing this contribution would not have been possibile without the active collaboration of many authors of the papers which have been quoted. It is therefore a pleasure to thank them for many discussions, and for specific clarifications, whose echoes can be found in the above pages. More specifically, the author would like to acknowledge the help of those who have read the manuscript, namely E. Vigezzi and P.F. Bortignon.

References

1. N. Lo Iudice, in *Perspectives on Theoretical Nuclear Physics*, proceedings of the "VII Convegno su Problemi di Fisica Nucleare Teorica", Cortona, 1998, edited by A. Fabrocini, G. Pisent and S. Rosati (Edizioni ETS, Pisa, 1999), p. 125.
2. P.F. Bortignon, in *Theoretical Nuclear Physics in Italy*, proceedings of the 8th Conference on Problems in Theoretical Nuclear Physics in Italy, Cortona,

2000, edited by G. Pisent, S. Boffi, L. Canton, A. Covello, A. Fabrocini and S. Rosati (World Scientific, Singapore, 2001), p. 3.
3. A. Gargano, in *Theoretical Nuclear Physics in Italy*, proceedings of the 9th Conference on Problems in Theoretical Nuclear Physics in Italy, Cortona, 2002, edited by S. Boffi, A. Covello, M. Di Toro, A. Fabrocini, G. Pisent and S. Rosati (World Scientific, Singapore, 2003), p. 23.
4. H. Bethe, *Sci. Am.* **189**, 58 (1953).
5. See, e.g., R. Machleidt and I. Slaus, *J. Phys.* **G27**, R69 (2001) and references therein.
6. P. Finelli, N. Kaiser, D. Vretenar, W. Weise, *Eur. Phys. J.* **A17**, 573 (2003); *Nucl. Phys.* **A735**, 449 (2004).
7. See the contribution of P. Finelli in these proceedings.
8. P. Ring, *Prog. Part. Nucl. Phys.* **C37**, 193 (1996).
9. R. Pezer, A. Ventura, D. Vretenar, *Nucl. Phys.* **A717**, 21 (2003).
10. B. Liu, V. Greco, V. Baran, M. Colonna and M. Di Toro, *Phys. Rev.* **C65**, 045201 (2002); V. Greco, V. Baran, M. Colonna, M. Di Toro, T. Gaitanos, H.H. Wolter, *Phys. Lett.* **B562**, 215 (2003).
11. S. Bogner, T.T.S. Kuo, L. Coraggio, A. Covello and N. Itaco, *Phys. Rev.* **C65**, 051301 (2002).
12. See the contribution of A. Gargano in these proceedings, and references therein.
13. L. Coraggio, N. Itaco, A. Covello, A. Gargano and T.T.S. Kuo, *Phys. Rev.* **C68**, 034320 (2003); L. Coraggio, A. Covello, A. Gargano, N. Itaco, T.T.S. Kuo and R. Machleidt, `nucl-th/0407003`.
14. A. Gargano, *Eur. Phys. J.* **A20**, 103 (2004); L. Coraggio, A. Covello, A. Gargano and N. Itaco, `nucl-th/0407002`.
15. P. Guazzoni, L. Zetta, A. Covello, A. Gargano, G. Graw, R. Hertenberger, H.-F. Wirth, M. Jaskola *Phys. Rev.* **C69**, 024619 (2004).
16. J.K. Hwang, A.V. Ramayya, J.H. Hamilton, Y.X. Luo, J.O. Rasmussen, C.J. Beyer, P.M. Gore, S.C. Wu, I.Y. Lee, C.M. Folden III, P.Fallon, P. Zielinski, K.E. Gregorich, A.O. Macchiavelli, M.A. Stoyer, S.J. Asztalos, T.N. Ginter, R. Donangelo, L. Coraggio, A. Covello, A. Gargano and N. Itaco, *Phys. Rev.* **C67**, 014317 (2003); J. Genevey, J.A. Pinston, H.R. Faust, R. Orlandi, A. Scherillo, G.S. Simpson, I.S. Tsekhanovich, A. Covello, A. Gargano, W. Urban, *Phys. Rev.* **C67**, 054312 (2003).
17. F. Andreozzi, N. Lo Iudice and A. Porrino, *J. Phys.* **G29**, 2319 (2003).
18. A. Covello, L. Coraggio, A. Gargano, and N. Itaco, `nucl-th/0310090`.
19. M. Baldo, A. Fiasconaro, H.Q. Song, G. Giansiracusa, U. Lombardo, *Phys. Rev.* **C65**, 017303 (2001).
20. M. Baldo, L. Lo Monaco, *Phys. Lett.* **B525**, 261 (2002).
21. W. Zuo, Caiwan Shen and U. Lombardo, *Phys. Rev.* **C67**, 037301 (2003).
22. M. Baldo, C. Maieron, *Phys. Rev.* **C69**, 014301 (2004).
23. S.C. Pieper and R.B. Wiringa, *Annu. Rev. Nucl. Part. Sci.* **51**, 51 (2001).
24. See the contribution of A. Kievsky in these proceedings.
25. S. Fantoni, A. Sarsa, K.E. Schmidt, *Phys. Rev. Lett.* **87**, 181101 (2001).
26. A. Sarsa, S. Fantoni, K.E. Schmidt, F. Pederiva, *Phys. Rev.* **C68**, 024308

(2003); F. Pederiva, A. Sarsa, K.E. Schmidt, S. Fantoni, *Nucl. Phys.* **A742**, 255 (2004); see also the contribution by F. Pederiva in these proceedings.
27. See the contribution of C. Bisconti in these proceedings.
28. M. Baldo, C. Maieron, P. Schuck, X. Viñas, *Nucl. Phys.* **A736**, 241 (2004).
29. S. Baroni, M. Armati, F. Barranco, R.A. Broglia, G. Colò, G. Gori and E Vigezzi, *J. Phys.* **G30**, 1353 (2004).
30. See: M. Bender, G.F. Bertsch and P.-H. Heenen, *Phys. Rev.* **C69**, 034340 (2004) and references therein.
31. E. Garrido, P. Sarriguren, E. Moya de Guerra, U. Lombardo, P. Schuck and H.J. Schulze, *Phys. Rev.* **C63**, 037304 (2001).
32. M. Baldo, M. Farine, U. Lombardo, E.E. Saperstein, P. Schuck and M.V. Zverev, *Eur. Phys. J.* **A18**, 17 (2003).
33. H. Kucharek and P. Ring, *Z. Phys.* **A339**, 23 (1991).
34. See the contribution of A. Drago in these proceedings.
35. F. Barranco, R.A. Broglia, G. Colò, G. Gori, E. Vigezzi and P.F. Bortignon, *Eur. Phys. J.* **A21**, 57 (2004).
36. F. Barranco, R.A. Broglia, G. Gori, E. Vigezzi, P.F. Bortignon, and J. Terasaki, *Phys. Rev. Lett.* **83**, 2147 (1999).
37. G. Gori, F. Barranco, E. Vigezzi and R.A. Broglia, *Phys. Rev.* **C69**, 041302(R) (2004).
38. F. Barranco, P.F. Bortignon, R.A. Broglia, G. Colò, E. Vigezzi *Eur. Phys. J.* **A11**, 385 (2001).
39. R.A. Broglia, K. Matsuyanagi, H. Sofia and A. Vitturi, *Nucl. Phys.* **A348**, 237 (1980).
40. M.B. Barbaro, R. Cenni, A. Molinari, M.R. Quaglia, `nucl-th/0306062`.
41. M.B. Barbaro, A. Molinari, F. Palumbo, M.R. Quaglia, `nucl-th/0304028`.
42. F. Palumbo, `nucl-th/0405045`.
43. J.M. Arias, C.E. Alonso, A. Vitturi, J.E. Garcìa-Ramos, J. Dukelsky, A. Frank, *Phys. Rev.* **C68**, 041302(R) (2003).
44. L. Fortunato and A. Vitturi, *J. Phys.* **G29**, 1341 (2003).
45. L. Fortunato and A. Vitturi, *J. Phys.* **G30**, 627 (2004).
46. S. Brant, D. Vretenar, A. Ventura, *Phys. Rev.* **C69**, 017304 (2004).
47. D.S. Delion, A. Insolia, R.J. Liotta, *Phys. Rev.* **C67**, 054317 (2003).
48. M. Bender, P.-H. Heenen, P.-G. Reinhard, *Rev. Mod. Phys.* **75**, 121 (2003).
49. V.I. Abrosimov, A. Dellafiore, F. Matera, *Nucl. Phys.* **A717**, 44 (2003); V.I. Abrosimov, O.I. Davidovskaya, A. Dellafiore, F. Matera, *Nucl. Phys.* **A727**, 220 (2003).
50. G. Giambrone, S. Scheit, F. Barranco, P.F. Bortignon, G. Colò, D. Sarchi, E. Vigezzi, *Nucl. Phys.* **A726**, 3 (2003).
51. See: P.F. Bortignon, R.A. Broglia, D.R. Bes, R. Liotta, *Phys. Rep.* **30**, 305 (1977), and references therein.
52. N. Lo Iudice and Ch. Stoyanov, *Phys. Rev.* **C69**, 044312 (2004).
53. V.G. Soloviev, *Theory of atomic nuclei: quasiparticles and phonons* (Institute of Physics Editor, Bristol, 1992).
54. G.F. Bertsch, P.F. Bortignon, R.A. Broglia, *Rev. Mod. Phys.* **55**, 287 (1983).
55. D. Sarchi, P.F. Bortignon, G. Colò, *Phys. Lett.* **B601**, 27 (2004).

56. J. Kvasil, N. Lo Iudice, R.G. Nazmitdinov, A. Porrino and F. Knapp, *Phys. Rev.* **C69**, 064308 (2004).
57. M. Fallot, Ph. Chomaz, M.V. Andrés, F. Catara, E.G. Lanza, J.A. Scarpaci, *Nucl. Phys.* **A729**, 699 (2003).
58. M. Grasso, F. Catara, and M. Sambataro, *Phys. Rev.* **C66**, 064303 (2002).
59. G. Colò and N. Van Giai, *Nucl. Phys.* **A731**, 15 (2004); G. Colò, N. Van Giai, J. Meyer, K. Bennaceur and P. Bonche, *Phys. Rev.* **C70**, 024307 (2004).
60. J. Kvasil, N. Lo Iudice, Ch. Stoyanov, P. Alexa, *J. Phys.* **G29**, 753 (2003).

LOW-MOMENTUM NUCLEON-NUCLEON POTENTIAL AND NUCLEAR STRUCTURE CALCULATIONS

A. GARGANO, L. CORAGGIO, A. COVELLO AND N. ITACO

Dipartimento di Scienze Fisiche, Università di Napoli Federico II, and Istituto Nazionale di Fisica Nucleare, Complesso Universitario di Monte S. Angelo, Via Cintia - I-80126 Napoli, Italy
E-mail: gargano@na.infn.it

A new approach for deriving the shell-model effective interaction from the free nucleon-nucleon (NN) potential is discussed. It consists in renormalizing the strong repulsive core contained in all modern NN potentials by constructing a low-momentum potential, $V_{\rm low-k}$, which is confined within a certain cutoff momentum Λ. Results of shell-model calculations performed within the framework of the $V_{\rm low-k}$ approach are compared with those obtained by using the usual G-matrix formalism and with experimental data. We also present some results of preliminary calculations with different modern NN potentials and discuss the choice of the cutoff momentum.

1. Introduction

An important result which has emerged during the last decade regards the practical value of realistic shell-model calculations. In fact, it has been shown that shell-model calculations with two-body effective interactions derived from modern NN potentials V_{NN} are able to give an accurate description of nuclear structure properties. This is well illustrated by a number of papers in which realistic shell-model calculations for nuclei with few valence particles or holes around doubly magic ^{100}Sn, ^{132}Sn, and ^{208}Pb have been performed.[1,2,3,4]

The first difficulty one is faced with in this kind of calculations is that all modern NN potentials contain a strongly repulsive core and therefore cannot be used directly in the derivation of the effective interaction, which is based on a perturbation expansion in terms of V_{NN}. In most of the realistic shell-model calculations to date this problem has been overcome by making use of the well-known Brueckner G-matrix method.

Recently, a new approach has been proposed in which the short-range repulsion of V_{NN} is renormalized by integrating out its high momentum components.[5,6] The resulting low-momentum potential is a smooth potential which preserves the physics of the original V_{NN} up to a certain cutoff momentum Λ and can be therefore used in nuclear structure calculations. The use of the $V_{\rm low-k}$ approach in realistic shell-model calculations is currently proving to be an advantageous alternative to the G-matrix method.[6,7,8] A relevant feature of this approach is that $V_{\rm low-k}$'s derived from different V_{NN}'s become close to each other for $\Lambda \sim$2 fm^{-1} or smaller.[9]

In this paper, we introduce $V_{\rm low-k}$ giving a brief description of its derivation. This is done in Sec. 2, where we also illustrate the preservation of the deuteron binding energy and phase shifts. A brief discussion is finally devoted to the dependence of $V_{\rm low-k}$ on the input potentials. In Sec. 3 some numerical applications of the $V_{\rm low-k}$ approach to shell-model calculations are presented. First, a comparison of the $V_{\rm low-k}$ and G-matrix results with experimental data is shown. Then, results obtained by using different modern NN potentials as input are reported and some comments on the choice of the cutoff momentum Λ are made.

However, before closing this section it is worthwhile to sketch our derivation of the shell-model effective interaction once the G matrix or the $V_{\rm low-k}$ has been calculated. The effective interaction $V_{\rm eff}$ is defined by

$$PH_{\rm eff}P\psi_\mu = P(H_0 + V_{\rm eff})P\psi_\mu = E_\mu P\psi_\mu \,, \tag{1}$$

where P denotes the projection operator onto the chosen shell-model space and $\mu = 1, 2 \cdots d$, with d denoting the dimension of this space. The unperturbed Hamiltonian is represented by $H_0 = T + U$, T being the kinetic energy and U an auxiliary potential introduced to define a convenient single-particle basis. This is usually chosen to be the harmonic oscillator potential. The eigenvalues E_μ are a subset of the original Hamiltonian, $H = H_0 + [V_{\rm low-k} - U]$ or $H = H_0 + [G - U]$, in the full space.

The $V_{\rm eff}$ of Eq. (1) may be derived by means of the folded-diagram expansion introduced by way of the time-dependent perturbation theory, as described in Ref. [10]. It can be written schematically as

$$V_{\rm eff} = \hat{Q} - \hat{Q}' \int \hat{Q} + \hat{Q}' \int \hat{Q} \int \hat{Q} - \hat{Q}' \int \hat{Q} \int \hat{Q} \int \hat{Q} + \dots \,, \tag{2}$$

where $\hat{Q}$ is in principle an infinite sum of irreducible and valence-linked diagrams in $V_{\rm low-k}$ (or G) and the integral sign stands for a generalized folding operation.[11] $\hat{Q}'$ is obtained from $\hat{Q}$ by removing terms of first order in the interaction. In our calculation of the $\hat{Q}$ box we include all diagrams up to second order and the computation of these diagrams is performed by including intermediate states composed of particle and hole states restricted to the two major shells above and below the Fermi surface. After $\hat{Q}$ is calculated, the series of Eq. (2) is summed to all orders by means of the Lee-Suzuki iteration method.[12]

It is worth mentioning that the effective interaction is usually derived for systems with two valence nucleons and therefore contains one- and two-body terms. It is customary, however, to use a subtraction procedure[13] so that only the two-body terms are retained. As regards the one-body terms, it is supposed that they are contained in the experimental single-particle energies which are generally used in shell-model calculations.

2. The low-momentum nucleon-nucleon potential

To introduce the low-momentum potential $V_{\rm low-k}$, let us start from the half-on-shell T matrix

$$T(k',k,k^2) = V_{NN}(k',k) + \wp \int_0^\infty q^2 dq V_{NN}(k',q) \frac{1}{k^2 - q^2} T(q,k,k^2)\,, \quad (3)$$

where k, k', and q stand for the relative momenta. We then define the effective low-momentum T matrix by

$$T_{\rm low-k}(p',p,p^2) = V_{\rm low-k}(p',p) + \wp \int_0^\Lambda q^2 dq V_{\rm low-k}(p',q) \frac{1}{p^2 - q^2} T_{\rm low-k}(q,p,p^2)\,, \quad (4)$$

where, as mentioned in the Introduction, Λ denotes the momentum space cutoff and $(p',p) \leq \Lambda$. The above $T_{\rm low-k}$ should satisfy the condition

$$T(p',p,p^2) = T_{\rm low-k}(p',p,p^2)\,, \quad (5)$$

which together with Eq. (4) defines the low-momentum potential $V_{\rm low-k}$. The calculation of this potential is performed by means of conventional effective interaction techniques (one of them is the above mentioned folded-diagram method). In such a way the preservation of the deuteron binding energy is guaranteed. It should be mentioned that the equivalence of a low-momentum potential derived through effective interaction theory and the $V_{\rm low-k}$ defined by Eqs. (4) and (5) is demonstrated in Refs. [6,14]. Clearly, the preservation of the half-on-shell T matrix by $V_{\rm low-k}$ implies that the phase shifts are preserved up to $E_{\rm lab} = 2\hbar^2\Lambda^2/M$, M being the nucleon mass.

Actually, $V_{\rm low-k}$ is derived by using the nonperturbative approach of Ref. [15], based on the Lee-Suzuki similarity transformation.[12] This consists in the construction of an iterative solution for $V_{\rm low-k}$ through matrix inversion techniques. It is worth mentioning that since the resulting potential is not Hermitian we also adopt the Hermitization procedure proposed in Ref. [15]. In a very recent paper[16] it is shown that different Hermitian $V_{\rm low-k}$'s obtained trough different Hermitization procedures are rather similar to each other and, in particular, they all preserve the deuteron binding energy and phase shifts of the original NN potential.

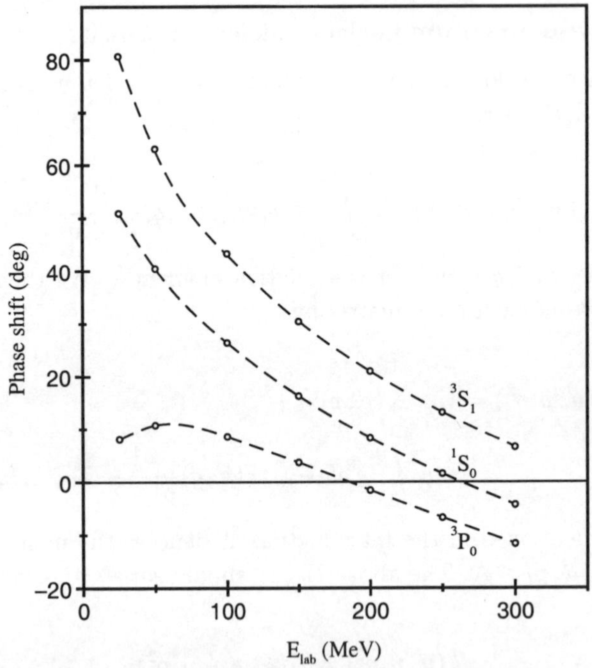

Figure 1. Comparison of phase shifts given by $V_{\rm low-k}$(dashed line) and V_{NN}(open circles) for the $^1S_0$, $^3S_1$, and $^3P_0$ channels. See text for details.

In Ref. [9] phase shifts obtained by a $V_{\rm low-k}$ derived from the CD-Bonn potential[17] using $\Lambda = 2.1\ \ {\rm fm}^{-1}$ are compared with those of the original V_{NN} in a number of cases. As an example, we show in Fig. 1 the neutron-proton $^1S_0$, $^3S_1$, and $^3P_0$ phase shifts as predicted by the full CD-Bonn potential and its $V_{\rm low-k}$ obtained using a cutoff momentum $\Lambda = 2.0\,{\rm fm}^{-1}$.

We have also carried out numerical checks for the deuteron binding energy. We have found that the binding energy given by $V_{\rm low-k}$ does not depend on the value of Λ and agrees very accurately with that given by the input V_{NN}.

Table 1. Calculated P_D's with different NN potentials and with the corresponding $V_{\rm low-k}$'s. See text for details.

	NijmII	AV18	CDB
Full potential	5.63	5.76	4.85
$V_{\rm low-k}$	4.32	4.37	4.04

In concluding this section, let us briefly comment on the dependence of the $V_{\rm low-k}$ on the input NN potential. It is worth mentioning that all modern NN potentials are on-shell equivalent, in the sense that they reproduce with equally high precision the NN scattering data and the deuteron properties. This does not mean, however, that these potentials are identical, having in fact quite different momentum-space matrix elements. A detailed comparison of $V_{\rm low-k}$'s derived from different NN potentials can be found in Refs. [9,18], where it was shown that diagonal matrix elements of different $V_{\rm low-k}$ are nearly identical for $\Lambda \lesssim 2$ fm^{-1}. Significant differences are found in some partial waves, as $^3F_2$, $^3D_3$, $^1F_3$, and $^3F_3$, which are explained[9] considering the different accuracy of the input V_{NN}'s in reproducing the phase shifts in such channels. In Ref. [9] some results regarding off-diagonal matrix elements are also reported. Here we only show how the differences in the D-state probabilities of the deuteron P_D predicted by various NN potentials are attenuated for the corresponding $V_{\rm low-k}$'s. This comparison is made in Table 1 for the three potentials Nijmegen II (NijmII),[19] Argonne v_{18} (AV18),[20] and CD-Bonn (CDB).[17] We see that the P_D's given by the full potentials range from 4.85 to 5.76% while after renormalization the difference reduces to 0.33.

3. Shell-model results

3.1. *Comparison of the G-matrix and $V_{\rm low-k}$ approaches*

A comparison of shell-model calculations performed within the framework of $V_{\rm low-k}$ and G-matrix formalisms has been presented in Refs. [6,7,8]. It has been shown that when using $V_{\rm low-k}$ the agreement between theory and experiment is as good as or even slightly better than that obtained by means of G matrix.

To illustrate this point, we consider here ^{134}Sn, having two neutrons in the 82-126 shell. In Fig. 2 we compare the calculated spectra obtained using the G-matrix and $V_{\rm low-k}$ formalisms with the experimental data. In both calculations we have taken the single-particle energies from experiment and adopted CD-Bonn as input NN potential. As for the $V_{\rm low-k}$, it has been derived for $\Lambda = 2.1\,{\rm fm}^{-1}$. This choice of Λ ensures that $V_{\rm low-k}$, as the original CD-Bonn potential, reproduces the empirical phase shifts up to $E_{\rm lab} = 350\,{\rm MeV}$.

The value of Λ one should use is an important issue. Several considerations on this point can be found in Ref. [6] while in Ref. [9] how $V_{\rm low-k}$ depends on the input potential is discussed in connection with the value of Λ. In a very recent paper[21] the ^{3}H, ^{3}He, and ^{4}He binding energies have been calculated with $V_{\rm low-k}$ for a wide range of momentum cutoffs with the aim to assess the role of the 3-body forces in the $V_{\rm low-k}$ approach. We feel, however, that no definite conclusions can be drawn at present and how to choose the cutoff momentum certainly requires further study. In the next section we show how results of shell-model calculations change with Λ for various modern NN potentials.

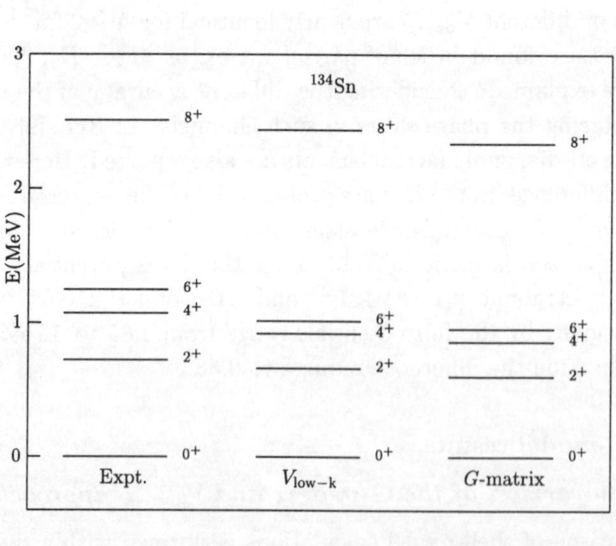

Figure 2. Experimental and calculated spectra for ^{134}Sn.

3.2. *Comparison for different modern NN potentials*

The spectrum of ^{130}Sn, with two holes in the 50-82 shell, has been calculated starting from the three NN potentials Nijmegen II, Argonne v_{18}, and CD-Bonn. In all cases use has been made of $V_{\text{low-k}}$ with the cutoff Λ ranging from 1.7 to 2.5 fm^{-1}. We have calculated the *rms* deviation σ[22] including the nine observed excited states up to 2.5 MeV and, in Fig. 3, we show its behavior as a function of Λ for the three potentials. We see that the curves relative to Nijmegen II and Argonne v_{18} practically overlap each other while that obtained starting from the CD-Bonn potential is rather different. In particular, CD-Bonn gives the lowest value of σ for $\Lambda \sim 1.8\,\text{fm}^{-1}$, while the minimum for the other two potentials corresponds to $\Lambda \sim 2.2\,\text{fm}^{-1}$. For these values of Λ we have found that not only the *rms* deviation is almost the same for the three potentials but they also give essentially the same energy for the various states. As an example, we report in Table 2 the ground-state energy of ^{130}Sn (relative to the doubly magic ^{132}Sn) and the excitation energy of the first three positive-parity yrast states.

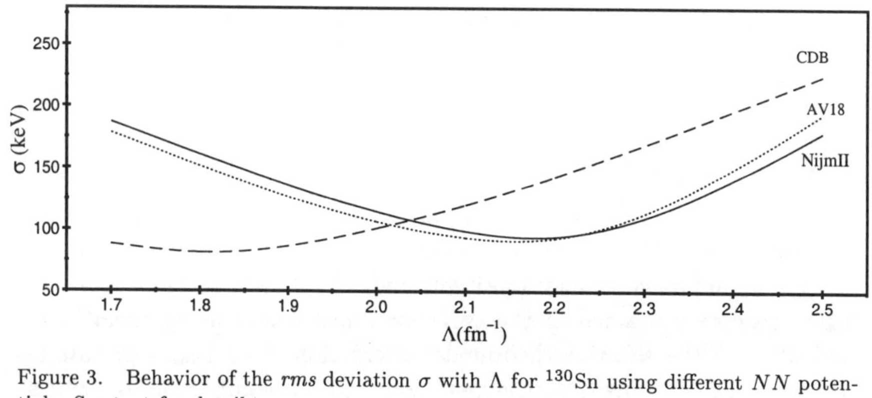

Figure 3. Behavior of the *rms* deviation σ with Λ for ^{130}Sn using different NN potentials. See text for details.

It is worth noting, however, that in all the three calculations moderate changes in the value of Λ do not change the quality of agreement between theory and experiment. For CD-Bonn, for instance, σ remains $\lesssim 100\,\text{keV}$ for $1.7\,\text{fm}^{-1} \lesssim \Lambda \lesssim 2.0\,\text{fm}^{-1}$. This is true also for the other two potentials for $2.0\,\text{fm}^{-1} \lesssim \Lambda \lesssim 2.3\,\text{fm}^{-1}$. Based on these findings, which seem to be confirmed by calculations in other mass regions, we may try to conclude that the shell-model results are not very sensitive to changes in Λ when its value is moved within a limited range, which we have found to depend on the input NN potential.

In concluding this section, we would like to briefly report on a point we are currently investigating. It regards the dependence of the "optimum" value of Λ by the number of intermediate states included in the calculation of the $\hat{Q}$ box. As mentioned in the Introduction, the calculations presented here have been performed including intermediate states composed of particle and hole states restricted to two major shell above and below the Fermi surface. We have now increased the number of intermediate states and we have found that a larger value of Λ seems to be needed to have the minimum *rms* deviation. Further calculations are in progress and will published in a forthcoming paper.

Table 2. Experimental energies (in MeV) of ^{130}Sn compared with calculated values for different NN potentials. See text for details.

	NijmII	AV18	CDB	Expt
E_{gs}	12.410	12.427	12.406	12.474
$E(2^+)$	1.462	1.448	1.433	1.221
$E(4^+)$	2.057	2.055	2.057	1.966
$E(6^+)$	2.227	2.214	2.240	2.257

4. Concluding remarks

In this paper we have discussed the low-momentum potential $V_{\rm low-k}$ within the framework of realistic shell-model calculations. We have shown that the $V_{\rm low-k}$ approach provides a simple and reliable way to renormalize the repulsive core contained in the NN potentials before using them in the derivation of the effective shell-model interaction. The $V_{\rm low-k}$ results are in very good agreement with experiment and are comparable to or even better than those obtained using the traditional G-matrix method.

We have performed shell-model calculations with $V_{\rm low-k}$ for different modern NN potentials. Here we have presented the results for ^{130}Sn obtained with Nijmegen II, Argonne v_{18}, and CD-Bonn and have also studied the behavior of the *rms* deviation as a function of the cutoff Λ. It has turned out that the predicted energies for ^{130}Sn depend only weakly on the NN potential used as input. In particular, we have found that the three potentials give practically the same results when using $\Lambda = 1.8\,\text{fm}^{-1}$ for CD-Bonn and $\Lambda = 2.2\,\text{fm}^{-1}$ for Nijmegen II and Argonne v_{18}. It should be stressed, however, that in all the three calculations the results are rather insensitive to moderate changes in the value of Λ.

Acknowledgments

This work was supported by the Italian Ministero dell'Istruzione, dell'Università e della Ricerca (MIUR).

References

1. A. Covello, L. Coraggio, A. Gargano, and N. Itaco, *Acta Phys. Pol.* **B32**, 871 (2001), and references therein.
2. L. Coraggio, A. Covello, A. Gargano, and N. Itaco, *Phys. Rev.* **C65**, 051306(R) (2002).
3. L. Coraggio, A. Covello, A. Gargano, N. Itaco, and T. T. S. Kuo, *Phys. Rev.* **C66**, 064311 (2002).
4. L. Coraggio, A. Covello, A. Gargano, and N. Itaco, *Phys. Rev.* **C70**, 034310 (2004).
5. S. Bogner, T. T. S. Kuo, and L. Coraggio, *Nucl. Phys.* **A684**, 432c (2001).
6. Scott Bogner, T. T. S. Kuo, L. Coraggio, A. Covello, and N. Itaco, *Phys. Rev.* **C65**, 051301(R) (2002).
7. A. Covello, L. Coraggio, A. Gargano, N. Itaco, and T. T. S. Kuo, in *Challenges of Nuclear Structure*, edited by A. Covello (World Scientific, Singapore, 2002), p. 129.
8. A. Covello, in *Proceedings of the International School of Physics "E. Fermi", Course CLIII*, edited by A. Molinari, L. Riccati, W. M. Alberico, and M. Morando (IOS Press, Amsterdam, 2003), p. 79.
9. S. K. Bogner, T. T. S. Kuo, and A. Schwenk, *Phys. Rep.* **386**, 1 (2003).
10. T. T. S. Kuo, S. Y. Lee, and K. F. Ratcliff, *Nucl. Phys.* **A176**, 62 (1971).
11. E. M. Krenciglowa and T. T. S. Kuo, *Nucl. Phys.* **A342**, 454 (1980).
12. K. Suzuki and S. Y. Lee, *Prog. Theor. Phys.* **64**, 2091 (1980).
13. J. Shurpin, T. T. S. Kuo, and D. Strottman, *Nucl. Phys.* **A408**, 310 (1983).
14. S. K. Bogner, A. Schwenk, T. T. S. Kuo, and G. E. Brown nucl-th/0111042.
15. F. Andreozzi, *Phys. Rev.* **C54**, 684 (1996).
16. Jason D. Holt, T. T. S. Kuo, and G. E. Brown, *Phys. Rev.* **C69**, 034329 (2004).
17. R. Machleidt, *Phys. Rev.* **C63**, 024001 (2001).
18. S. K. Bogner, T. T. S. Kuo, A. Schwenk, D. R. Entem, and R. Machleidt, *Phys. Lett.* **B576**, 265 (2003).
19. V. G. J. Stoks, R. A. M. Klomp, C. P. F. Terheggen, and J. J. de Swart, *Phys. Rev.* **C49**, 2950 (1994).
20. R. B. Wiringa, V. G. J. Stoks, and R. Schiavilla, *Phys. Rev.* **C51**, 38 (1995).
21. A. Nogga, S. K. Bogner, and A. Schwenk, nucl-th/0405016.
22. We define $\sigma = \{(1/N_d)\sum_i[E_{exp}(i) - E_{calc}(i)]^2\}^{1/2}$, where N_d is the number of data.

RECENT RESULTS IN CBF THEORY FOR HEAVY-MEDIUM NUCLEI *

C. BISCONTI AND G. CO'

Dipartimento di Fisica, Università di Lecce
and,
Istituto Nazionale di Fisica Nucleare, sez. di Lecce
Lecce, Italy

F. ARIAS DE SAAVEDRA

Departamento de Física Moderna, Universidad de Granada, Spain

A. FABROCINI

Dipartimento di Fisica, Università di Pisa
and,
Istituto Nazionale di Fisica Nucleare, sez. di Pisa
Pisa, Italy

A formalism based on the Correlated Basis Functions (CBF) theory is developed to describe nuclei with $N \neq Z$ in a jj coupling scheme. The binding energies and the density distributions of doubly magic nuclei have been calculated by solving the Fermi Hypernetted Chain (FHNC) integral equations in the single operator chain approximation (SOC). The realistic Argonne v_8' two-nucleon potential has been used in the application to the ^{40}Ca, ^{48}Ca and ^{208}Pb nuclei.

The aim of this contribution is to report on the progresses done in the framework of a project [1-5] aimed to apply the CBF theory to finite nuclear systems. The starting point of the theory is the variational principle,

$$\delta E[\Psi] = \delta \left[\frac{< \Psi|H|\Psi >}{< \Psi|\Psi >} \right] = 0 \tag{1}$$

with the following ansatz for the expression of the many-body wave functions Ψ:

$$\Psi(1,2,...,A) = G(1,2,...A)\,\Phi(1,2,...,A) \tag{2}$$

*Presented by C. Bisconti

The search of the minumum of the energy functional is done by making variations of the many-body correlation operator $G(1,2,...,A)$ and of the single particle wave functions forming the Slater determinant $\Phi(1,2,...,A)$. The many-body correlation operator is supposed to be described by a product of two-body operators F_{ij}:

$$G(1,2...,A) = S\prod_{i<j} F_{ij} \tag{3}$$

where the operator S symmetrizes the product. The two-body correlations F_{ij} have an operatorial dependence analogous to that of the modern nucleon-nucleon interactions. We consider F_{ij} of the form:

$$F_{ij} = \sum_{p=1,8} f^p(r_{ij})\, O^p_{ij} \tag{4}$$

where the involved operators are:

$$O^{p=1,8}_{ij} = [1, \boldsymbol{\sigma}_i \cdot \boldsymbol{\sigma}_j, S_{ij}, (\mathbf{L}\cdot\mathbf{S})] \otimes [1, \boldsymbol{\tau}_i \cdot \boldsymbol{\tau}_j] \tag{5}$$

In the above equation $\boldsymbol{\sigma}$ and $\boldsymbol{\tau}$ indicate the usual spin and isospin Pauli matrices, and S_{ij} the tensor operator.

The evaluation of the many-variables integrals necessary to calculate the energy functional (1) is done by using the integral summation technique known as Fermi hypernetted chain (FHNC) [6], originally developed for infinite systems. The FHNC equations allows the sum of a set of infinite classes of Mayer-like diagrams resulting from the cluster expansion of Eq. (1). The use of state dependent correlations, as those of Eq. (4), requires special attention because the various correlation operators do not commute. In our calculations this difficulty is handled by considering, in addition to all the scalar correlations, only those diagrams having chains of correlation functions containing a single operator with $p > 1$. This is the so-called single operator chain (SOC) approximation [7].

In the past, we have applied this theory to describe the ground state of doubly closed shell nuclei by using two-body potentials and correlations containing operator terms up to the tensor components [1,2,3,4], and also a three-body interaction [5]. These calculations have been limited to the ^{16}O and ^{40}Ca nuclei since the formalism was developed in ls coupling scheme and for single particle wave functions equal for both protons and neutrons. Formally this situation is very similar to that of the symmetric nuclear matter.

In this report we present the first results obtained by extending the FHNC/SOC scheme for nuclei with different wave functions for protons and neutrons and in jj coupling representation. The results of these calculations have been done by using the realistic v_8' Argonne two-body potential, but without any three-body force, which, in any case, it will be soon implemented. For this reason, the results we show here, should not be considered fully realistic. In any case, already at this stage, they give interesting informations on the nuclear structure.

The separated treatment of protons and neutrons requires that the correlations should be considered separately in their spin and isospin part:

$$F_{ij} = \sum_{k=1}^{3}\sum_{l=0}^{1} f_{2k-1+l}(r_{ij})O_{ij}^{2k-1+l} = \sum_{l=0}^{1}(\vec{\tau}_i \cdot \vec{\tau}_j)^l \sum_{k=1}^{3} f_{2k-1+l}(r_{ij})P_{ij}^k \quad (6)$$

Another, in principle more obvious, consequence, is that the FHNC/SOC equations should be almost triplicated to consider one- and two-body densities of proton-proton, neutron-neutron, and isospin mixed type. In jj coupling these equations should be further extended to distinguish between two different types of statistical correlations, with parallel or antiparallel spin.

As already stated, the calculations have been done by using the v_8' reduction of the Argonne v_{18} two-body potential [8]. The operator structure of this potential is given in Eq. (5). The channels up to $p = 6$ have been treated within the FHNC/SOC formalism without any approximation, while the contributions of the two spin-orbit channels have been calculated by using perturbation theory [5,7].

In the correlations the two spin-orbit channels have been neglected. The results have been obtained by using a fixed set of single particle wave functions and searching for the minimum by modifying the correlations. The two-body correlation functions, have been generated by solving a set of coupled Euler-Lagrange equations. The search of the minimum has been done by changing only two parameters: the healing distance d_t of the two tensor channels, and another healing distance d_c related to all the other channels. These are the nucleonic relative distances where the various two-body correlation functions reach their asymptotic values, one for $p = 1$ and zero for all the other cases.

We show in Tab. 1 the results obtained for the ^{40}Ca, ^{48}Ca and ^{208}Pb nuclei. The single particle wave functions have been generated by using a Woods-Saxon potentials whose parameters have been taken from the literature [3]. For each nucleus, in addition to the full calculation, whose results are indicated in the F_6 columns, we also made calculations by using only the first four central channels of the correlations. These are the results

Table 1. Contributions to the binding energies per nucleon for the three nuclei considered. All the quantites are expressed in MeV.

	F_4	F_6	F_4	F_6	F_4	F_6
	^{40}Ca	^{40}Ca	^{48}Ca	^{48}Ca	^{208}Pb	^{208} Pb
T/A	38.61	38.18	36.36	36.21	36.93	36.21
T_{jj}/A	0.008	0.006	0.16	0.11	0.15	0.09
V_8/A	-51.81	-47.11	-49.29	-45.29	-54.17	-47.63
V_{8jj}/A	-0.0002	-0.0003	0.01	0.11	0.02	0.02
V_C/Z	3.78	3.92	2.55	2.67	9.58	10.45
V_{Cjj}/Z	0.018	-0.0012	-0.016	0.008	-0.01	-0.01
E_{ee}/A	0.47	0.46	0.45	0.44	0.34	0.32
E/A	-11.56	-6.73	-10.99	-7.15	-12.97	-6.87
E_{exp}		-8.55		-8.66		-7.80

shown by the F_4 columns. The rows of the table indicate the contributions of the various terms to the binding energy per nucleon: T is the kinetic energy, V_8 the nuclear interaction term, V_C the Coulomb term (here divided by the number of protons), and finally E_{ee} is the contribution of a relevant elementary diagram that should be added to the FHNC/SOC calculations [1] to properly satisfy the various sum rules. All the terms with the jj labels indicate the contribution due to the antiparallel spin densities.

A first remark, is that the f_4 calculations produce more binding than the F_6 ones. The tensor terms of the correlation reduce the binding as it is shown by the V_8 row. The kinetic energies have similar values in all the calculations, and also the Coulomb terms are not affected by the tensor correlation.

In Fig. 1 we compare the correlation functions of the three nuclei considered. In the upper panel the scalar correlation functions, are shown, while in the other panel the tensor-isospin correlation functions, $p = 6$ channel in Eq.(5), are compared. The scalar correlation functions are rather similar for all the three nuclei considered, they heal at 1.4 fm. The tensor-isospin correlations heal at larger distances and show a strong dependence on the nucleus. The healing for the ^{208}Pb is larger than that for the other two nuclei.

Another interesting remark comes from observing the results obtained by the same type of calculation in the various nuclei. Kinetic energies and nuclear energies clearly show saturation properties while the repulsive Coulomb contribution increases with increasing proton number. The comparison between the Coulomb terms of the two calcium isotopes show a relatively large difference. This is surprising, but it demonstrates that the presence of a different number of neutrons modifies the proton densities, as

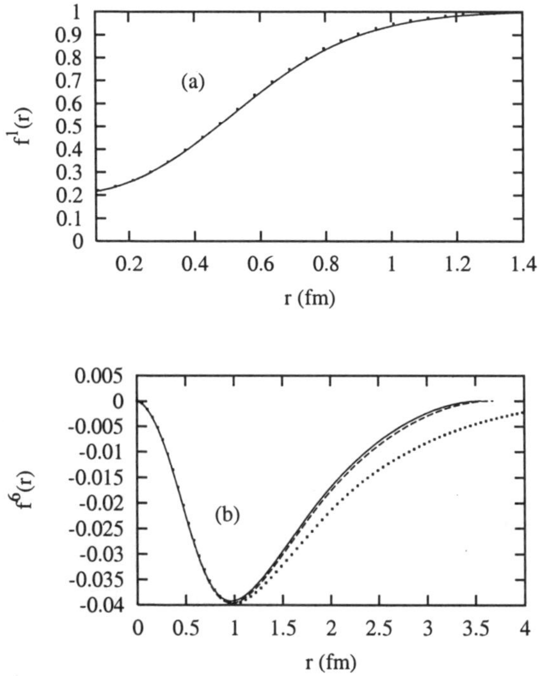

Figure 1. The f^1,f^6 correlation functions. The ^{40}Ca,^{48}Ca,^{208}Pb correlations are indicated with solid , dashed and dotted lines respectively.

it is shown in Fig. 2.

A further remark about the results shown in Tab. 1 regards the contribution of the antiparallel spin terms, the jj contributions. In general, these contributions are rather small and, as expected, they become even smaller in ^{40}Ca where all the spin-orbit partners levels are saturated.

To conclude, we may say that the extension of the FHNC/SOC formalism to treat nuclei with N$\neq$Z in jj coupling scheme has been successful. This has allowed us to perform calculations of binding energies and density distributions of medium-heavy nuclei with neutron exess such as ^{48}Ca and ^{208}Pb, by using a realistic two-body potential. To the best of our knowledge this is the first microscopic calculation for the ^{208}Pb nucleus. As already mentioned, a fully realistic calculation should include also a three-body in-

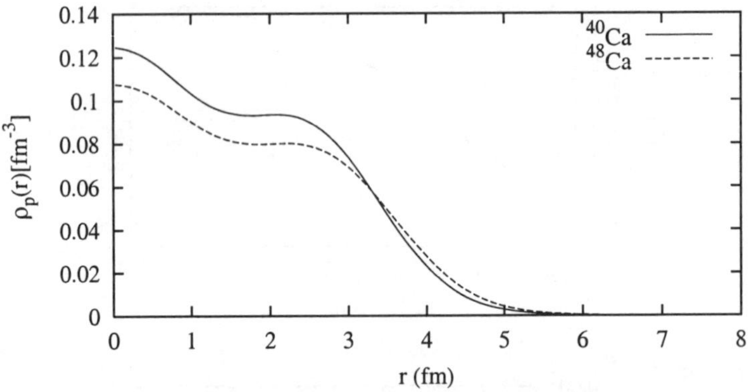

Figure 2. Proton densities of the ^{40}Ca and ^{48}Ca nuclei.

teraction. The work in this direction has quite advanced and we plan to obtain results in short time.

References

1. G. Co', A. Fabrocini, S. Fantoni, I. E. Lagaris, *Nucl. Phys.* **A549**, 439 (1992).
2. G. Co', A. Fabrocini, S. Fantoni, *Nucl. Phys.* **A568**, 73 (1994).
3. F. Arias de Saavedra, G. Co', A. Fabrocini, S. Fantoni, *Nucl. Phys.* **A605**, 359 (1996).
4. A. Fabrocini, F. Arias de Saavedra, G. Co', P. Folgarait, *Phys. Rev.* **C57**, 1668 (1998).
5. A. Fabrocini, F. Arias de Saavedra, G. Co', *Phys. Rev.* **C61**, 1668 (2000).
6. S. Rosati, in *From nuclei to particles,* Proc. Int. School E. Fermi, course LXXIX, ed. Molinari(1979).
7. V.R. Pandharipande, R.B. Wiringa, *Rev. Mod. Phys.* **51**, 821 (1979).
8. R.B. Wiringa, V.G.J. Stoks and R. Schiavilla, *Phys. Rev.* **C51**, 38 (1995).

AUXILIARY FIELD DIFFUSION MONTE CARLO CALCULATION OF PROPERTIES OF OXYGEN ISOTOPES

S. GANDOLFI AND F. PEDERIVA

Dipartimento di Fisica, Università di Trento,
and INFN Gruppo Collegato di Trento
via Sommarive, 14
I-38050 Povo, Trento (Italy)

S. FANTONI

S.I.S.S.A, International School of Advanced Studies,
Trieste (Italy)

K.E. SCHMIDT

Department of Physics and Astronomy, Arizona State University,
Tempe, AZ (USA)

The Auxiliary Field Diffusion Monte Carlo method has been applied to simulate oxygen isotopes in the $1D_{5/2}$ shell using a realistic nucleon-nucleon interactions, which include tensor, spin-orbit and three-body forces. The closed shell ^{16}O has been replaced by the self–consistent potential obtained by solving Hartree-Fock equations with the effective interaction Skyrme I, and AFDMC was used only on the external neutrons. We report results for energies of the ground-state and of the first excited states.

1. INTRODUCTION

In the recent years the study of the structure of the neutron matter has regained interest due to the strong connections with astrophysical problems which require a detailed knowledge of the behavior of the condensed neutron systems. On the other hand the creation of exotic nuclei near the drip line opened interesting questions about the role of the terms of the N-N interaction in granting or preventing the stability of neutron rich nuclei. The Green's Function Monte Carlo method proved to be extremely efficient in computing nuclear properties. Recently the Auxiliary Field Diffusion Monte Carlo (AFDMC) was introduced[1], which was employed to study the equation of state and other properties of neutron matter[2] and neutron droplets[3]. In this paper we present the application of AFDMC to the study of the binding energy of neutron–rich oxygen isotopes. The adopted scheme consists in replacing the closed shell ^{16}O-core with a single particle potential, generated by Hartree-Fock calculations using Skyrme forces[4]. In this way we can perform AFDMC calculations on the external neutrons. In this paper we first describe general features of the AFDMC algorithm, followed by results for the ground state energies of oxygen isotopes 18, 19, 20, 21 and 22, and the excited states of the isotope 18 and 21.

2. HAMILTONIAN

Properties of neutron drops and of oxygen isotopes are computed starting from a nonrelativistic Hamiltonian of the following form:

$$\hat{H} = -\sum_i \frac{\hbar}{2m}\nabla_i^2 + \sum_i V_{ext}(\vec{r}_i) + \sum_{i<j} v_{ij} + \sum_{i<j<k} V_{ijk}$$

The one-body potential V_{ext} depends on system studied. For the oxygen isotopes we employed the self consistent well generated in a Hartree-Fock calculation for ^{16}O using the Skyrme I effective interaction. The sum runs therefore only over the off-shell neutrons. The two–body NN interaction considered is in the class of the Urbana-Argonne v_l potentials :

$$v_l = \sum_{i<j}\sum_{p=1}^{l} v_p(r_{ij})O^{(p)}(i,j), \tag{1}$$

truncated to include only the following 8 operators (v_8'):

$$O^{p=1,8}(i,j) = (1, \vec{\sigma}_i \cdot \vec{\sigma}_j, S_{ij}, \vec{L}_{ij} \cdot \vec{S}_{ij}) \times (1, \vec{\tau}_i \cdot \vec{\tau}_j) \tag{2}$$

where the operator $S_{ij} = 3\vec{\sigma}_i \cdot \hat{r}_{ij}\vec{\sigma}_j \cdot \hat{r}_{ij} - \vec{\sigma}_i \cdot \vec{\sigma}_j$ is the tensor operator and $\vec{L}_{ij} = -\imath\hbar\vec{r}_{ij} \times (\vec{\nabla}_i - \vec{\nabla}_j)/2$ and $\vec{S}_{ij} = \hbar(\vec{\sigma}_i + \vec{\sigma}_j)/2$ are the relative angular momentum and the total spin for the pair ij. For neutrons $\vec{\tau}_i \cdot \vec{\tau}_j = 1$, and we are left with an isoscalar potential.

The v_8' potential is a simplified version of the v_{18} potential, having the same isoscalar parts of v_{18} in all S and P waves, as well as in the $^3D_1$ channel and its coupling to the $^3S_1$. It is semirealistic in the sense that it does not fit the Nijmegen N–N data at a confidence level of $\chi^2/N_{data} \sim 1$, as v_{18} does. However, the difference between v_{18} and v_8' is rather small for densities smaller or of the order of the nuclear matter equilibrium density $\rho_0 = 0.16 fm^{-3}$, and it can be safely treated perturbatively. The v_8' potential should be considered as a *realistic* homework potential, and it has been used in a number of calculations on light nuclei, symmetric nuclear matter, neutron matter and spin polarized neutron matter.

3. METHOD

The Auxiliary Field Diffusion Monte Carlo, is an extension of the standard Diffusion Monte Carlo method that can be employed for general spin– and isospin– dependent interactions. One of the major problems that prevents the use of standard DMC techniques in nuclear problems with an even moderately large number of nucleons consists in the fact that operators quadratic in spin and isospin that appear in the Green's function used for projecting the ground–state wavefunction would imply summing over all the possible spin and isospin states of the system. The terms in this sum grow exponentially with the number of nucleons, and so does the computer time required for simulations.

In the AFDMC algorithm the quadratic dependence on the spin and isospin operators is taken care of by sampling a set of auxiliary variables, which allow to linearize the Green's function by means of an Hubbard–Stratonovich transformations. A detailed discussion of the method can be found in Ref. 2 and 5.

As an example, the v_6' two–body potential can be separated into a spin–independent and a spin–dependent part:

$$V = V^{SI} + V^{SD} \; ,$$
$$V^{SD} = \sum_{i,j} \sigma_{i\alpha} A_{i\alpha;j\beta} \sigma_{j\beta} \; ,$$

where the elements of the matrix A are given by the proper combinations of the components v_p in Eq. (1). Latin indices, like i and j, are used to for particles, while the greek ones, like α and β, refer to the Cartesian components of the operators. We use the summation convention that all repeated greek indices are summed from 1 to 3.

Because $A_{i\alpha;i\beta} = 0$ the $3N$ by $3N$ matrix A has real eigenvalues and eigenvectors, defined by:

$$\sum_j A_{i\alpha;j\beta} \psi_n^{j\beta} = \lambda_n \psi_n^{i\alpha} \; . \tag{3}$$

The spin–dependent potential can therefore be written in terms of such eigenvalues and eigenvectors in the following form:

$$V^{SD} = \frac{1}{2}\sum_n \left[\sum_{i,j} \sigma_{i\alpha}\psi_n^{i\alpha}\lambda_n\psi_n^{j\beta}\sigma_{j\beta}\right] \ . \tag{4}$$

If one defines new N-body spinorial operators

$$O_n = \sum_i \sigma_{i\alpha}\psi_n^{i\alpha} \ , \tag{5}$$

the spin–dependent potential becomes

$$V^{SD} = \frac{1}{2}\sum_{n=1}^{3N} \lambda_n O_n^2 \ . \tag{6}$$

In the short–time limit we can decompose the imaginary time propagator of the diffusion process, which projects the ground state out of a trial wavefunction in the following way:

$$e^{-H\Delta\tau} \sim e^{-T\Delta\tau} e^{-V_c\Delta\tau} e^{-V^{SD}\Delta\tau} \ , \tag{7}$$

where $V_c = \sum V_{ext}(r_i) + V^{SI}$ is the spin independent part of the interaction. The propagation accounting for the kinetic and V_c operators gives rise to the usual drift–diffusion scheme of DMC. The spin–dependent two–body potential part $e^{-V^{SD}\Delta\tau}$ is handled by making use of the following Hubbard–Stratonovich transformation

$$e^{-\frac{1}{2}\lambda_n O_n^2\Delta\tau} = \frac{1}{\sqrt{2\pi}}\int_{-\infty}^{+\infty} dx_n \, e^{-\frac{x_n^2}{2} - \sqrt{-\lambda_n\Delta\tau}x_n O_n} \ , \tag{8}$$

with

$$e^{-V^{SD}\Delta\tau} \sim \prod_n e^{-\frac{1}{2}\lambda_n O_n^2\Delta\tau} \ , \tag{9}$$

where the commutators amongst the O_n are neglected, which requires to keep the time step $\Delta\tau$ small enough.

In Eq.(8) the quadratic dependence on the spin operators is transformed into a linear expression which corresponds to a rotation in the spin space. For each eigenvalue $\psi_n^{i\alpha}$ a value of x_n is sampled, and the current spinor value for each particle is multiplied by the set of matrices given by the transformation in Eq.(8).

The spin-orbit and three-body potentials can be treated within the same scheme, see details in Ref. 1, 2, 3 and 5.

The AFDMC algorithm is implemented as usual, with a propagation in imaginary time of a population of walkers $|R, S\rangle$ according to the propagator in eq. (7) with the standard drift-diffusion procedure. In addition one has to sample the x_n auxiliary variables given in Eq.(8) to rotate the spinors. After that all the weight factors are computed, they are combined to evaluate a new value of $\langle\Psi|R, S\rangle$.

The importance function is composed by a Jastrow term computed by FHNC/SOC at saturation density for the neutron matter, and by a sum of Slater determinant built from numerical single-particle wavefunctions obtained by the Hartree-Fock simulation of the oxygen core. Only states in the $1D_{5/2}$ shell were considered in this work. The determinants are combined in order to obtain eigenstates of J^2. The obtained configurations include at most three Slater determinants. No variational optimization has been performed for the importance function, which has been used also for projecting the eigenenergy.

In order to overcome the exponential growth of the variance due to the sign problem, the usual constrained-path approximation was employed, i.e. walkers reaching a point in the coordinates-spin space for which the determinantal part of the importance function changes sign are destroyed. This artificial boundary condition introduces a dependence of the results on the nodal surface of the wavefunction.

4. RESULTS

In table 1 and in figure 1 we report differences between the ground-state energies for the oxygen isotopes we had studied.

As it can be seen there is a quite good agreement with experimental values, except for the isotope ^{19}O. In this case the shell is filled by three neutrons, and therefore it is half-closed. With this configuration the effects of the constrained path approximation might be amplified. A large discrepancy might indeed indicate that the determinantal part of the wavefunction has to be deeply modified. Two possible sources of a bad nodal structure might be the need of including effects of deformation (meaning that eigenstates of J^2 do not describe correctly the ground state nodal surface), and enhanced effects of pairing, which are not explicitly included in the present importance function. The absolute values of the energies are instead quite far from the experimental results. In particular for $^{18}\mathrm{O}$

Table 1. Ground-state energy calculated with AFDMC. The differences between the isotopes studied are reported. All the energies are express in MeV.

	E_{AFDMC}	E_{exp}
$E_{^{19}O} - E_{^{18}O}$	-1.61(7)	-3.957
$E_{^{20}O} - E_{^{18}O}$	-11.32(8)	-11.564
$E_{^{21}O} - E_{^{18}O}$	-15.53(9)	-15.370
$E_{^{22}O} - E_{^{18}O}$	-22.44(7)	-22.219

the AFDMC ground state energy is -18.04(2) MeV, versus an experimental value of -12.188 MeV. The fact that energy differences are nevertheless very close to the experimental ones indicate that the correction due to the use of an external field replacing the ^{16}O amounts to a value which is nearly constant for all the isotopes series.

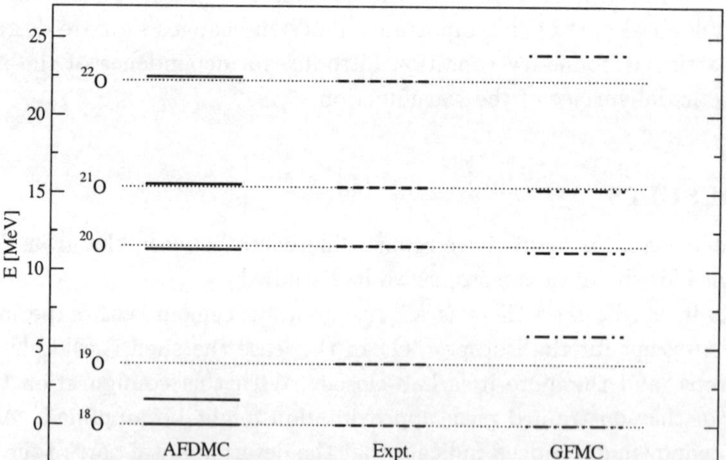

Figure 1. Outline of differences between energies of each isotope studied. All the energies are expressed in MeV. All values are referred to the energy of the isotope ^{18}O. The third column represent GFMC calculations of Ref. 6

We also performed calculations of the low-lying excited states of the ^{18}O and ^{21}O isotopes. In particular, we computed the energies for the 2^+ and 4^+ states in ^{18}O. The symmetry of the state is fixed a priori by the trial wave-function which is built to be a correct eigenstate of J.

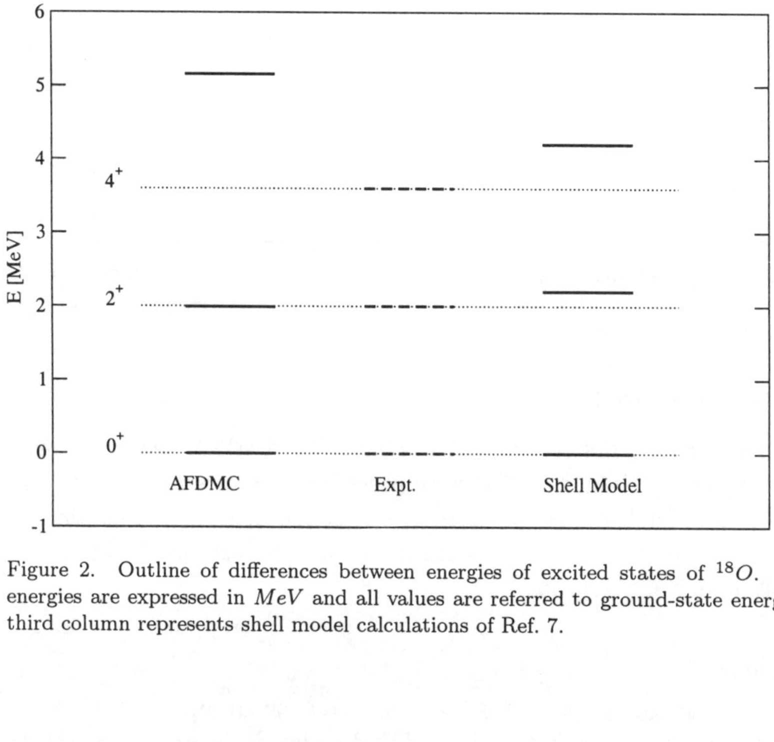

Figure 2. Outline of differences between energies of excited states of ^{18}O. All the energies are expressed in MeV and all values are referred to ground-state energy. The third column represents shell model calculations of Ref. 7.

The constrained path approximation and the fact that the importance functions for the ground– and excited–states are mutually orthogonal ensures the convergence of the energy to the eigenvalue of the excited state. We report our results in figure 2. As it can be seen, there is a very good agreement for the 2^+ state, but the difference between 4^+ and 0^+ seems to be too large. This is probably due again to the constrained path approximation. On the other hand, results obtained for the $1/2^+$ and $3/2^+$ states of ^{21}O are in very good agreement with experimental data, as shown in figure 3.

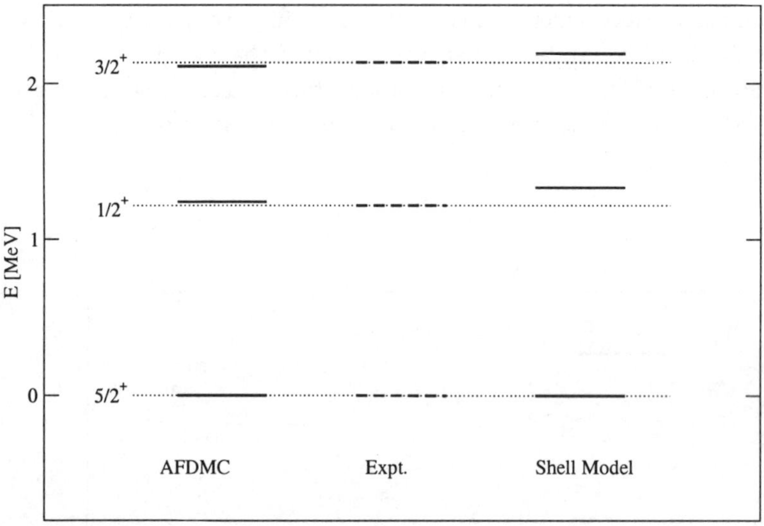

Figure 3. Outline of differences between energies of excited states of ^{21}O. All the energies are expressed in MeV and all values are referred to the ground-state energy. The third column represents shell model calculations of Ref. 8.

5. CONCLUSIONS

We reported on the first attempt of applying the AFDMC method to the study of neutron rich nuclei. The same procedure can be straightforwardly extended to other isotopes, possibly closer to the drip line. Inclusion of pairing effects and general improvements of the importance function are currently under investigation.

References

1. K. E. Schmidt and S. Fantoni, *Phys. Lett. B* **446**, 93 (1999)
2. A. Sarsa, S. Fantoni, K. E. Schmidt and F. Pederiva, *Phys. Rev. C* **68**, 024308 (2003)
3. F. Pederiva, A. Sarsa, K. E. Schmidt and S. Fantoni, *Nucl. Phys.* **A742**, 255 (2004).
4. D. Vautherin and D. M. Brink, *Phys. Rev. C* **5**, 626 (1972)
5. S. Fantoni, A. Sarsa and K. E. Schmidt, *Prog. Part. Nucl. Phys.* **44**, 63 (2000)
6. S. Y. Chang, J. Morales, Jr., V. R. Pandharipande, D. G. Ravenhall, J. Carlson, Steven C. Pieper, R. B. Wiringa and K. E. Schmidt, *nucl-th/0401016* (2004)
7. I. Morrison et al., *Phys. Rev. C* **17**, 1485 (1978)
8. M. Stanoiu et al., *Phys. Rev. C* **69**, 034312 (2004)

NUCLEAR DENSITY FUNCTIONAL CONSTRAINED BY LOW-ENERGY QCD*

PAOLO FINELLI, NORBERT KAISER, WOLFRAM WEISE
Physik Department, Technische Universität München, D-85747 Garching, Germany

DARIO VRETENAR
Physics Department, University of Zagreb, 10000 Zagreb, Croatia

We have developed a relativistic point-coupling model of nuclear many-body dynamics constrained by the low-energy sector of QCD. The effective Lagrangian is characterized by density-dependent coupling strengths determined by chiral one- and two-pion exchange (with single and double delta isobar excitations) and by large isoscalar background fields that arise through changes of the quark condensate and the quark density at finite baryon density. The model has been tested in the analysis of nuclear ground-state properties along different isotope chains of medium and heavy nuclei. The agreement with experimental data is comparable with purely phenomenological predictions. The built-in QCD constraints and the explicit treatment of pion exchange restrict the freedom in adjusting parameters and functional forms of density-dependent couplings. It is shown that chiral pionic fluctuations play an important role for nuclear binding and saturation mechanism, whereas background fields of about equal magnitude and opposite sign generate the effective spin-orbit potential in nuclei.

1. Introduction

In this work we would like to investigate the connection between QCD, its symmetry breaking patterns, and the nuclear many-body problem. Usual nuclear structure approaches [1] are consistent with the symmetries of QCD (in particular chiral symmetry) but only functional forms of the interaction terms can be determined. Model parameters cannot be constrained at the level of accuracy required for a quantitative analysis of structure data; they can only be estimated with the Naive Dimensional Analysis [2].

*Work supported in part by BMBF and GSI

The approach we propose is based on the following ingredients:

(1) The presence of large isoscalar background fields which have their origin in the in-medium changes of the scalar quark condensate and of the quark density.
(2) The nucleon-nucleon interaction is described by one- and two-pion exchange (with medium insertions and delta isobar excitations), in combination with Pauli blocking effects.

The first point has a clear connection with QCD sum rules at finite density [3], in which large nucleon self-energies naturally arise in the presence of a filled Fermi sea of nucleons.
The second point is motivated by the observation that, at the nuclear matter level, the nucleon Fermi momentum k_f, the pion mass m_π and the $\Delta - N$ mass difference represent comparable scales [4]. Pionic degrees of freedom are threfore included explicitly (through density-dependent couplings), in contrast to the phenomenological relativistic models, in which effects of iterated one-pion and two-pion exchange are treated implicitly through an *effective* scalar field [5].

2. The model

The model is defined by the Lagrangian density[a]

$$\mathcal{L} = \mathcal{L}_{free} + \mathcal{L}_{4f} + \mathcal{L}_{der} + \mathcal{L}_{coul} \tag{1}$$

with the four terms specified as follows:

$$\mathcal{L}_{free} = \bar{\psi}[i\gamma_\mu\partial^\mu - M_N]\psi \tag{2}$$

$$\mathcal{L}_{4f} = -\frac{1}{2}G_S(\hat{\rho})(\bar{\psi}\psi)(\bar{\psi}\psi) - \frac{1}{2}G_V(\hat{\rho})(\bar{\psi}\gamma_\mu\psi)(\bar{\psi}\gamma^\mu\psi) \tag{3}$$

$$-\frac{1}{2}G_{TS}(\hat{\rho})(\bar{\psi}\vec{\tau}\psi)\cdot(\bar{\psi}\vec{\tau}\psi) - \frac{1}{2}G_{TV}(\hat{\rho})(\bar{\psi}\vec{\tau}\gamma_\mu\psi)\cdot(\bar{\psi}\vec{\tau}\gamma^\mu\psi) \tag{4}$$

$$\mathcal{L}_{der} = -\frac{1}{2}D_S(\partial_\nu\bar{\psi}\psi)(\partial^\nu\bar{\psi}\psi) - \frac{1}{2}D_V(\partial_\nu\bar{\psi}\gamma_\mu\psi)(\partial^\nu\bar{\psi}\gamma^\mu\psi) \tag{5}$$

$$-\frac{1}{2}D_{TS}(\partial_\nu\bar{\psi}\vec{\tau}\psi)\cdot(\partial^\nu\bar{\psi}\vec{\tau}\psi) - \frac{1}{2}D_{TV}(\partial_\nu\bar{\psi}\vec{\tau}\gamma_\mu\psi)\cdot(\partial^\nu\bar{\psi}\vec{\tau}\gamma^\mu\psi) \tag{6}$$

$$\mathcal{L}_{coul} = -eA^\mu\bar{\psi}\frac{1+\tau_3}{2}\gamma_\mu\psi - \frac{1}{4}F_{\mu\nu}F^{\mu\nu} \,. \tag{7}$$

This Lagrangian is understood to be formally used in the mean-field approximation, with fluctuations encoded in density-dependent couplings $G_i(\hat{\rho})$.

[a] more details about density-dependent hadron field theory can be found in [6]

Their functional dependence will be determined from finite-density QCD sum rules and in-medium chiral perturbation theory

$$G_i(\hat{\rho}) = G^{(0)}(\hat{\rho}) + G^{(\pi)}(\hat{\rho}) \,, \tag{8}$$

where the index i labels all isospin-Lorentz structures of Eq. (1). The variation of the Lagrangian with respect to $\bar{\psi}$, leads to the single-nucleon Dirac equation

$$[\gamma_\mu(i\partial^\mu - V^\mu - V_R^\mu) - (M+S)]\psi = 0. \tag{9}$$

In addition to the usual self-energies V^μ and S, the density-dependence of the vertex functions produces the rearrangement contribution[b]

$$V_R^\mu = u^\mu \left(\frac{1}{2}\frac{\partial G_S}{\partial \hat{\rho}}\rho_s^2 + \frac{1}{2}\frac{\partial G_{TS}}{\partial \hat{\rho}}\vec{\rho}_s \cdot \vec{\rho}_s + \frac{1}{2}\frac{\partial G_V}{\partial \hat{\rho}} j^\nu j_\nu + \frac{1}{2}\frac{\partial G_{TV}}{\partial \hat{\rho}}\vec{j}^\nu \cdot \vec{j}_\nu \right) . \tag{10}$$

The inclusion of this additional term ensures energy-momentum conservation and thermodynamical consistency. For a complete treatment of the density dependent point-coupling model, the reader is referred to [7].

3. Interactions

3.1. *Self-energies from QCD sum rules*

In leading order, which should be valid below and around saturated nuclear matter, the condensate part of the scalar self-energy

$$\Sigma_S^{(0)} = -\frac{8\pi^2}{\Lambda_B^2}[\langle \bar{q}q\rangle_\rho - \langle \bar{q}q\rangle_0] = -\frac{8\pi^2}{\Lambda_B^2}\frac{\sigma_N}{m_u+m_d}\rho_s \,, \tag{11}$$

is expressed in terms of nucleon sigma-mass term ($\sigma_N = \langle \bar{N}|m_q\bar{q}q|N\rangle$) and the quark masses. At the same order, the time-component of the vector self-energy reads

$$\Sigma_V^{(0)} = \frac{64\pi^2}{3\Lambda_B^2}\langle q^\dagger q\rangle_\rho = \frac{32\pi^2}{\Lambda_B^2}\rho \,, \tag{12}$$

where the quark baryon density is related to that of the nucleons by $\langle q^\dagger q\rangle_\rho = \frac{3}{2}\rho$. In both cases, $\Lambda_B \simeq 1$ GeV is the characteristic scale (the Borel mass), which approximately separates the perturbative and non-perturbative energy domains.

[b] the four-velocity u^μ is defined as $(1-\mathbf{v}^2)^{-1/2}(1,\mathbf{v})$ ($\mathbf{v}=0$ in the rest-frame of the nuclear system)

For typical values of the nucleon sigma mass term σ_N ($\simeq$ 50 MeV [8]) and $m_u + m_d$ ($\simeq$ 12 MeV at a renormalization scale of 1 GeV [9]), the in-medium QCD sum rules predict large scalar and vector self energies of about equal magnitude ($\simeq 300 - 400$ MeV in agreement with relativistic phenomenological models [1]), and opposite in sign. Neglecting corrections from higher order condensates, these estimates have large uncertainties and the error in the ratio $\Sigma_S^{(0)}/\Sigma_V^{(0)} \simeq -1$ is about 20%.

Given the self-energies arising from the condensate background, the corresponding equivalent point-coupling strenghts $G_{S,V}^{(0)}$ are simply determined by

$$G_S^{(0)} = \frac{\Sigma_S^{(0)}}{\rho_s} \quad \text{and} \quad G_V^{(0)} = \frac{\Sigma_V^{(0)}}{\rho} . \tag{13}$$

At leading order the condensate terms of the couplings are constant and do not contribute to the rearrangement self-energy.

3.2. *Self-energies from in-medium chiral perturbation theory*

In recent years the nuclear matter problem has been extensively studied in the framework of in-medium chiral perturbation theory. The calculations have been performed to three-loop order in the energy density and include one-pion exchange Fock term, one-pion iterated exchange and irreducible two-pion exchange terms with medium insertions and delta isobar excitations [10]. By adjusting the coupling constants of few NN contact terms, and encoding short-range effects not resolved at relevant scales, to the properties of the empirical saturation point of isospin-symmetric nuclear matter ($\bar{E}_0 = -16$ MeV and $\rho_0 = 0.16$ fm^{-1}), several aspects of the problem have been successfully investigated: the symmetric and asymmetric nuclear matter equation of state (EOS), single-particle properties, the low-temperature behaviour, and the connection with non-relativistic nuclear energy density functionals [10,11].

The resulting nucleon self-energies are expressed as expansions in powers of the Fermi momentum k_f. The expansion coefficients are functions of k_f/m_π, the dimensionless ratio of the two relevant small scales:

$$\Sigma^{CHPT} = f(k_f/m_\pi)\frac{k_f^3}{M_N^2} \quad \text{where} \quad f(k_f/m_\pi) = \sum_k c_k \left(\frac{k_f}{m_\pi}\right)^k . \tag{14}$$

The density-dependence of the strength parameters is determined by equating the point-coupling self-energies in the single-nucleon Dirac equation (9) with those calculated using in-medium chiral perturbation theory[c]:

$$G_S^{(\pi)}\rho_s = \Sigma_S^{CHPT}(k_f) \tag{15}$$

$$G_V^{(\pi)}\rho + V_R^{(\pi)} = \Sigma_V^{CHPT}(k_f) \tag{16}$$

$$G_{TS}^{(\pi)}\rho_s^3 = \Sigma_{TS}^{CHPT}(k_f) \tag{17}$$

$$G_{TV}^{(\pi)}\rho^3 = \Sigma_{TV}^{CHPT}(k_f). \tag{18}$$

In Fig. (1) the resulting equation of state of isospin-symmetric nuclear matter is compared with the CHPT nuclear matter EOS [10]. The agreement is satisfactory, and the small difference can be attributed to the approximations involved (i.e. the momentum dependence is neglected).

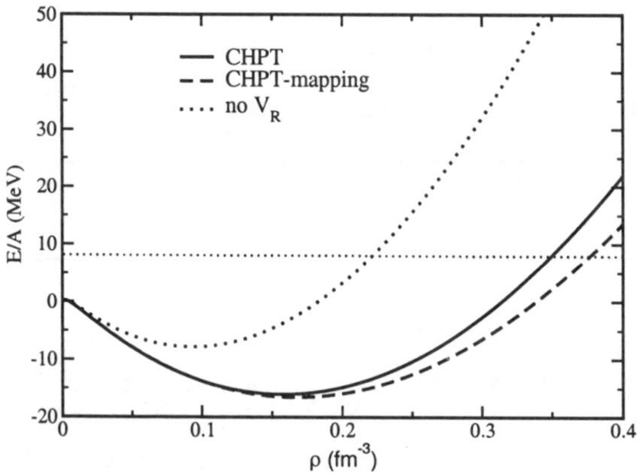

Figure 1. Binding energies for symmetric nuclear matter as a function of baryon density. The solid curve (CHPT) is the EOS calculated by using in-medium CHPT. The EOS displayed by the dashed curve (CHPT-mapping) is obtained when the CHPT nucleon self-energies are mapped on the self-energies of the relativistic point-coupling model with density-dependent couplings. The dotted curve denotes the latter EOS when rearrangement terms are neglected.

4. Results

In this section the relativistic point-coupling model will be applied to the description of ground-state properties of finite nuclei. The density-dependent coupling strenghts will be determined in a *least-squares fit* to observables

[c]for the definition of ρ, ρ_s, ρ_3, ρ_{s3} the reader is referred to [7]

($\mathcal{O}$) of a rather large set of nuclei along the valley of β-stability (from ^{16}O to ^{214}Pb):

$$\chi^2 = \sum_{nuclei} \left(\sum_{observ.} \left| \frac{\langle \mathcal{O}_{th} - \mathcal{O}_{exp} \rangle}{weights} \right|^2 \right). \tag{19}$$

In Fig. 2 the overall agreement is shown in comparison with two standard relativistic meson-exchange parametrization: NL3 [12], with explicit higher order self-interaction terms for the σ-meson, and DD-ME1 [13], with phenomenological density-dependent couplings strenghts. The interaction

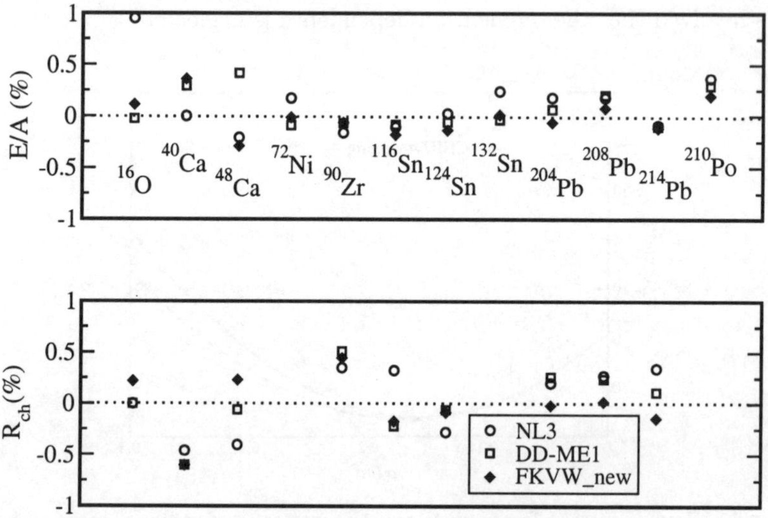

Figure 2. Relative errors for the observables binding energies (upper panel), and charge radii (lower panel), in comparison with the results of NL3 and DD-ME1.

determined by the *least-squares fit* procedure contains pionic fluctuations on top of the large background fields. Nonetheless, it is interesting to analyze the role of pionic fluctuations, without the presence of background fields, and how these fields modify the single-particle spectra in finite systems. In Fig. 3 we display the single proton levels in ^{40}Ca without and with background fields. It is obvious that the the spin-orbit potential

$$V_{s.o} \simeq \frac{1}{2M^2} \left(\frac{1}{r} \frac{\partial}{\partial r} (V^0(r) - S(r)) \right) \mathbf{l} \cdot \mathbf{s} \,, \tag{20}$$

represents a short-range effect (see also [10]), determined by the condensate structure of QCD. The isospin-dependent part of the interaction has been

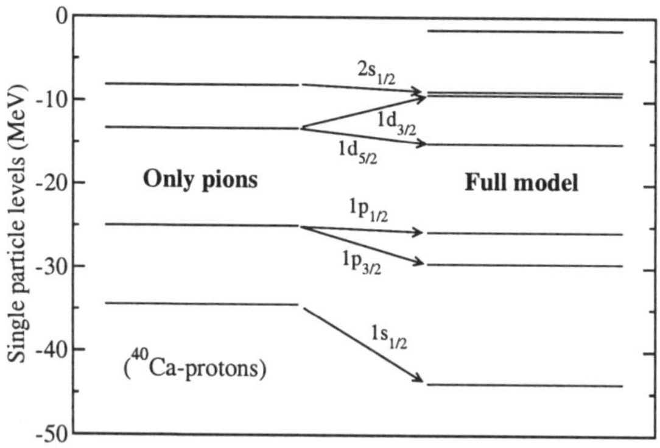

Figure 3. Single proton levels for ^{40}Ca with only pionic fluctuations (left panel) and with the inclusion of background fields (right panel). In the latter case the degeneracy of the p- and d-levels is removed.

studied in the description of the neutron radii. Standard relativistic mean-field calculations sistematically overestimate the difference between neutron and proton radii [1]. It turns out that the density dependence of the isovector channel of the interaction is crucial in order to reproduce these observables, as also shown in [13]. In Fig. 4 we show the calculated values of $r_n - r_p$ for the

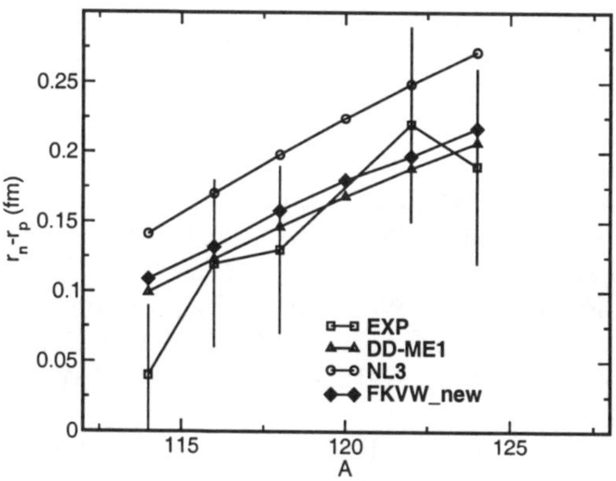

Figure 4. Predictions (diamonds) for the differences between neutron and proton radii of Sn isotopes, in comparison with NL3 (empty circles), DD-ME1 (empty triangles) predictions and experimental data (empty squares) .

Sn isotope chain in comparison with experimental data [14]. The agreement is excellent, and it is a clear indication that the isovector channel can be successfully described by pionic fluctuations.

5. Conclusions

It has been demonstrated that an approach to nuclear dynamics constrained by the low-energy sector of QCD provides a quantitative description of properties of nuclear matter and finite nuclei.

References

1. M. Bender, P.-H. Heenen, and P.-G. Reinhard, Rev. Mod. Phys. **75** (2003) 121, P. Ring, Prog. Part. Nucl. Phys. **37** (1996) 193, B. D. Serot and J. D. Walecka, Int. J. Mod. Phys. E **6** (1997) 515.
2. J. J. Rusnak and R. J. Furnstahl, Nucl. Phys. A **627** (1997) 495 .
3. T. D. Cohen, R. J. Furnstahl and D. K. Griegel, Phys. Rev. Lett. **67** (1991) 961, R. J. Furnstahl, D. K. Griegel and T. D. Cohen, Phys. Rev. C **46** (1992) 1507, T. D. Cohen, R. J. Furnstahl, D. K. Griegel and X. m. Jin, Prog. Part. Nucl. Phys. **35** (1995) 221.
4. W. Weise, Chiral Dynamics and the Hadronic Phase of QCD, Proc. of the International School of Physics Enrico Fermi From Nuclei and their Constituents to Stars, Varenna (2002), A. Molinari et al., eds., IOS Press, Amsterdam (2003), p. 473-529.
5. R. J. Furnstahl and B. D. Serot, Comments Nucl. Part. Phys. **2** (2000) A23 .
6. C. Fuchs, H. Lenske and H. H. Wolter, Phys. Rev. C **52** (1995) 3043, S. Typel and H. H. Wolter, Nucl. Phys. A **656** (1999) 331, T. Niksic, D. Vretenar, P. Finelli and P. Ring, Phys. Rev. C **66**, 024306 (2002).
7. P. Finelli, N. Kaiser, D. Vretenar and W. Weise, Eur. Phys. J. A **17** (2003) 573, P. Finelli, N. Kaiser, D. Vretenar and W. Weise, Nucl. Phys. A **735** (2004) 449.
8. J. Gasser, H. Leutwyler and M. E. Sainio, Phys. Lett. B **253** (1991) 252.
9. A. Pich and J. Prades, Nucl. Phys. Proc. Suppl. **86** (2000) 236, B. L. Ioffe, Phys. Atom. Nucl. **66** (2003) 30 [Yad. Fiz. **66** (2003) 32].
10. N. Kaiser, S. Fritsch and W. Weise, Nucl. Phys. A **697** (2002) 255, S. Fritsch, N. Kaiser and W. Weise, arXiv:nucl-th/0406038.
11. N. Kaiser, S. Fritsch and W. Weise, Nucl. Phys. A **700** (2002) 343, S. Fritsch, N. Kaiser and W. Weise, Phys. Lett. B **545** (2002) 73, S. Fritsch and N. Kaiser, Eur. Phys. J. A **17** (2003) 11, N. Kaiser, S. Fritsch and W. Weise, Nucl. Phys. A **724** (2003) 47.
12. G. A. Lalazissis, J. Konig and P. Ring, Phys. Rev. C **55** (1997) 540 .
13. T. Niksic, D. Vretenar, P. Finelli and P. Ring, Phys. Rev. C **66**, 024306 (2002).
14. A. Krasznahorkay *et al.*, Phys. Rev. Lett. **82**, 3216 (1999).

QUARK GLUON PLASMA AND RELATIVISTIC HEAVY ION COLLISIONS

F. BECATTINI
Università di Firenze and INFN Sezione di Firenze
Via G. Sansone 1, I-50019, Sesto F.no (Firenze)
E-mail: becattini@fi.infn.it

In this paper I summarize major recent achievements in the search of Quark Gluon Plasma and I review theoretical research activity carried out in Italy. The main focus will be on phenomenology of relativistic heavy ion collisions.

1. Introduction

Quark Gluon Plasma (QGP) is a new state of matter whose existence is predicted, by the generally accepted theory of strong interactions, Quantum-Chromo-Dynamics (QCD), to occur at sufficiently high temperatures and/or baryon densities. In this state of matter quarks and gluons are deconfined, i.e. they can freely move over distances much larger than a typical hadron size. This peculiar QCD prediction follows from a property of the theory known as asymptotic freedom: the interaction of the fundamental fields become weaker, and the mean free paths of particles become longer, as the energy density increases. Therefore, for a strongly interacting system, there should be a critical temperature (at vanishing chemical potentials) separating the phase where quarks and gluons are deconfined from a hadron gas phase. A major theoretical issue is whether this critical temperature defines a first-order, second-order phase transition, or rather a smooth crossover. Recent lattice calculations [1] point to a crossover transition at low baryon densities up to a temperature of $\simeq 170$ MeV at vanishing baryon density, whilst there are indications of a first-order transition at higher baryon densities and lower temperatures (see fig. 7 later). What seems by now established is that, at least for $\mu_B = 0$, the crossover for deconfinement and chiral symmetry restoration occur at the same temperature (see fig. 1).

The quest for Quark Gluon Plasma started more than 20 years ago.

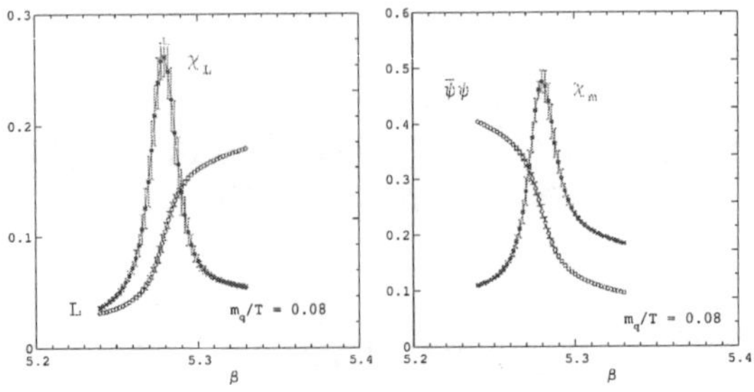

Figure 1. Left: Polyakov loop expectation value and temperature derivative (χ_L) as a function of the lattice coupling $\beta = 6/g^2$ which is monotonically related to the temperature T. Right: the chiral condensate $\langle\bar{\psi}\psi\rangle$ and the negative of its derivative (χ_m) [2].

It was soon proposed that QGP could have been produced in high energy collisions of heavy nuclei. The main reason of colliding large nuclei is that, in order to obtain a real deconfinement, the region of space at high energy density is achieved should be much larger than a typical hadron size. Therefore, for a fixed nucleon beam energy, it seems more likely to produce a deconfined phase of quarks and gluons in collisions of heavy nuclei (several fm's of diameter) rather than, e.g., in proton-proton collisions. A rich experimental programme of heavy ion collisions at several centre-of-mass energies was then initiated; from few GeV (Au-Au collisions at BNL-AGS) to an intermediate range of 6-17 GeV (Pb-Pb collisions at CERN-SPS) up $O(100)$ GeV at BNL-RHIC, started in 2000. The next big step will be at CERN in 2007, where collisions of Pb nuclei will be carried out in LHC at $\sqrt{s_{NN}} \approx 5$ TeV. It is now widely believed that the threshold for QGP production has been overcome at RHIC and probably at SPS (see fig. 2). The main issue of the whole experimental programme is how to detect the QGP, i.e. to find a signature of its formation in the final observable particles produced by the collision. Indeed, most of the effort of theorists in this field has been spent into the quest of the most powerful signatures, besides the description of the collision itself.

Because of the complexity of the process of collision of two nuclei at high energy and the lack (or impossibility) of QCD first-principle calculations at the energy scale of interest, most phenomenology in this field has to resort to various models. These models apply to different stages of the process, but some of them have significant overlap and are therefore in competition to better reproduce the experimental data. As yet, phenomenological models,

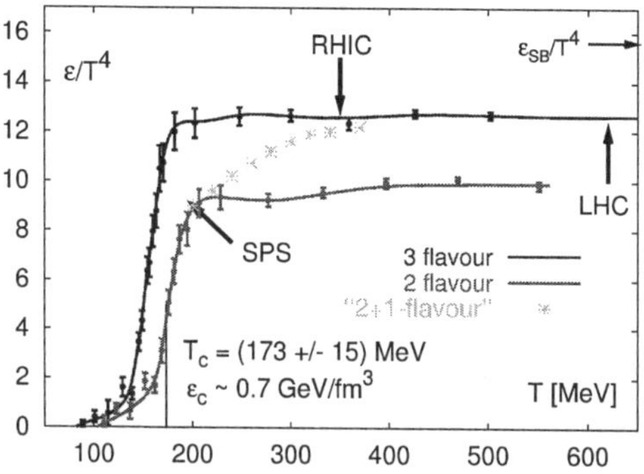

Figure 2. Energy density in units of T^4 for QCD with two and three dynamical quark flavours [3]. The curves labelled "2 flavour" and "3 flavour" were calculated for two and three light quark flavours with mass $m_q = 0.4T$. The "2+1 flavour" indicates the calculated curve for two light quark with $m_q = 0.4T$ and one with $m_q = T$. The critical temperature turns out to be 173±15 MeV. Also shown the energy density achieved in various heavy ion collisions calculated with a hydrodynamical model.

either based on QCD or more general physical schemes (like statistical, transport and hydrodynamical model) are indispensable tools to interpret the data and probe the possible formation of QGP.

2. Highlights from experiments

The experiments of heavy ion collisions have produced over a decade a tremendous amount of data which improved dramatically the understanding of the collision process. Some of the observations are directly related to proposed QGP signatures and those are certainly the most intriguing for experimentalists, yet many other measurements and observations have a special interest independently of their relevance to QGP discovery. Here I summarize the most outstanding according to general wisdom and author's taste.

2.1. *Jet quenching*

It was proposed many years ago [4] that early produced high p_T partons giving rise to jet formation in pp collisions should behave differently in heavy ion collisions owing to the surrounding medium. The idea is that these hard-scattering partons undergo peculiar energy loss in the Quark

Gluon Plasma which is otherwise absent. If sufficiently large, this energy loss finally prevents hard partons from being detected as high p_T jets at all. Therefore, the high p_T jet rate in heavy ion collisions is expected to be smaller than in pp collisions at the same energy (a phenomenon called *jet quenching*). The main source of energy loss has was identified later [5] as bremsstrahlung in the deconfined medium and several expressions of the relevant dE/dx calculated [6]. These studies triggered much interest in the field of heavy ion collisions and RHIC experiments have been designed to detect hadronic jets in order to reveal this interesting feature.

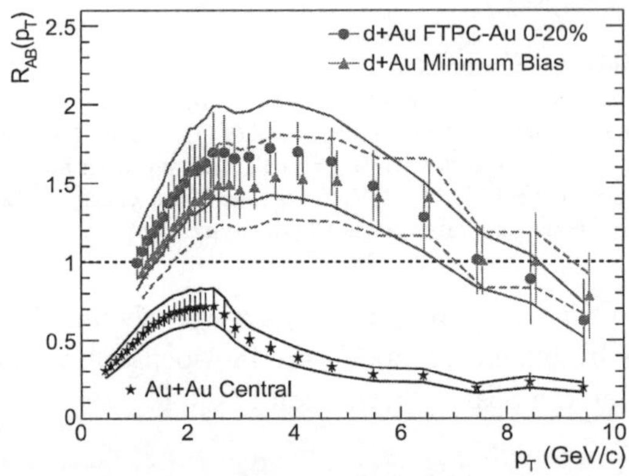

Figure 3. The nuclear modification factor R_{AA} for charged hadrons measured in D-Au and central Au-Au collisions at $\sqrt{s}_{NN} = 200$ GeV [7].

RHIC experiments have shown that the high p_T particle yield is indeed suppressed with respect to an extrapolation from pp collisions based on the Glauber model. In fact, the so-called *nuclear modification factor*:

$$R_{AA}(p_T) = \frac{\dfrac{dN_{AA}(p_T)}{dp_T}}{N_c \dfrac{dN_{pp}(p_T)}{dp_T}} \tag{1}$$

turns out to be less than 1 over the high p_T range for many particle species (see fig. 3) in central Au-Au collisions at RHIC at $\sqrt{s}_{NN} = 130$ and 200 GeV unlike in peripheral collisions; the number of collisions N_c is estimated by using Glauber's model. This result has a very simple geometrical interpretation: in central collisions the hot central region where QGP is formed is larger, so most hard-scattered partons have to traverse a relatively large region before emerging from the medium and manifest themselves as hadron jets. On the other hand, in peripheral collisions, the region occupied by the deconfined medium is smaller and energy loss mechanism is less effective. Altogether, one expects less jet quenching in peripheral than central collisions and this is just in agreement with the data.

This inclusive measurement has been confirmed by a more specific one, where the suppression of back-to-back jets have been observed by triggering high p_T jets. It can be seen from fig. 4 that the high momentum particle flow in the direction opposite to that of the triggered jet ($\phi = 0$) does not show any bump in central Au-Au collisions, unlike in pp collisions and D-Au collisions. This is a rather striking indication that the back-scattered parton associated to an identified hard parton could not emerge as a high energy jet, as it normally does in pp collisions, thus it must have lost its energy at some point. Jet emission in Au-Au seems to occur only from the surface of the hot fireball.

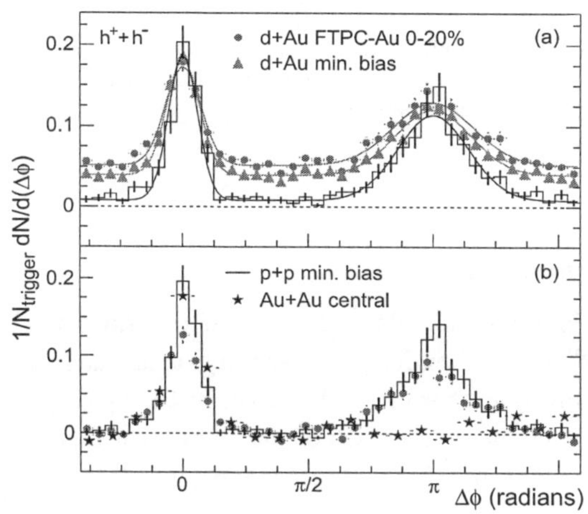

Figure 4. Azimuthal angular correlations in min. bias pp, D+Au and central Au-Au collisions relevant to an identified high p_T jet at $\phi = 0$ [7].

It has been argued, on the basis of the Colour-Glass-Condensate theory (see later), that these effects could stem from the initial state, namely the suppression of low x nuclear gluon structure functions rather than being related to final-state QGP formation. However, this possibility has been ruled out at RHIC by means of D-Au collisions: if a modification of gold nuclear structure function was the actual responsible of jet quenching, this effect should be observed also in D-Au, but this was not the case: jet quenching seems to be a final state effect. Even though many specific features and details are still to be understood, if confirmed, this is likely to be a good piece of evidence in favour of Quark Gluon Plasma.

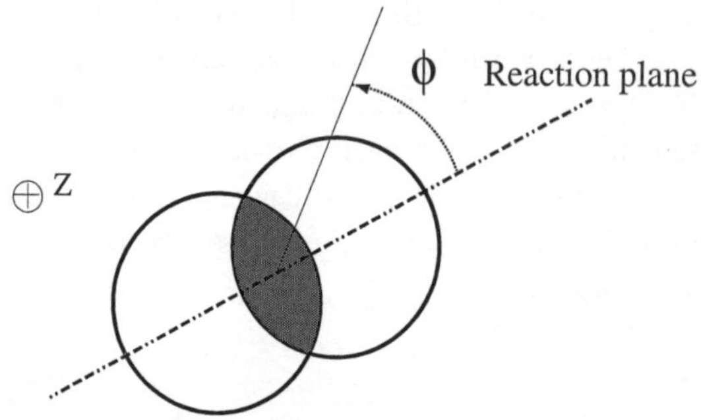

Figure 5. Geometry of the peripheral heavy ion collisions: the direction of the initial momenta of colliding nuclei is orthogonal to the page. The reaction plane is defined by the beam axis and the impact parameter vector (the distance vector between centres of the nuclei).

2.2. *Hydrodynamical flow*

One of the most popular models in heavy ion collisions is the hydrodynamical model. The basic underlying assumption is that matter reaches local thermal equilibrium at some early stage, in the QGP phase. For this pciture to make sense, mean free path of particles must be consistently less than the size of the system. Hydrodynamical calculations try to reproduce measured transverse momentum spectra at low p_T and require as input the equation of state and initial conditions. The actually used equation of states are phenomenological ones mimicking lattice QCD results at high temperature and hadron-resonance gas at temperatures lower than critical one. Mostly adopted initial conditions include a density profile calculated from the geometrical overlap of the two incoming nuclei and longitudinal flow profile according to Bjorken's scaling hypothesis with vanishing transverse flow. The free parameters, to be fitted to the data, are the initial energy and baryon density, the initial equilibration time and the freeze-out temperature, when particle cease to scatter elastically. Most calculations (see table 1) indicate an early equilibration time, less than 1 fm/c both at SPS and RHIC energies, with an initial temperature well above the critical value of ≈ 170 MeV (see arrows in fig. 2).

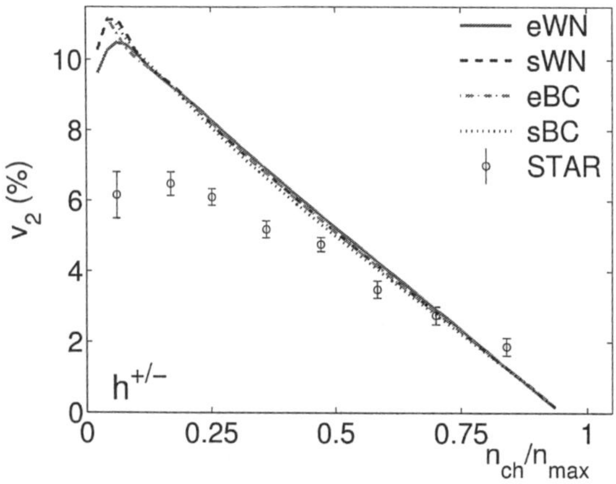

Figure 6. p_T-integrated elliptic flow parameter v_2 as a function of centrality for Au-Au collisions at $\sqrt{s}_{NN} = 130$ GeV. The curves are hydrodynamical calculation for different parametrizations of the initial transverse density profile [8].

Of special interest is the so-called *elliptic flow*, which shows as an anisotropy of differential azimuthal particle momentum spectrum in the reaction plane (see fig. 5) in peripheral collisions. Because of the non-zero impact parameter in peripheral collisions, the overlapping region of the two incoming nuclei has an initial almond shape, unlike in central collisions where it is spherical. If local thermal equilibrium sets in early, pressure gradient are significantly higher in the reaction plane than in the orthogonal direction and this drives an enhancement of the collective flow along the reaction plane, resulting in an enhancement of particle momentum. The magnitude of the elliptic flow is usually gauged with the second coefficient of the Fourier expansion of the $dN/dp_T d\varphi$ spectrum as a function of φ for fixed p_T, the so-called $v_2(p_T)$. This parameter shows a very interesting behaviour in the data (see fig. 6), namely it attains the highest possible value according to ideal fluid hydrodynamical calculations for nearly central collisions and stays below the model value for very peripheral collisions. Necessary conditions for reaching high values of v_2 are an early equilibration time of produced matter and quasi-ideal fluid behaviour (this has been shown in detail in dedicated studies). The fact that matter behaves like an ideal fluid with almost no viscosity implies that even at the highest initial temperatures, in the QGP phase there are strong non-perturbative interactions which make mean free paths very small. Therefore, the naive expectation of a gas of weakly interacting particles does not apply and one should rather think of a strongly interacting fluid (s-QGP).

The indication from hydrodynamics is altogether quite fascinating: if its interpretation is correct, a state of thermally equilibrated matter is produced in the early stage of the collision at a temperature above the critical one. This might be properly called a Quark Gluon Plasma.

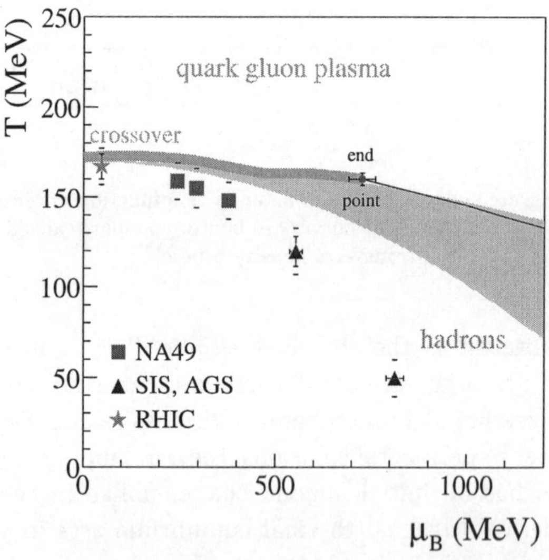

Figure 7. QCD phase diagram with chemical freeze-out points extracted with statistical model analysis [9] of hadron abundances over a centre-of-mass energy range 2-130 GeV [10]. The curve and shaded band are the estimates of deconfinement phase transition from lattice QCD [1] Also shown the estimated critical end-point.

2.3. *Particle yields and strangeness enhancement*

The abundances of particle emitted from the source are in agreement with those expected from a thermal source at a decoupling temperature (chemical freeze-out) of $\simeq$ 160 MeV at SPS energies and above. In fig. 7 the different chemical freeze-out temperatures and baryon-chemical potentials from AGS to RHIC are shown. Unfortunately, this is not a compelling case for Quark Gluon Plasma because the same behaviour is observed in elementary collisions [11] as well: particle yields are in agreement with those predicted by a statistical model in many collisions. Whether this is an indication of a genuine statistical behaviour or simply a mimicking of a thermodynamical behaviour due to peculiar hadronization dynamics (*phase space dominance*) is still under investigation [12]. What seems to be a peculiarity of heavy ion collisions with respect to elementary collisions as far as bulk hadron production is concerned is the enhanced strangeness yield, which was predicted long ago as a possible signature of QGP formation [13].

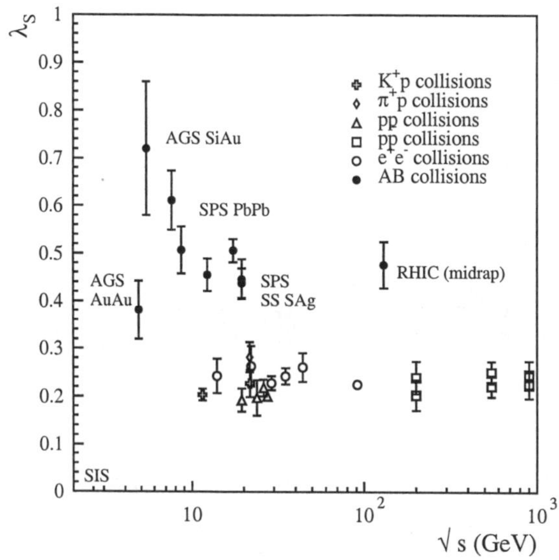

Figure 8. Wroblewski parameter $\lambda_S = 2\langle s\bar{s}\rangle / (\langle d\bar{d}\rangle + \langle u\bar{u}\rangle)$, i.e. the ratio between newly produced strange and light quark-antiquark pair in high energy elementary and heavy ion collisions [14]. Primary valence quarks yields are estimated by using the statistical model fits.

This is better seen in the Wroblewski parameter λ_S, namely the ratio between newly produced $\bar{\mathrm{s}}$ and $\bar{\mathrm{d}}, \bar{\mathrm{u}}$ valence quarks in the collision. This ratio is approximately constant in elementary collisions over two orders of magnitude of centre-of-mass energy, but it is as twice as high in heavy ion collisions (see fig. 8). It is very difficult to explain this enhancement in terms of "conventional" hadronic physics (i.e. post-hadronization rescattering) [8], although the increasing size of the system might explain it partly within a statistical model scenario.

An intriguing signal in this regard has been observed by NA49 experiment, which measured the K^+/π^+ ratio over the range of energy 6-17 GeV and found an anomalous peak around $\simeq$ 7 GeV (see fig. 9). This spike cannot be explained with smooth interpolations of the statistical model [15]. There is a specific model [16] relating this peak to the onset of deconfinement transition and, even more, with the proximity of the chemical freeze-out point in the $T - \mu_B$ plane (see fig. 7) to the end-point of the QCD first order transition line, which could give rise to enhanced fluctuations of the $\bar{\mathrm{s}}/\bar{\mathrm{d}}$ ratio [17]. All this matter is still under investigation.

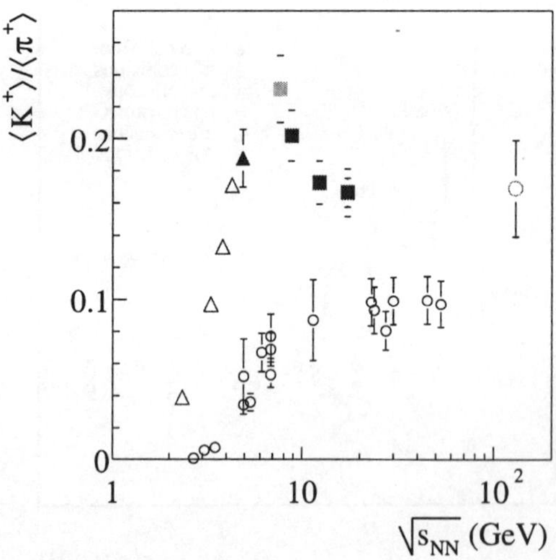

Figure 9. K^+/π^+ ratio (after strong decays) as a function of $\sqrt{s}_{NN}$ measured in central Au-Au and Pb-Pb heavy ion collisions [18].

3. Recent activity in Italy

The interest of theorists in heavy ion physics in Italy is rapidly growing. Presently, there are several groups who have been working in heavy ion collisions and related topics over the past few years. Besides those who participate in dedicated projects, there are other theorists whose work can be linked to this field. Since I cannot give a comprehensive summary of all the research work carried out during past two years, I will confine myself to summarize scientific work pertaining to phenomenology.

3.1. *Saturation model*

At low x and sufficiently large Q^2, the gluon structure functions $G(x, Q^2)$, i.e. the number of gluons, become large. Yet, the growth of G cannot continue indefinitely at low x because of unitarity constraint. If gluon density is very high, the gluonic field may be treated as a classical one whose source are valence quarks. In this regime the field is intense, but the coupling constant α_S is small. The perturbative evolution equation are to be modified, become non-linear and this entails a saturation of the structure functions at scales $\Lambda^2_{QCD} < Q^2 < Q^2_s(x)$, where $Q_s(x)$ is defined as *saturation scale.* This particular regime is called "Colour Glass Condensate" (GCG). The scale $Q^2_s(x)$ at which saturation sets in depends on the colliding particles, and it is higher for large nuclei (with a linear dependence on the number of nucleons participating in the collision). The unintegrated gluon distribution $\varphi(x, k_T)$ shows a typical pQCD-behaviour $1/k^2_T$ at $k_T > Q_s(x)$ and $\approx \log[Q^2_s(x)/k^2_T]$ otherwise.

One of the typical predictions of the saturation model was concerned with pseudorapidity distributions in Au-Au collisions at RHIC energies ($\sqrt{s} = 130$ and 200 GeV, see fig. 10). Here, one assumes strict local parton-hadron duality (i.e.the proportionality between kinematical distribution before and after hadronization) and that particle production is dominated by gluon-gluon scattering. For more details, see for instance refs. [19]. The model is quite succesful in reproducing both the dependence on centre-of-mass energy and on centrality [19,20]. Despite the agreement with the data, it is still unclear why the multiple physical processes occurring between initial partonic scattering and final particle production should not affect the rapidity spectra. In fact, there is now enough evidence for hydrodynamical flow in measured spectra at low p_T and CGC has indeed been used as initial condition for further hydrodynamical evolution, showing a remarkable difference between input and output distribution [21]. This makes difficult

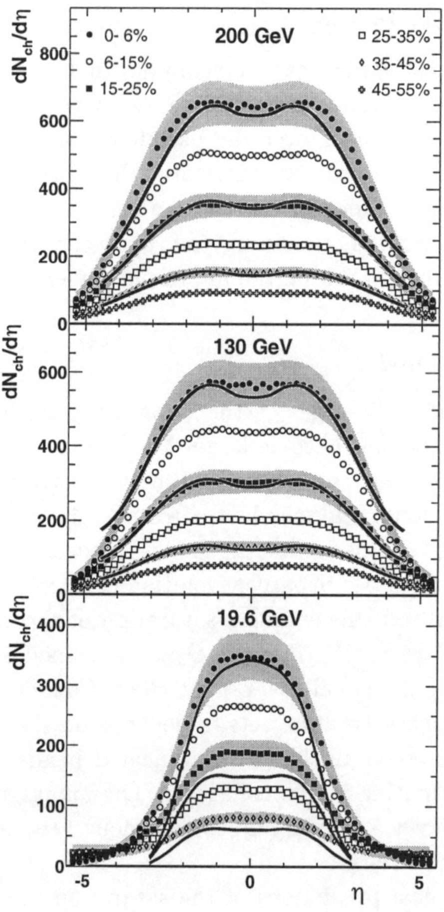

Figure 10. Pseudorapidity distributions of charged particle in Au-Au collisions at different centre-of-mass energies compared with the saturation model calculations (from [20]).

to understand the striking agreement between calculated distributions at perturbative partonic level and at the level of final charged particles, after lots of physical processes and a long decay chain.

3.2. *Cronin effect*

The study of the Cronin effect in pA collisions is a very interesting one because of its relation with jet quenching studies in Au-Au and D-Au collisions. Jet quenching in Au-Au (see previous section), as defined by Eq. (1), is not seen in D-Au collisions (see fig. 3). It is however necessary to fully understand all the features of this spectrum to establish that some anomalous phenomenon occurs in Au-Au. It can be seen already from fig. 3 that in D-Au production of particles at intermediate p_T (between 2 and 8 GeV) is higher than expected from simple geometrical scaling (i.e. a superposition of NN collisions) as calculated in the Glauber model. This broadening - mainly observed in proton-nucleus collisions - is known as *Cronin effect* and it is possibly the result of several mechanisms (modification of intrinsic k_T gluon structure functions, multiple partonic collisions, shadowing etc.). The involved p_T are large enough to allow the use of perturbative QCD and many of these effects can be studied quantitatively. Recent studies [22] show that almost all features of measured spectra in D-Au at RHIC can be accomodated within a "traditional" perturbative QCD picture without invoking saturation and CGC.

17

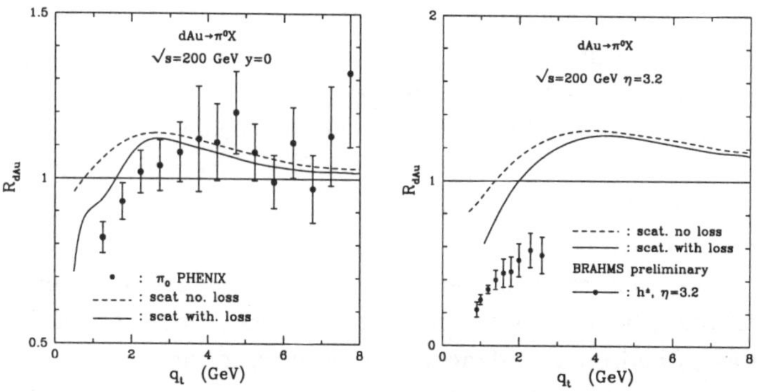

Figure 11. Calculated nuclear modification factor in D-Au→ π^0 X compared with the data at $\sqrt{s}_{NN} = 200$ GeV. The dashed line refers to exact kinematics and solid line to approximate kinematics. Left: $\eta = 0$. Right: $\eta = 3.2$ [23].

The italian collaboration led by D. Treleani has studied the effect of exact kinematical constraints (overall conservation of energy and momentum) in the perturbative calculation of partonic p_T spectra in proton-nucleus collisions. The model is based on a Glauber-eikonal approach and has built-in geometrical shadowing and multiple collisional broadening. For more details on the formalism and methods, see ref. [23]. The authors have been able to point out that the kinematical constraint may have some considerable effect on the spectrum broadening especially at LHC in p-Pb collisions at $\sqrt{s} \simeq 8$ TeV. They have also compared the measured π^0 transverse momentum spectrum at midrapidity in D-Au collisions at RHIC, finding good agreement with the data (see fig. 11). The agreement is much worse at large rapidity, yet the covered p_T range of the measurements hardly attains 3 GeV, a region where perturbative calculations are not reliable. On the other hand, other calculations [24] find better consistence.

3.3. *Statistical model*

The statistical model has been long used to determine the bulk properties of the source at the final stage of its evolution, i.e. when hadrons stop interactions. The measured hadron abundances turn out to be in very good agreement with those expected from a thermal source at varying temperatures and baryon chemical potentials from AGS energies ($\sqrt{s}_{NN} \simeq 5$ GeV) to RHIC energies. As an example of the quality of this agreement, see fig 12. A long-standing issue is whether this is a result of an actual equilibration process at hadronic level through inelastic collisions or the outcome of the hadronization process itself, since hadron multiplicities are in very good agreement with statistical model predictions in elementary collisions as well [11]. There is an ongoing debate on how this comes about and probably more detailed studies are needed to verify whether this statistical-like behaviour is genuine or just mimicked (phase space dominance) by the data.

In a recent study by the author [9], different proposed versions of the statistical model picture have been tested against mainly NA49 measured multiplicities, in Pb-Pb collisions at centre-of-mass energies between 7 and 17 GeV. Fits of hadron abundances have been performed with or without extra strangeness suppression parameter γ_S, with or without extra light-quark suppression parameter γ_q and with strangeness correlation volume fixing $\gamma_S = 1$. Furthermore, a two component model has been studied where the hadron production is assumed to stem from the superposition of particles emerging from single NN collisions and from a central fully equilibrated ($\gamma_S = 1$) fireball. These studies indicate that both the model with γ_S and the two component model are consistent with the data, whereas the strangeness correlation volume picture is disfavoured. The data sample is not rich and accurate enough to prove the need of a further extra parameter γ_q. Finally, the use of midrapidity yields at SPS [25] has been proved to artificially enhance the yield of strange particle yields. As far as

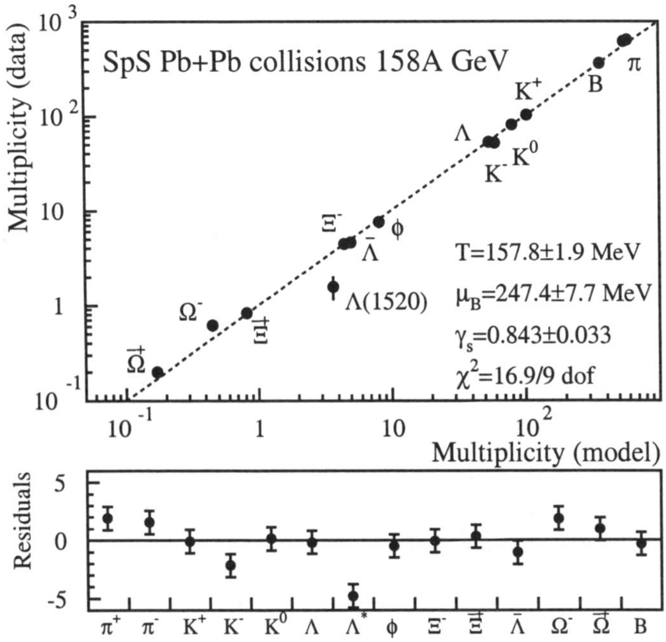

Figure 12. Top panel: measured versus fitted multiplicities in the statistical model for Pb-Pb collisions at $\sqrt{s}_{NN}$ = 17.2 GeV. Also quoted the best-fit parameters. Bottom panel: residual distribution [9].

strangeness production is concerned, previous finding have been confirmed, with an additional anomalous peak at $\sqrt{s} = 7.4$ GeV which is the reflection of a peak in the K^+/π^+ ratio (see fig. 9).

Many more studies on the statistical model (microcanonical calculations, fluctuations, tests at low energy) are in progress, which should lead to a big development of its scope and a clearer understanding of its meaning.

3.4. *J/ψ suppression*

Maiani *et al.* [26] have studied the J/ψ suppression pattern found by the experiment NA50 at SPS at $\sqrt{s}_{NN} = 17.2$ GeV. The goal of this work was to probe the existence of an anomalous suppression of the J/ψ beside that calculated in a purely hadronic scenario, where the dissociation reaction

$$M + J/\psi \to D + \bar{D} \tag{2}$$

M being a meson, is the dominant mechanism. The absorption by the nuclear medium, i.e through reactions with nucleons is also taken into account by using the measurements by NA50 in pA collisions. The general scheme is of "traditional" type: all J/ψ's are produced at the very beginning of the collision and they have thereafter to traverse a medium, which can be either hadronic or partonic (QGP). In the latter case, one expects melting of the J/ψ in the plasma [27]. In the former case, J/ψ absorption can be calculated by assuming, for instance, that the hadronic medium is a hadron-resonance gas at some temperature and that J/ψ undergoes collisions with all of the meson species. The energy density which allows to determine the temperature of the hadron gas is determined on the basis of a plausible geometrical picture of the collision. The dissociation (2) cross section is calculated for all mesons of the lowest lying nonets (pseudoscalar and vector).

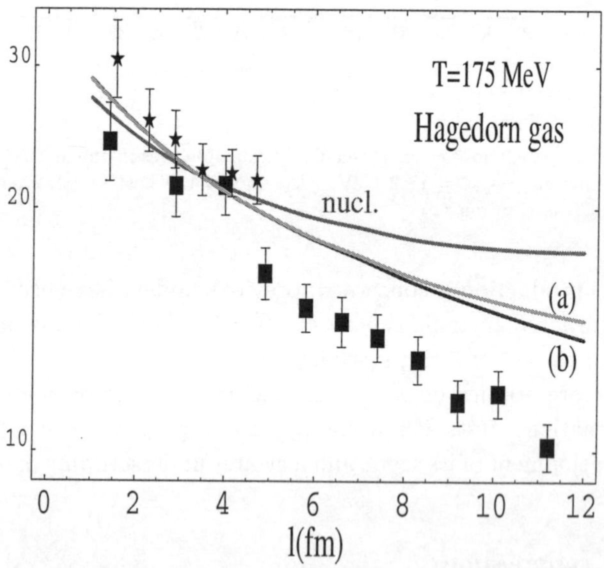

Figure 13. Calculated absorption curves of J/ψ in an ideal hadron-resonance gas compared with the data in S-U and Pb-Pb collisions [26]. The top curve is the simple nuclear absorption, the two lower curves refer to the absorption due to meson dissociation with (b) and without (a) geometrical effects.

Good agreement is found with the data when the hadron gas includes only these low lying mesons. Conversely, an anomalous suppression shows up when modifying the equation of state of the hadron-resonance gas to include an infinitely large number of species (Hagedorn states) because the temperature decreases (see fig. 13). On the other hand, the authors point out that the dissociation of J/ψ with the included heavy resonances has been neglected. Though the single contribution to the absorption is tiny, the very large number of mesonic states could make the overall effect significant. More studies are under way.

3.5. *Transport models*

This is a traditional approach to ultrarelativistic heavy ion collisions. The idea is to achieve a complete description of the process, from the onset of the collision to the emission of final particles through the simulation of a sequence of microscopic (binary) collisions or decays. These kinetic or transport models have several implementations in codes like HIJING, URQMD etc. In this framework, the group in LNS has recently studied [28] the effect of Fermi statistics on particle production. They found that a suppression as large as 10% of final particle yields due to Pauli blocking at early partonic level may show up at very high energy (LHC). At lower energy this effect is much lower, due to the reduced density. As yet, the model cannot reproduce satisfactorily the measured particle multiplicities. However, several important physical mechanisms are still to be introduced and this may lead to an improvement of the agreement.

3.6. *Others*

There is much more work relevant to this field which has been carried out during the past few years by italian collaborations and which would be definitely worth being presented and discussed more in detail. I wish to apologize with the authors for not doing this here. This does not mean that their work is less important, it is just a matter of taste of the author and lack of space.

However, I would like to quote studies of the equation of state of the hadron gas and QGP in the non-extensive Tsallis statistics [29], which shows significant differences with respect to the "classical" one; of the QCD phases with an effective Nambu-Jona Lasinio model with non-vanishing isospin chemical potential [30] (see also [31]); of the pure gauge SU(3) first-order phase transition with a gluon condensate evaporation model [32].

4. Conclusions

The field of ultrarelativistic heavy ion physics has grown in importance during the last years driven by an increasing theoretical and experimental effort. Although much has been learnt about the relevant physics and many evidencies of the formation of a peculiar, strongly interacting state of matter have been found, especially at RHIC, a clearcut proof of QGP is still missing [33]. The confidence on QGP discovery has to rest on the comparison with different models which, though very promising, have not yet developed into a mature, fully predictive, theoretical framework. As yet, many questions are still to be answered and much more studies are needed to achieve this goal.

The interest of theorists in this field is rapidly increasing in Italy, as witnessed by an expanding activity (e.g. GISELDA and FI31 projects), and it is expected that it will continue to do so in the near future.

5. Acknowledgments

The author would like to thank the organizers of this conference for their invitation and for having provided a stimulating environment.

References

1. F. Karsch, *Nucl. Phys.* **A698**, 199c (2002); Z. Fodor, S. D. Katz, *JHEP* **03** 014 (2002).
2. F. Karsch, E. Laermann, *Phys. Rev.* **D50**, 6954 (1994); Z. Fodor, S. D. Katz, *JHEP* **03** 014 (2002).
3. F. Karsch, E. Laermann, A. Peikert, *Nucl. Phys.* **B83-84**, 390 (2000).
4. J. D. Bjorken, Fermilab-Pub-82/59-THY (1982), unpublished.
5. M. Gyulassy and M. Plumer, *Phys. Lett.* **B243**, 432 (1990); M. Gyulassy and X. N. Wang, *Nucl. Phys.* **B420**, 583 (1994); X. N. Wang, M. Gyulassy and M. Plumer, *Phys. Rev.* **D51**, 3436 (1995).
6. R. Baier, *Nucl. Phys.* **A715**, 209c (2003) and references therein.
7. J. Adams *et al.*, STAR coll., *Phys. Rev. Lett.* **91**, 072304 (2003).
8. U. Heinz, hep-ph/0407360 and references therein.
9. F. Becattini *et al.*, *Phys. Rev.* **C69**, 024905 (2004).
10. M. Van Leeuwen *et al.*, NA49 coll., *Nucl. Phys.* **A715**, 161c (2003).
11. F. Becattini *Z. Phys.* **C69**, 485 (1996); F. Becattini, Proc. of *Universality features in multihadron production and the leading effect* p. 74 (1997); F. Becattini, U. Heinz, *Z. Phys.* **C76**, 269 (1997); F. Becattini, G. Passaleva, *Eur. Phys. J.* **C23**, 551.
12. Hormuzdiar J *et al.*, *Int. J. Mod. Phys.* **E12**, 649 (2003); R. Stock, *Phys. Lett.* **B456**, 277 (1999); V. Koch, *Nucl. Phys.* **A715**, 108 (2003); F. Becattini, hep-ph/0410403, to appear in the Proc. of *Focus on multiplicity*.

13. J. Rafelski, B. Müller, *Phys. Rev. Lett.* **48**, 1066 (1982).
14. F. Becattini, *J. Phys.* **G28**, 1553 (2002).
15. J. Cleymans *et al.*, hep-ph/0411187.
16. M. Gazdzicki, M. Gorenstein, *Acta Phys. Polon.* **B30**, 2705 (1999).
17. R. Stock, hep-ph/0404125.
18. M. Gazdzicki, *J. Phys.* **G30**, S1073 (2004).
19. D. Kharzeev, E. Levin and M. Nardi, *Nucl. Phys.* **A730**, 448 (2004); D. Kharzeev and M. Nardi, *Phys. Lett.* **B507**, 121 (2001).
20. M. Nardi, *Hadronic multiplicities at RHIC and LHC*, to appear in the Proc. of *Focus on multiplicity.*
21. T. Hirano, Y. Nara, *Nucl. Phys.* **A743**, 305 (2004); K. Golec-Biernat, talk given at *The future of high energy collisions*, Kielce (Poland) October 14-17 2004.
22. G. G. Barnafoldi, G. Papp, P. Levai and G. Fai, *J. Phys.* **G 30**, S1125 (2004) and references therein.
23. E. Cattaruzza and D. Treleani, *Phys. Rev.* **D69**, 094006 (2004);
24. G. Papp, talk given at *The future of high energy collisions*, Kielce (Poland) October 14-17 2004.
25. P. Braun-Munzinger, I. Heppe and J. Stachel, *Phys. Lett.* **B465**, 15 (1999).
26. L. Maiani, F. Piccinini, A. D. Polosa and V. Riquer, hep-ph/0408150; L. Maiani, F. Piccinini, A. D. Polosa and V. Riquer, *Nucl. Phys.* **A741**, 273 (2004).
27. T. Matsui, H. Satz, *Phys. Lett.* **B178**, 416 (1986).
28. B. H. Sa, A. Bonasera, *Phys. Rev.* **C70**, 034904 (2004).
29. A. Drago, A. Lavagno, P. Quarati,*Physica* **A344**, 472 (2004).
30. A. Barducci *et al.*, *Phys. Rev.* **D69**, 096004 (2004).
31. M. Di Toro *et al.*, nucl-th/0210052.
32. A. Drago, M. Gibilisco, C. Ratti, *Nucl. Phys.* **A742**, 165 (2004).
33. STAR Collaboration White Paper: *Are we there yet? The STAR collaboration's critical evaluation of the evidence regarding formation of a Quark Gluon Plasma in RHIC collisions* (2004) unpublished.

THERMODYNAMICS OF THE TWO-COLOUR NJL MODEL*

C. RATTI AND W. WEISE
Physik-Department, Technische Universität München,
D-85747 Garching, Germany
E-mail: Claudia_Ratti@ph.tum.de

We investigate two-flavour and two-colour QCD at finite temperature and chemical potential in comparison with a corresponding Nambu and Jona-Lasinio model. By minimizing the thermodynamic potential of the system, we confirm that a second order phase transition occurs at a value of the chemical potential equal to half the mass of the chiral Goldstone mode. For chemical potentials beyond this value the scalar diquarks undergo Bose condensation and the diquark condensate is nonzero. We evaluate the behaviour of the chiral condensate, the diquark condensate, the baryon charge density and the masses of scalar diquark, antidiquark and pion, as functions of the chemical potential. Very good agreement is found with lattice QCD ($N_c = 2$) results. We also compare with a model based on leading-order chiral effective field theory.

1. Introduction

The study of full QCD at finite baryon density is still a formidable challenge, due to the limitations of standard Monte Carlo simulations when applied to systems at finite chemical potential (for recent results see [1,2]). Present developments are aimed at improved strategies [3] to deal with the fact that the determinant of the Euclidean Dirac operator becomes complex at finite chemical potential. One response to this situation has been to start from simpler QCD-like theories with additional antiunitary symmetries that guarantee the Fermion determinant to be real at non-zero chemical potential and therefore allow the study of such theories on the lattice. Examples of such explorations include QCD with two colours and fundamental quarks and QCD with an arbitrary number of colours and adjoint quarks [4].

In two-colour QCD, diquarks can form colour singlets which are the baryons of the theory. The lightest baryons and the lightest quark-antiquark excitations (pions) have a common mass, m_π, and this spectrum

*This work is supported in part by INFN and BMBF

determines the properties of the ground state for small chemical potentials. General arguments [5] predict a phase transition from the vacuum to a state with finite baryon density at a critical chemical potential μ_c, which is the lowest energy per quark that can be realized by an excited state of the system. This state is populated by light diquarks, and one expects $\mu_c = m_\pi/2$. The Bose-Einstein condensation of diquarks, with nonzero baryon number, can be interpreted as baryon charge superconductivity.

In our paper [6] we investigate the relationship between $N_c = 2$ QCD and a corresponding Nambu and Jona-Lasinio (NJL) model [7,8,9,10,11]. Similar models have already been used to study the QCD colour superconductivity phase with two [12,13,14,15,16,17] and three flavours [18,19,20] (for a recent review see [21]). The specific aim of our work is to test the effectiveness of the NJL model, with its dynamically generated quasiparticles, in reproducing the thermodynamics of two-colour QCD, and to compare our results quantitatively with those obtained from recent lattice computations. We study the behaviour of the chiral and diquark condensates, and of the baryon density, as functions of temperature and chemical potential. As further applications we evaluate the pion, diquark and antidiquark masses, as functions of the chemical potential. We compare our results to lattice data and also to the predictions from chiral effective field theory.

2. Two colour NJL model

We consider as a starting point the Lagrangian

$$\mathcal{L} = \bar{\psi}(x)\left(i\gamma^\mu\partial_\mu - \mathbf{m_0}\right)\psi(x) - G_c\sum_{a=1}^{3} J_\mu^a(x)\, J_a^\mu(x)\,, \tag{1}$$

with a four point interaction that represents the local coupling between colour currents $J_\mu^a = \bar{\psi}\gamma_\mu t^a\psi$ involving the quark fields ψ and the $SU(2)_{colour}$ generators t_a with $tr(t^a t^b) = 2\delta_{ab}$. Here G_c is an effective coupling strength of dimension (length)^2 and $\mathbf{m_0}$ is the diagonal current quark mass matrix.

It is convenient to rewrite the interaction between quarks, by Fierz transformation, in terms of the colour singlet pseudoscalar/scalar quark-antiquark and scalar diquark channels. The resulting Lagrangian reads

$$\begin{aligned}
\mathcal{L}_{NJL} &= \bar{\psi}\left(i\gamma^\mu\partial_\mu - \mathbf{m_0}\right)\psi + \mathcal{L}_{q\bar{q}} + \mathcal{L}_{qq} + \text{(colour triplet terms)}, \qquad (2)\\
\mathcal{L}_{q\bar{q}} &= \frac{G}{2}\left[\left(\bar{\psi}\psi\right)^2 + \left(\bar{\psi}i\gamma_5\vec{\tau}\psi\right)^2\right],\\
\mathcal{L}_{qq} &= \frac{H}{2}\left(\bar{\psi}i\gamma_5\tau_2 t_2 C\bar{\psi}^T\right)\left(\psi^T C i\gamma_5\tau_2 t_2\psi\right)
\end{aligned}$$

where G and H are constants which describe quark-antiquark and quark-quark interactions, respectively, t_a are Pauli matrices in colour space and τ_i are Pauli matrices in flavour (isospin) space. We have introduced the charge conjugation operator for fermions:

$$C = i\gamma_0\gamma_2. \tag{3}$$

The coupling constants G and H in the Lagrangian (2) are uniquely fixed by this procedure. One obtains

$$G = H = \frac{3}{2}G_c \tag{4}$$

Starting from the Lagrangian (2) and using standard bosonization techniques, we introduce the auxiliary scalar (σ), pseudoscalar triplet[a] ($\vec{\pi}$) and diquark (Δ, Δ^*) fields, thus obtaining the following equivalent Lagrangian in the colour singlet sector:

$$\begin{aligned}\tilde{\mathcal{L}} = \bar{\psi}\left(i\gamma^\mu\partial_\mu - \mathbf{m_0} + \sigma + i\gamma_5\vec{\tau}\cdot\vec{\pi}\right)\psi - \frac{1}{2}\Delta^*\psi^T C\gamma_5\tau_2 t_2\psi \\ + \frac{1}{2}\Delta\bar{\psi}\gamma_5\tau_2 t_2 C\bar{\psi}^T - \frac{\sigma^2+\vec{\pi}^2}{2G} - \frac{|\Delta|^2}{2H}.\end{aligned} \tag{5}$$

3. Parameter fixing

The three parameters of the model are the "bare" quark mass m_0, a loop-momentum cutoff Λ and the coupling strength $G = H$. Even if we are considering the $N_c = 2$ NJL model, we choose to reproduce the known chiral physics in the hadronic sector.

For this reason, we fix those parameters through the constraints imposed by the pion decay constant and the chiral (quark) condensate: The current quark mass m_0 is fixed from the Gell-Mann, Oakes, Renner (GMOR) relation. In the chiral limit, $m_0 = 0$ and $m_\pi = 0$.

Table 1. Parameter set used in this work, and the corresponding physical quantities.

	$\lvert\langle\bar{\psi}_u\psi_u\rangle\rvert^{1/3}$	f_π	m_π
NJL	$\left[-4miN_c\int\frac{d^4p}{(2\pi)^4}\frac{\theta(\Lambda^2-\vec{p}^{\,2})}{p^2-m^2+i\epsilon}\right]^{1/3}$	$\sqrt{-4im^2N_c\int\frac{d^4p}{(2\pi)^4}\frac{\theta(\Lambda^2-\vec{p}^{\,2})}{(p^2-m^2+i\epsilon)^2}}.$	$\sqrt{\frac{-m_0\langle\bar{\psi}\psi\rangle}{f_\pi^2}}.$
"exp."	(240±20) MeV	92.4 MeV	139.6 MeV
Th.	259 MeV	89.6 MeV	139.3 MeV
	Λ [GeV]	$G = H$[GeV^{-2}]	m_0[MeV]
	0.78	10.3	4.5

[a]Isovectors such as the pion field are denoted by $\vec{\pi}$.

4. Results at finite T and μ

We now extend the NJL model to finite temperature T and chemical potentials μ using the Matsubara formalism. We consider the isospin symmetric case, with an equal number (and therefore a single chemical potential) of u and d quarks. The quantity to be minimized at finite temperature is the thermodynamic potential, which reads:

$$\Omega\left(T,\mu\right) = -4\int \frac{d^3p}{(2\pi)^3}\left[2T\ln\left(1+\exp\left(-\frac{E^+}{T}\right)\right)+\right. \tag{6}$$
$$\left.+\,2T\ln\left(1+\exp\left(-\frac{E^-}{T}\right)\right)+\left(E^+ + E^-\right)\right]\theta\left(\Lambda^2-\vec{p}^{\,2}\right)+\frac{\sigma^2}{2G}+\frac{|\Delta|^2}{2H}.$$

where we have defined $E^{\pm} = \sqrt{\left(\epsilon^{\pm}\right)^2+|\Delta|^2}$, with $\epsilon^{\pm} = \epsilon \pm \mu$, $\epsilon = \sqrt{\vec{p}^{\,2}+m^2}$, $m = m_0 - \langle\sigma\rangle = m_0 - G\langle\bar{\psi}\psi\rangle$.

The mean values for the σ and Δ fields are determined by minimizing the thermodynamic potential. Their behaviour as a function of the chemical potential is shown in Fig. 1 in comparison with the lattice data taken from ref. [22]

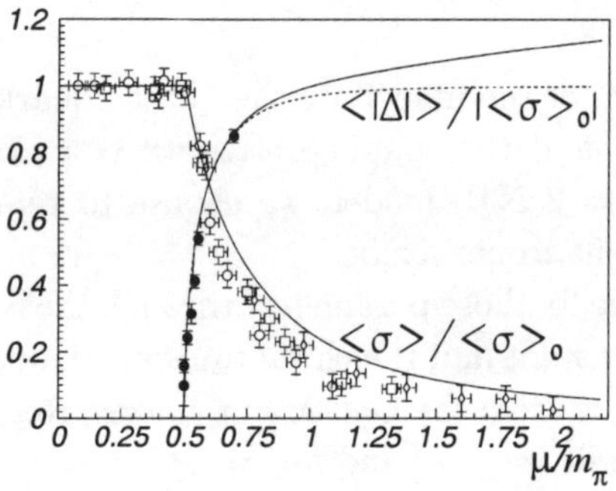

Figure 1. Scaled $\langle\sigma\rangle$ and $\langle|\Delta|\rangle$ as a function of the chemical potential at $T = 0$: our results (solid lines) are compared to the lattice data taken from ref. [22]. The different symbols (open circles, squares and diamonds) for the chiral condensate correspond to different values for the quark masses. The dashed lines are the predictions from chiral effective field theory [4].

An interesting quantity is the baryonic density

$$\rho = -\frac{\partial\Omega(T,\mu)}{\partial\mu}. \tag{7}$$

The lattice data of ref. [22] show a scaled baryonic density defined as

$$\tilde{\rho} = \frac{\rho}{4N_f f_\pi^2 m_\pi}. \tag{8}$$

Fig. 2 presents our results for the scaled baryonic density (8) as a function of the chemical potential at zero temperature, in comparison with the lattice data for the same quantity.

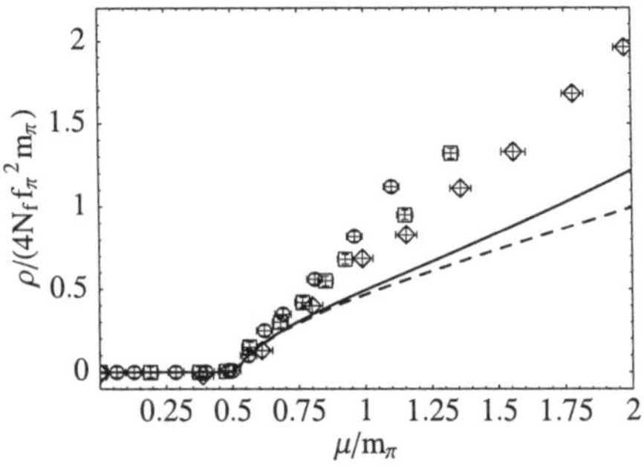

Figure 2. Scaled baryonic density as a function of the chemical potential at $T = 0$ (Continuous line). The lattice data are taken from ref. [22]. The different symbols correspond to different values for the quark masses. The dashed line is the prediction from chiral effective field theory [4].

5. Pion and scalar diquark properties

In order to evaluate the masses of the bosonic fields, we expand the effective action in a power series of the meson and diquark fields around their mean field values. The second-order term of this expansion, $S^{(2)}_{eff}$, identifies the mass spectrum of mesons and diquarks. Mixing terms arise, at $\mu > \mu_c$, between the σ, Δ and Δ^* fields: these terms are proportional to $|\Delta|$, and the mixing occurs because the presence of a nonzero diquark condensate spontaneously breaks the baryon number symmetry. The mass matrix turns out to have the following form:

$$M = \begin{pmatrix} \frac{\partial^2 S^{(2)}_{eff}}{\partial \vec{\pi}^2} & 0 & 0 & 0 \\ 0 & \frac{\partial^2 S^{(2)}_{eff}}{\partial \sigma^2} & \frac{\partial^2 S^{(2)}_{eff}}{\partial \sigma \partial \Delta} & \frac{\partial^2 S^{(2)}_{eff}}{\partial \sigma \partial \Delta^*} \\ 0 & \frac{\partial^2 S^{(2)}_{eff}}{\partial \Delta \partial \sigma} & \frac{\partial^2 S^{(2)}_{eff}}{\partial \Delta^2} & \frac{\partial^2 S^{(2)}_{eff}}{\partial \Delta \partial \Delta^*} \\ 0 & \frac{\partial^2 S^{(2)}_{eff}}{\partial \Delta^* \partial \sigma} & \frac{\partial^2 S^{(2)}_{eff}}{\partial \Delta^* \partial \Delta} & \frac{\partial^2 S^{(2)}_{eff}}{\partial \Delta^{*2}} \end{pmatrix}, \tag{9}$$

and the masses of the various modes are found by solving the equation

$$\det(M) = 0. \tag{10}$$

Evidently the pion fields do not mix with the others, while the σ, the diquark and the antidiquark fields mix in the phase with $|\Delta| \neq 0$.

The behaviour of the scaled pion mass as a function of the chemical potential is shown in Fig. 3 (a), in comparison to the lattice data. The pion mass increases linearly with the chemical potential at $\mu > \mu_c$. This behaviour was anticipated in the calculations by Kogut *et al.* [4], as indicated by the dashed line in Fig. 3 (a). Our result is in very good agreement with both the lattice data and the predictions using the leading-order chiral effective Lagrangian.

Next, consider the other two bosonic modes of the theory: the scalar diquark and its antidiquark. The behaviour of their masses at finite chemical potential is shown in Fig. 3 (b) in comparison to the pion mass: at $\mu = 0$ they are all degenerate, as predicted on the basis of general arguments, but they behave in different ways as the chemical potential increases: for $\mu < \mu_c = m_\pi^{(0)}/2$ the pion, which does not carry baryon charge, is not affected by μ, while the diquark and antidiquark masses are shifted according to their baryon number $B = \pm 1$. They follow in fact the behaviour observed also in chiral effective field theory [4]:

$$m_\Delta = m_\pi - 2\mu, \qquad\qquad m_{\Delta^*} = m_\pi + 2\mu. \tag{11}$$

For $\mu > \mu_c$, the appearance of the diquark condensate spontaneously breaks the baryon number symmetry. The scalar modes (diquark, antidiquark and sigma) get mixed. The new eigenmodes are linear combinations of the original quasiparticle states. By solving eq. (10) we find the masses of the new orthogonal modes. One of them, which we denote by $\tilde{\Delta}$, is massless and can be identified with the true Goldstone boson of the theory, corresponding to the spontaneous breaking of the baryon number ($U(1)$) symmetry. The other two modes are massive. One of them, which we denote by $\tilde{\Delta}^*$, follows the behaviour derived in the paper by Kogut *et al.*:

$$m_{\tilde{\Delta}^*} = 2\mu\sqrt{1 + 3\,(m_\pi/2\mu)^4}. \tag{12}$$

6. Chiral limit

In the chiral limit $m_0 \to 0$ ($m_\pi \to 0$), and at $\mu = 0$, the thermodynamic potential (6) (with $G = H$) is a function only of $\sigma^2 + |\Delta|^2$. This is a natural outcome once the relation between the coefficients G and H is fixed through the Fierz transformation of the colour current-current interaction (see eq. (4)). As a result, Ω is invariant under the rotation which connects the chiral and the diquark condensate along the circle $\sigma^2 + |\Delta|^2$=const.

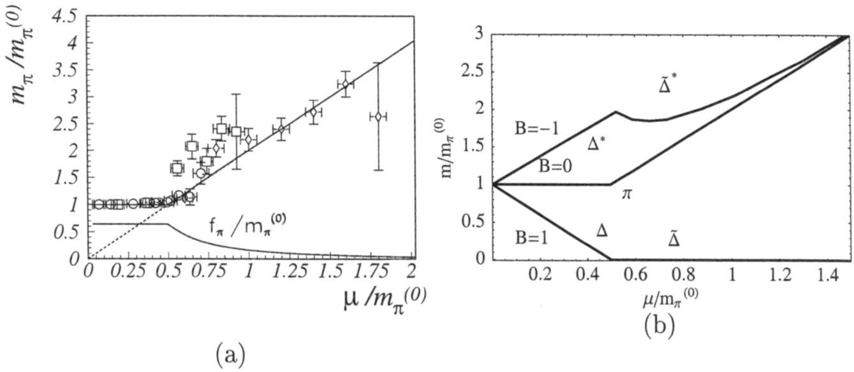

Figure 3. (a) Scaled pion mass as a function of $\mu/m_\pi^{(0)}$ at $T = 0$ (continuous line). The lattice data are taken from ref. [22] and have been rescaled in order to show dimensionless quantities. The different symbols correspond to different values for the quark masses. The dashed line is $m_\pi = 2\mu$, as predicted in leading-order chiral effective field theory [4]. Also shown is the (scaled) pion decay constant $f_\pi/m_\pi^{(0)}$ and its evolution with increasing μ. (b) Spectrum of pions and diquarks/antidiquarks as a function of the (scaled) chemical potential at zero temperature.

Because of this symmetry, the chiral condensate is indistinguishable from the diquark condensate for $m_0 = \mu = 0$, so that a state with finite $\langle\sigma\rangle$ can always be transformed into a state with finite $\langle|\Delta|\rangle$ and $\langle\sigma\rangle = 0$. The phases with spontaneously broken chiral and baryon number symmetries are degenerate in this limit. As soon as the chemical potential takes a finite value, the favourable phase is the one with a non-zero diquark condensate and zero chiral condensate.

Consider next the pion and diquark masses in the chiral limit and their variations with increasing chemical potential. The chiral condensate is always equal to zero in this limit. Consequently, the $\tilde{\Delta}$ mode is a true Goldstone boson and its mass is always equal to zero, while the $\tilde{\Delta}^*$ and pion masses are degenerate. Explicit symmetry breaking by a finite chemical potential lets these masses scale as $m_{\tilde{\Delta}^*} = m_\pi = 2\mu$. The degeneracy of $\tilde{\Delta}^*$ and π is removed as soon as a small non-zero quark mass m_0 is introduced. This also gives a finite mass to the $\tilde{\Delta}$ mode, which is again equal to zero above $\mu_c = m_\pi^{(0)}/2$.

7. Conclusions

The starting point of our work is the assumption that gluon dynamics can be integrated out and reduced to a local interaction between quark colour currents. By Fierz rearrangement, this implies a one-to-one correspondence between interactions in colour singlet quark-antiquark and diquark channels (the Pauli-Gürsey symmetry).

The resulting spontaneous (dynamical) symmetry breaking pattern identifies pseudoscalar Goldstone bosons (pions) and scalar diquarks as the thermodynamically active quasiparticles. The successful comparison with $N_c = 2$ lattice data indicates that this simple NJL model does indeed draw a remarkably realistic picture of the quasiparticle dynamics emerging from $N_c = 2$ QCD.

We confirm that a diquark condensate develops at chemical potentials $\mu > \mu_c = m_\pi/2$. The correlated evolution of the chiral and diquark condensates with increasing μ, as observed in $N_c = 2$ lattice QCD, is very well reproduced. It appears that modelling the low-energy dynamics of $N_c = 2$ QCD is already done surprisingly well when using just a colour current-current interaction with a single strength parameter.

References

1. S. Muroya, A. Nakamura, C. Nonaka, and T. Takaishi, Prog. Theor. Phys. **110**, 615 (2003).
2. F. Karsch, K. Redlich, and A. Tawfik, Phys. Lett. **B571**, 67 (2003).
3. Z. Fodor and S. D. Katz, JHEP **0203**, 014 (2002).
4. J. B. Kogut *et al.*, Nucl. Phys. **B582**, 477 (2000).
5. M. A. Halasz *et al.*, Phys. Rev. **D58**, 096007 (1998).
6. C. Ratti and W. Weise, Phys. Rev. **D70**, 054013 (2004), hep-ph/0406159.
7. Y. Nambu and G. Jona-Lasinio, Phys. Rev. **122**, 345 (1961).
8. Y. Nambu and G. Jona-Lasinio, Phys. Rev. **124**, 246 (1961).
9. U. Vogl and W. Weise, Prog. Part. Nucl. Phys. **27**, 195 (1991).
10. S. P. Klevansky, Rev. Mod. Phys. **64**, 649 (1992).
11. T. Hatsuda and T. Kunihiro, Phys. Reports **247**, 221 (1994).
12. J. Berges and K. Rajagopal, Nucl. Phys. **B538**, 215 (1999).
13. K. Langfeld and M. Rho, Nucl. Phys. **A660**, 475 (1999).
14. T. M. Schwarz, S. P. Klevansky and G. Papp, Phys. Rev. **C60**, 055205 (1999).
15. M. Buballa, J. Hosek, and M. Oertel, Phys. Rev. **D65**, 014018 (2002).
16. M. Buballa, J. Hosek, and M. Oertel, Phys. Rev. Lett. **90**, 182002 (2003).
17. D. Blaschke, M. K. Volkov, and V. L. Yudichev, Eur. Phys. J. **A17**, 103 (2003).
18. F. Gastineau, R. Nebauer and J. Aichelin, Phys. Rev. **C65**, 045204 (2002).
19. M. Buballa and M. Oertel, Nucl. Phys. **A703**, 770 (2002).
20. F. Neumann, M. Buballa, and M. Oertel, Nucl. Phys. **A714**, 481 (2003).
21. M. Buballa, hep-ph/0402234, Submitted to Phys. Reports.
22. S. Hands, I. Montvay, L. Scorzato, and J. Skullerud, Eur. Phys. J. **C22**, 451 (2001).

MESON CORRELATION FUNCTIONS IN HOT QCD

A. BERAUDO

Dipartimento di Fisica Teorica dell'Università di Torino and
Istituto Nazionale di Fisica Nucleare, Sezione di Torino,
via P.Giuria 1, I-10125 Torino, Italy

The temporal pseudoscalar meson correlation function in a QCD plasma is investigated in a range of temperatures above T_c of experimental interest. Only the flavour-singlet channel is considered and the imaginary time formalism is employed for the finite temperature calculations. The behaviour of the meson spectral function and of the temporal correlator is first studied in the HTL approximation replacing the free thermal quark propagators with the HTL resummed ones. This procedure satisfactory describes the soft fermionic modes, but its application to the propagation of hard quarks is not reliable. An improved version of the so-called NLA scheme, which allows a better treatment of the hard fermionic modes, is then proposed. The impact of the improved NLA on the pseudoscalar temporal correlator is investigated.

1. Thermal meson correlation function

We consider the following current operator, carrying the quantum numbers of a meson in the flavour-singlet channel (there is no matrix mixing the flavours):

$$J_M(-i\tau,\mathbf{x}) = \bar{q}(-i\tau,\mathbf{x})\Gamma_M q(-i\tau,\mathbf{x}) \ , \tag{1}$$

where $\Gamma_M = 1,\gamma^5,\gamma^\mu,\gamma^\mu\gamma^5$ for the scalar, pseudoscalar, vector and pseudovector channels, respectively. We next define the fluctuation operator $\widetilde{J}_M$ as

$$\widetilde{J}_M(-i\tau,\mathbf{x}) = J_M(-i\tau,\mathbf{x}) - \langle J_M(-i\tau,\mathbf{x})\rangle \ . \tag{2}$$

The chief quantity we address is the *thermal meson 2 point correlation function*:

$$\begin{aligned} G_M(-i\tau,\mathbf{x}) &= \langle \widetilde{J}_M(-i\tau,\mathbf{x})\widetilde{J}^\dagger_M(0,\mathbf{0})\rangle \\ &= \langle J_M(-i\tau,\mathbf{x})J^\dagger_M(0,\mathbf{0})\rangle - \langle J_M(-i\tau,\mathbf{x})\rangle\langle J^\dagger_M(0,\mathbf{0})\rangle \\ &= \frac{1}{\beta}\sum_{n=-\infty}^{+\infty}\int\frac{d^3p}{(2\pi)^3}e^{-i\omega_n\tau}e^{i\mathbf{p}\cdot\mathbf{x}}\chi_M(i\omega_n,\mathbf{p}) \quad , \end{aligned} \tag{3}$$

with $\tau \in [0, \beta = 1/T]$ and $\omega_n = 2n\pi T$ $(n = 0, \pm 1, \pm 2 \ldots)$.
It is convenient to adopt the following spectral representation for the meson propagator in momentum space:

$$\chi_M(i\omega_n, \mathbf{p}) = -\int_{-\infty}^{+\infty} d\omega \frac{\sigma_M(\omega, \mathbf{p})}{i\omega_n - \omega} \quad \Rightarrow \quad \sigma_M(\omega, \mathbf{p}) = \frac{1}{\pi} \mathrm{Im}\, \chi_M(\omega + i\eta, \mathbf{p}) \,. \tag{4}$$

Hence one can write the meson correlation function in the mixed representation, in terms of the meson spectral function σ_M, as follows:

$$G_M(-i\tau, \mathbf{p}) = -\frac{1}{\beta} \sum_{n=-\infty}^{+\infty} e^{-i\omega_n \tau} \int_{-\infty}^{+\infty} d\omega \frac{\sigma_M(\omega, \mathbf{p})}{i\omega_n - \omega} \,, \tag{5}$$

which, performing the summation over the Matsubara frequencies, becomes

$$G_M(-i\tau, \mathbf{p}) = \int_0^{+\infty} d\omega \; \sigma_M(\omega, \mathbf{p}) \frac{\cosh(\omega(\tau - \beta/2))}{\sinh(\omega\beta/2)} \,. \tag{6}$$

2. Free spectral functions

We start by revisiting the computation of the free meson spectral functions for the different channels. We use the short-hand notation $(\tau, \mathbf{x})$ for $(t = -i\tau, \mathbf{x})$. From the definitions given in Eqs. (1,2) and applying Wick's theorem one easily gets:

$$\langle \tilde{J}_M(\tau, \mathbf{x}) \tilde{J}_M^\dagger(0, \mathbf{0}) \rangle = \mathrm{Tr}(\Gamma_M S_F(\tau, \mathbf{x}) \gamma^0 \Gamma_M^\dagger \gamma^0 S_F(\beta - \tau, -\mathbf{x})) \,, \tag{7}$$

$S_F(\tau, \mathbf{x})$ being the free thermal quark propagator. Hence in Fourier space the mesonic 2 point correlation function reads (we omit the subscript M):

$$\chi(i\omega_l, \mathbf{p}) = -2N_c \frac{1}{\beta} \sum_{n=-\infty}^{+\infty} \int \frac{d^3k}{(2\pi)^3} \mathrm{Tr}[\Gamma S_F(i\omega_n, \mathbf{k}) \gamma^0 \Gamma^\dagger \gamma^0 S_F(i\omega_n - i\omega_l, \mathbf{k} - \mathbf{p})], \tag{8}$$

where the mesonic frequency is $\omega_l = 2l\pi T$, while in the quark propagators $\omega_n = (2\mathrm{n}+1)\pi T$. The overall factor $2N_c$ comes from the trace over the light flavours and colours.
Globally, in the case $\mathbf{p} = \mathbf{0}$, the results for the different channels can be summarized in the compact expression:

$$\sigma_M^{\mathrm{free}}(\omega, \mathbf{0}) = \frac{N_c}{4\pi^2} \theta(\omega - 2m) \sqrt{1 - \left(\frac{2m}{\omega}\right)^2}\, \omega^2 \tanh(\omega/4T) \left(a + b\left(\frac{2m}{\omega}\right)^2\right) , \tag{9}$$

where the coefficient (a,b) turn out to be given respectively by (1,-1), (1,0), (-2,-1), (-2,3) in the scalar, pseudoscalar, vector and pseudovector channel[a].

3. Meson spectral function in the HTL approximation

In this section we evaluate the meson spectral function in the HTL approximation for the pseudoscalar channel. This amounts to replacing the free thermal quark propagators with HTL resummed ones which can be cast in the form:

$$ {}^{\star}S(i\omega_n,\mathbf{p}) = -\frac{\gamma^0-\gamma\cdot\hat{\mathbf{p}}}{2}\int_{-\infty}^{+\infty} d\omega \frac{\rho_+(\omega,p)}{i\omega_n-\omega} - \frac{\gamma^0+\gamma\cdot\hat{\mathbf{p}}}{2}\int_{-\infty}^{+\infty} d\omega \frac{\rho_-(\omega,p)}{i\omega_n-\omega}\,, \tag{10} $$

having introduced the quark spectral functions [1,2,3,4]

$$ \rho_\pm(\omega,k) = \frac{\omega^2-k^2}{2m_q^2}[\delta(\omega-\omega_\pm)+\delta(\omega+\omega_\mp)] + \beta_\pm(\omega,k)\theta(k^2-\omega^2)\,, \tag{11} $$

with

$$ \beta_\pm(\omega,k) = -\frac{m_q^2}{2}\frac{\pm\omega-k}{\left[k(-\omega\pm k)+m_q^2\left(\pm 1-\frac{\pm\omega-k}{2k}\ln\frac{k+\omega}{k-\omega}\right)\right]^2+\left[\frac{\pi}{2}m_q^2\frac{\pm\omega-k}{k}\right]^2}\,. \tag{12} $$

In the above $m_q = gT/\sqrt{6}$ is the so-called *thermal gap mass* of the quark. The HTL spectral functions in Eq. (11) consist of two pieces: a pole term and a continuum term, corresponding to the Landau damping of a quark propagating in the bath. In the time-like domain $\omega > k$ the dispersion relation $\omega_+(k)$ corresponds to the propagation of a quasi-particle with chirality and helicity eigenvalues of the same sign. On the other hand $\omega_-(k)$ describes the propagation of an excitation, referred to as *plasmino*, with negative helicity over chirality ratio. Both these excitations are undamped at this level of approximation.
The pseudoscalar vertex receives no HTL correction. Hence one simply replaces, in Eq.(8), the free thermal quark propagators with HTL resummed ones (which we denote with a star), obtaining:

$$ \chi^{\mathrm{ps}}(i\omega_l,\mathbf{p}) = 2N_c\frac{1}{\beta}\sum_{n=-\infty}^{+\infty}\int\frac{d^3k}{(2\pi)^3}\mathrm{Tr}[\gamma^5\,{}^{\star}S(i\omega_n,\mathbf{k})\gamma^5\,{}^{\star}S(i\omega_n-i\omega_l,\mathbf{k}-\mathbf{p})]\,. \tag{13} $$

[a]These coefficients do not coincide with the ones of Refs. [3,4], due to our choice of working with ordinary Dirac matrices satisfying the anti-commutation relation $\{\gamma^\mu,\gamma^\nu\}=2g^{\mu\nu}$, $g^{\mu\nu}$ being the Minkowskian metric tensor.

Making use of the spectral representation (10) of the quark propagator, in the case $\mathbf{p} = 0$ one obtains [3,4,5]:

$$\sigma^{\rm ps}(\omega, \mathbf{0}) = \frac{2N_c}{\pi^2}(e^{\beta\omega} - 1) \int_0^{+\infty} dk\, k^2 \int_{-\infty}^{+\infty} d\omega_1 \int_{-\infty}^{+\infty} d\omega_2 \tilde{n}(\omega_1)\tilde{n}(\omega_2)$$
$$\delta(\omega - \omega_1 - \omega_2)[\rho_+(\omega_1, k)\rho_+(\omega_2, k) + \rho_-(\omega_1, k)\rho_-(\omega_2, k)] \,. \tag{14}$$

Inserting then into Eq. (14) the explicit expression for $\rho_\pm$ given in Eq. (11), one finds, as first pointed out in Ref. [1], that the HTL meson spectral function consists of the sum of three terms: pole-pole (pp), pole-cut (pc) and cut-cut (cc).

4. Beyond HTL

In this section we attempt to improve upon the HTL result for $\sigma^{\rm ps}(\omega)$ quoted in the previous section. For this purpose we observe that, in making the convolution of the two fermionic propagators, one has to integrate over *all the scale of momenta* (hard and soft). Now, while the HTL approximation is supposed to dress correctly the propagation of the *soft modes*, this is not so for what concerns the *hard modes.* Thus, by replacing naively the free thermal quark propagators in Eq. (8) with HTL resummed ones, one treats incorrectly the contribution to the integral arising from hard momenta. This is analogous to what happens when one tries to evaluate thermodynamical quantities for the QGP, like the entropy and the baryon density, in a pure HTL approximation. As pointed out in Ref. [6], in HTL one gets the right contribution of order g^2 to such quantities, but *only* part of the g^3 term (actually this one is strictly a non-perturbative contribution, being non analytical in $\alpha_s = g^2/4\pi$), namely the contribution arising from the soft modes. To get the remaining part of the g^3 contribution one has to evaluate the correction $\delta\Sigma_\pm(\omega = \pm k)$ to the self energy of a normal quark mode (with hard momentum) stemming from its interaction with soft gluons (described by HTL resummed propagators). It is possible to account for this effect in an effective way, through a correction to the HTL quark asymptotic mass $\hat{m}_\infty = gT/\sqrt{3}$ reading [6,7,8]:

$$\delta m_\infty^2 = -\frac{1}{2\pi}\, g^2\, C_f\, T\, \hat{m}_D\,, \tag{15}$$

where C_f is the SU(3) Casimir operator and (in the case of zero chemical potential) the Debye screening mass in the HTL approximation is

$$\hat{m}_D = \sqrt{\frac{N_c}{3} + \frac{N_f}{6}}\, gT \, . \tag{16}$$

In the Next to Leading Approximation (NLA) the quark asymptotic mass is given by the solution of the "self-consistent like" equation:

$$m_\infty^2 = \hat{m}_\infty^2 - \frac{1}{2\pi} g^2 C_f \sqrt{N_c + \frac{N_f}{2}}\, T\, m_\infty \, . \tag{17}$$

The strategy followed in Ref. [6,7,8] was to introduce a cutoff Λ at an intermediate scale of momenta, keeping the HTL approximation for momenta below the cutoff and adopting the NLA asymptotic mass above. A reasonable choice for the cutoff is:

$$\Lambda = \sqrt{2\pi T \hat{m}_D} \quad , \tag{18}$$

which represents the geometric mean of the spacing between the Matsubara frequencies (hard scale) and the Debye screening mass in HTL (soft scale). Here we adopt this strategy for the evaluation of meson correlation functions. Actually, in order to allow for a smooth transition between the soft and the hard regime, we introduce two additional momenta Λ_1 and Λ_2 defined as follows:

$$\omega_+(\Lambda_1) = \sqrt{\Lambda^2 + m_\infty^2} \tag{19a}$$

$$\omega_+(\Lambda) = \sqrt{\Lambda_2^2 + m_\infty^2} \, . \tag{19b}$$

Hence we guess for the NLA quark spectral function the following expression:

$$\rho_\pm^{\rm NLA}(\omega, k) = \begin{cases} \rho_\pm(\omega, k) & \text{if } k < \Lambda_1 \\ \cos^2(\alpha(k))\rho_\pm(\omega, k) + \sin^2(\alpha(k))\delta(\omega \mp \epsilon_k^{\rm NLA}) & \text{if } \Lambda_1 < k < \Lambda_2 \\ \delta(\omega \mp \epsilon_k^{\rm NLA}) & \text{if } k > \Lambda_2 \end{cases} \tag{20}$$

with

$$\alpha(k) = \frac{\pi}{2} \cdot \frac{k - \Lambda_1}{\Lambda_2 - \Lambda_1} \, . \tag{21}$$

For further details on the motivations for this ansatz see Ref. [5].

Replacing then the HTL quark spectral functions entering into Eq. (14)

with the above given ansatz, one gets the following NLA expression for the pseudo-scalar meson spectral function (for $\omega > 0$):

$$\sigma^{\rm ps}_{\rm NLA}(\omega,\mathbf{0}) = \frac{2N_c}{\pi^2}(e^{\beta\omega}-1)\int_0^{+\infty} dk\, k^2 \int_{-\infty}^{+\infty} d\omega_1 \int_{-\infty}^{+\infty} d\omega_2 \tilde{n}(\omega_1)\tilde{n}(\omega_2)$$
$$\delta(\omega-\omega_1-\omega_2)[\rho_+^{\rm NLA}(\omega_1,k)\rho_+^{\rm NLA}(\omega_2,k) + \rho_-^{\rm NLA}(\omega_1,k)\rho_-^{\rm NLA}(\omega_2,k)]\,. \quad (22)$$

In conformity with Eq. (6), one can next obtain the associated zero momentum temporal correlator $G^{\rm ps}_{\rm NLA}(\tau,\mathbf{0})$ whose behaviour will be later shown for different values of the temperature.

5. Numerical results

Here we report our numerical findings for the pseudoscalar meson correlator in the HTL and NLA approximations. They are displayed in Figs. (1-4). In Fig. (1), where the so-called Van Hove singularities are clearly standing out, we recover the results of Ref.[4]. In Figs. (2-4) we explore where and how the NLA predictions differ from the HTL ones.

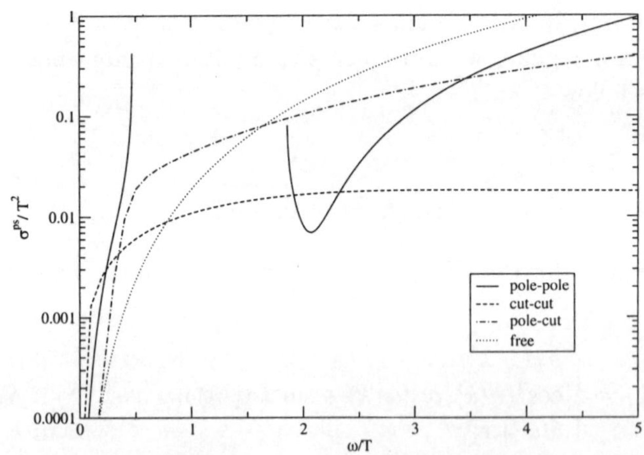

Figure 1. The various contributions (pole-pole, pole-cut and cut-cut) to the dimensionless spectral function of a pseudoscalar meson $\sigma^{\rm ps}/T^2$ at zero spatial momentum as a function of $x = \omega/T$ in the HTL scheme. The plot is given for a value of T such that $g(T) = \sqrt{6}$, entailing $m_q = T$. In the pole-pole contribution one recognizes the occurrence of the Van Hove singularities at $x = 0.47$ and at $x = 1.856$. Also plotted is the free spectral function.

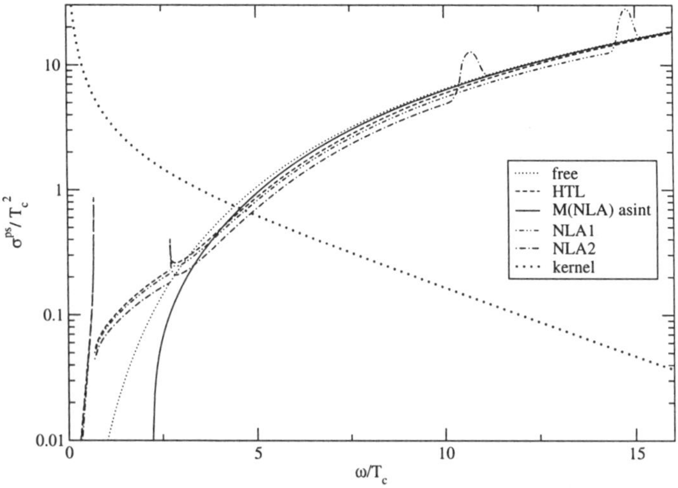

Figure 2. Behaviour of the zero momentum pseudoscalar spectral function σ^{ps}/T_c^2 versus ω/T_c in different approximations: free result, HTL, NLA and quarks endowed with a thermal mass $m_{\rm NLA}$. NLA1 corresponds to the choice $\Lambda = \sqrt{2\pi T \hat{m}_D}$, NLA2 to $\Lambda = \sqrt{\pi T \hat{m}_D}$. The plot refers to $T = 2T_c$.

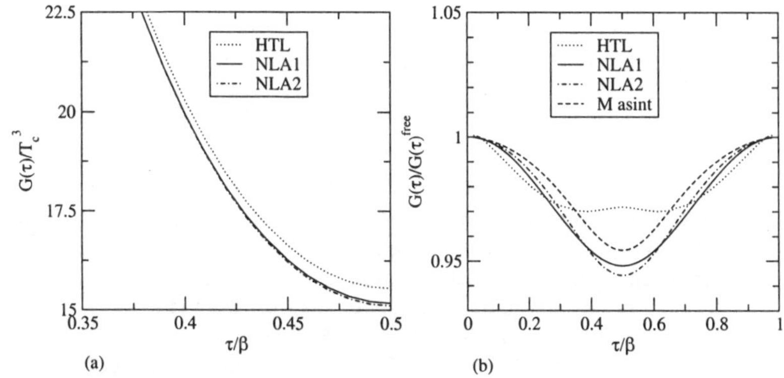

Figure 3. (a): Behaviour of $G(\tau)/T_c^3$ vs τ/β. (b): Behaviour of $G(\tau)/G^{\rm free}(\tau)$ vs τ/β. NLA1 corresponds to $\Lambda = \sqrt{2\pi T \hat{m}_D}$, NLA2 corresponds to $\Lambda = \sqrt{\pi T \hat{m}_D}$. In panel (b) we also display the result obtained in the case of quarks endowed with a thermal mass $m = m_{\rm NLA}$. The curves are given for $T = 2T_c$.

6. Conclusions

We have explored the predictions of the NLA framework on the thermal meson correlation functions in the deconfined phase of QCD, thus extending past work of Blaizot et al. [6,7,8] who successfully investigated, employing

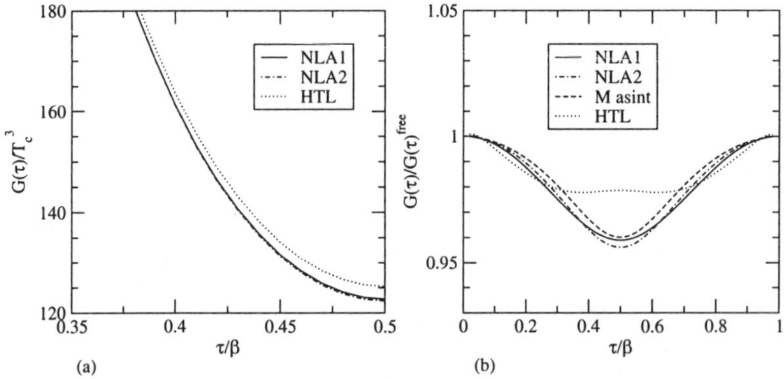

Figure 4. The same as in Fig. (3), but for $T = 4T_c$.

the NLA scheme, the thermodynamical properties of the QGP.
Actually what we have proposed is a variant of the NLA scheme, smoothly matching the soft and hard momenta regimes.
With a more careful treatment of the contribution to the meson correlation function arising from the hard quark modes we have still found that the NLA does not change dramatically the HTL results. In particular the peculiar behaviour of the meson spectral function for soft energies (Van-Hove singularities) remains (almost) unaltered.

References

1. E. Braaten, R.D. Pisarski and T.C. Yuan, Phys.Rev.Lett. **64** (1990), 2242.
2. M. Le Bellac, Thermal Field Theory, Cambridge University Press, 1996.
3. M.G. Mustafa, M.H. Thoma , Pramana **60** (2003), 711.
4. F. Karsch, M.G. Mustafa, M.H. Thoma, Phys. Lett. B**497** (2001), 249.
5. W.M. Alberico, A. Beraudo, A. Molinari, hep-ph/0411346.
6. J.P. Blaizot and E. Iancu, Phys. Rev. D**63** (2001)
7. J.P. Blaizot, E. Iancu, A. Rebhan, Phys. Lett. B**523** (2001), 143.
8. J.P. Blaizot, E. Iancu, A. Rebhan, Phys. Lett. B**470** (1999), 181.

NUCLEAR ASTROPHYSICS

ALESSANDRO DRAGO

Dipartimento di Fisica, Università di Ferrara and INFN, Sezione di Ferrara,
Via Paradiso, 12 – 44100 Ferrara, Italy
E-mail: drago@fe.infn.it

The activity of the Italian nuclear physicists community in the field of Nuclear Astrophysics is reported. The researches here described have been performed within the project "Fisica teorica del nucleo e dei sistemi a multi corpi", supported by the Ministero dell'Istruzione, dell'Università e della Ricerca.

1. Introduction

In the last years the research in Nuclear Astrophysics all over the world has received a significant boost, mainly due to the abundance of data obtained from various types of satellites, as well as from the new possibilities opened by the detection of neutrinos in laboratory experiments. Moreover, a new era is starting, concerning the search of gravitational waves, since powerful detectors are now fully operational.

In Italy, the nuclear physicists community has been extremely reactive to the suggestions coming from the observational data. Important researches have been developed concerning the structure of compact stellar objects, the study of various nuclear reactions of astrophysical importance, the search of possible connections between gravitational wave signals and the structure of compact stars. Moreover, new possible applications of neutrino detectors have been explored, opening the possibility of using neutrinos as a tool to investigate the inner structure of the Earth.

In this review I will present the main results in all these areas, and in particular I will try to emphasize the more and more strict collaboration between the various research groups participating to the project "Fisica teorica del nucleo e dei sistemi a multi corpi". The collaboration between the various components of the nuclear physicists community is particularly important in a research field like Astrophysics, which is characterized by its multidisciplinarity.

In the following I will first discuss the researches connected with the astrophysics of Compact Stars (CS). All the aspects of CSs have been touched by the theoretical research in Italy, from the analysis of the properties of the crust and in particular the estimate of the specific heat of the inner superfluid crust, to the study of neutrino propagations in Hadronic Matter (HM), to sophisticated calculations of the Equation Of State (EOS) of beta-stable HM, to the possible formation of Quark Matter (QM) inside the CS. In particular, concerning this last problem, several questions have been investigated, from the effect of the presence of QM on the mass and radius of a CS, to the computation of the viscosity of a hybrid hadron-quark star, to the analysis of the possible scenarios of formation of QM inside the CS and the search of various dynamical signals of the quark deconfinement process. For instance, the formation of QM could help Supernovae (SN) to explode, or could be at the origin of at least some of the so-called Gamma Ray Bursts (GRB), or could contribute to generating a significant kick to SN remnants or, finally, could be associated with the emission of bursts of Gravitational Waves.

In the following I will also shortly discuss a few other topics not related to CSs. In particular I will mention the studies concerning nuclear reactions, where important advances have been made both in the few body technique and in a new approach based on applying non-standard statistics to the study of several astrophysical processes. Finally, I will mention a new and very promising application of neutrino detection to the study of Geophysics.

2. Astrophysics of Compact Stars

The data accumulating from the new X-ray satellites are at the origin of the rapid development of this sector. Another strong impulse to the theoretical investigation came from the discovery of the possibility that a diquark condensate can form in the inner part of CSs and from the huge variety of phenomenological implications of this possibility. In the following I will report on the main research activities in this area.

2.1. *Crust of Compact Stars*

The structure of the crust of a CS is in itself a very interesting topic since it plays a crucial role connecting the inner part of the star, which is difficult to investigate in a direct way, to the exterior of the star. In particular the relation between the temperature of the so-called inner crust and the temperature of the exterior is of extreme importance. The relation between

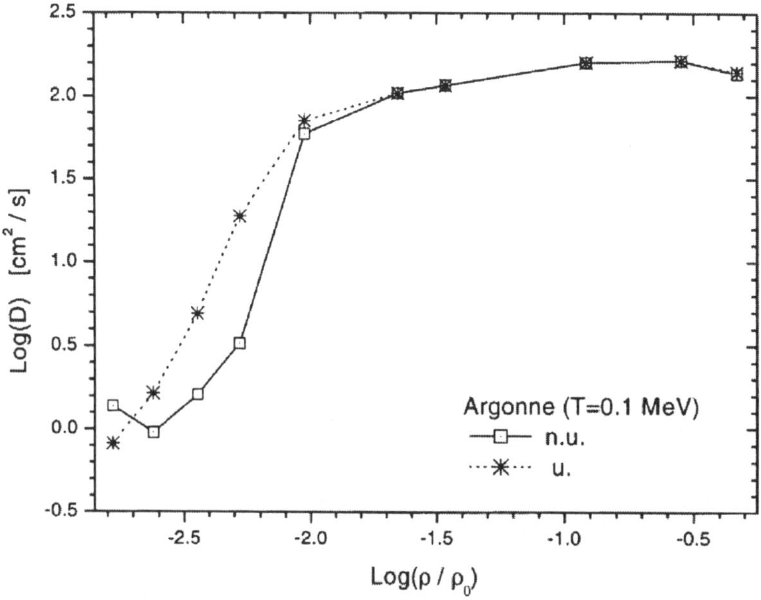

Figure 1. Thermal diffusivity for inner crust matter at T=0.1 MeV with the Argonne interaction for pairing. The solid line represents the case of nonuniform neutron matter with nuclear impurities, while the dashed line is the standard uniform neutron matter. From Ref. [1].

these two temperatures, which has been investigated many times in the past, has been discussed in a few papers by the Milano and Catania groups [1,2,3,4]. In particular, the question at the center of the analysis of Ref. [1] concerns how the presence of neutron superfluidity in the inner crust, where a gas of unbound neutrons permeates a Coulomb lattice of neutron-rich nuclei, will affect the thermal properties of this region. To estimate the superfluid gap the authors solved Hartree-Fock-Bogoliubov equations in 10 different Wigner-Seitz cells. In this way they have been able to compute the specific heat of the superfluid neutrons in each cell. The main interest in computing the heat capacity C_V lies on its relation with thermal diffusivity $D = k/C_V$, where k is the thermal conductivity. Integrating on the thermal diffusivity, one gets the diffusion time through the inner crust layer, having thickness $R_{\rm shell}$, as

$$t_{\rm diff} = \gamma \int_0^{R_{\rm shell}} \frac{{\rm d}R^2}{D(r)} \,.$$

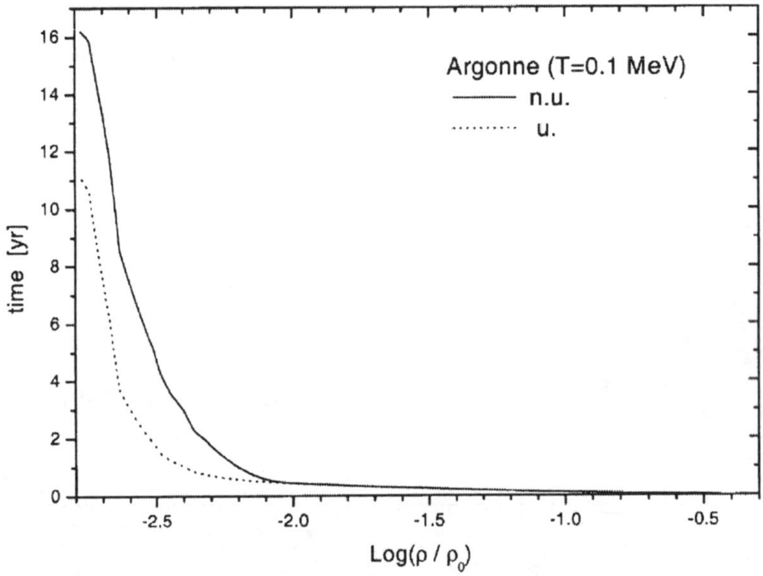

Figure 2. Diffusion time along the inner crust for nonuniform (solid line) and uniform (dashed line) neutron matter. From Ref. [1].

In Figs.1,2 the thermal diffusivity and the diffusion time are shown. It is clear that the low density region can be considered as a "bottleneck" and that the precise evaluation of $t_{\rm diff}$ is crucial to estimate the cooling time of a CS. Therefore the main result of Ref. [1], namely the effect of nuclear impurities on the diffusion time, is important for the cooling history of the star, particularly in the case of the so-called rapid cooling scenario, in which the temperature of the star rapidly drops during the first ten years of its life.

In order to investigate the pairing correlations, a detailed knowledge is necessary of the effective NN interaction, taking into account the modifications of this force due to the nuclear medium. A way to incorporate these effects is to introduce three-body forces, which are known to be crucial, e.g., for reproducing the saturation properties of nuclear matter [5,6]. In Ref. [4], the effect of a microscopic three-body force on the proton and neutron superfluidity in the $^1S_0$ channel in β-stable neutron matter is investigated. The authors found that, while the three-body force has only a limited influence on the neutron gap, it strongly suppresses the proton gap (see Fig. 3). Moreover, the density region for the proton superfluid phase is considerably shrunken as compared to the pure two-body force prediction. This result is important for the cooling of CSs, since the $^3P_2$ proton gap is strongly suppressed by relativistic effects and, therefore, the main suppression of the neutrino production comes from the proton gap.

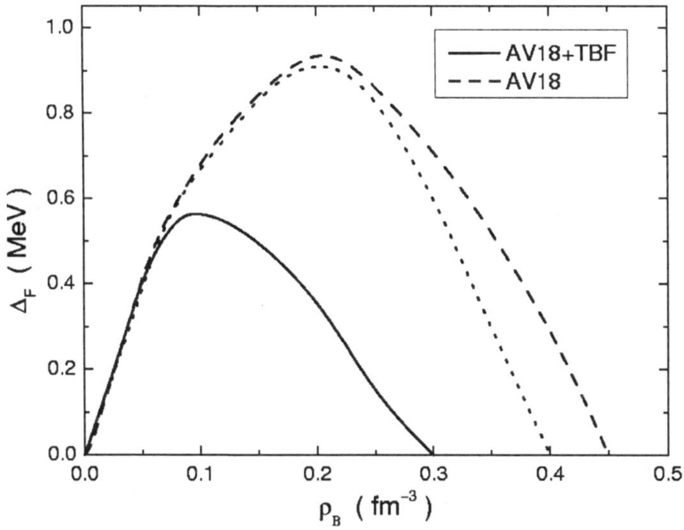

Figure 3. Proton $^1S_0$ gap as a function of the baryon density. The solid line takes into account three-body force effects. The dashed curve is obtained using the AV18 two-body force only. The dotted line corresponds to using the three-body force in the calculation of the proton fraction, but neglecting this force in the gap equation. From Ref. [4].

2.2. *Neutrino mean free path in Neutron Stars*

A crucial information in evaluating the thermal evolution of a CS in its early life is the estimate of the neutrino mean free path inside the newly formed CS. The precise estimate of this quantity is based on the evaluation of the so-called spin susceptibility of neutron matter, which depends on the details of the adopted nuclear EOS. The Catania and the Pisa groups have investigated this quantity using realistic EOSs [7,8]. In particular, they have studied the effects of short and long range correlations on neutrino transport. A result of these analysis is that it is important to take into account both neutrino scattering via neutral currents and also neutrino absorption via charge exchange processes. Moreover, absorption process dominates. In Figs. 4 and 5 it is easy to appreciate the effect of the residual interaction which enhances the neutrino mean free path and of three-body forces which magnify this effect.

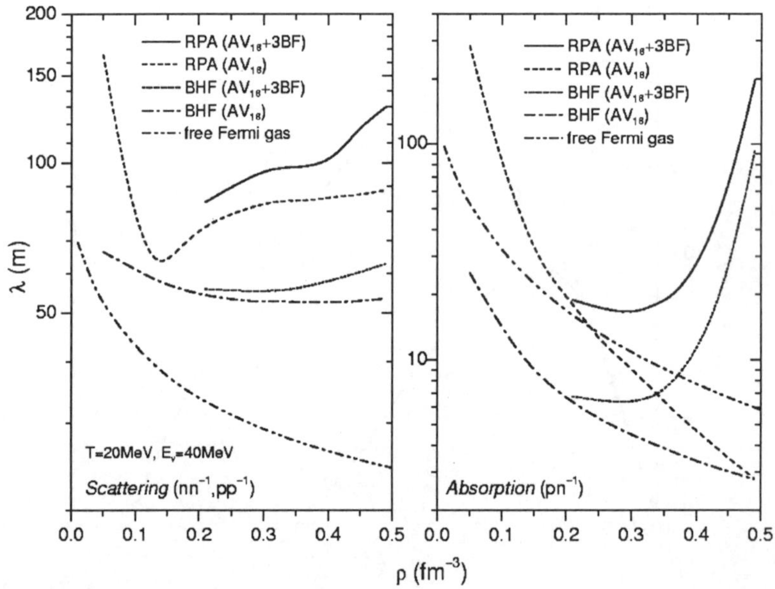

Figure 4. Neutrino mean free path for scattering (left panel) and absorption (right panel). From Ref. [7].

2.3. *Equation of State of β-stable hadronic matter*

One of the most important ingredient in several calculations of the structure of CSs is clearly the hadronic EOS, as distinguished from the quark EOS to be discussed later. Although the problem of evaluating the hadronic EOS has been at the center of an incredibly large number of works in the past, the study of β-stable matter is still far from having reached a definitive conclusion, also due to the absence of direct information from laboratory experiments. In recent years, two groups from Catania have discussed this important topic. In particular, in Ref. [9] the effect of three body forces on the mass-radius relation of a neutron star has been considered. It has been shown (see Fig.6) that the maximum mass of a purely nucleonic CS can be significantly increased when three body forces are taken into account. This result is particularly interesting at the light of the difficulties in reaching

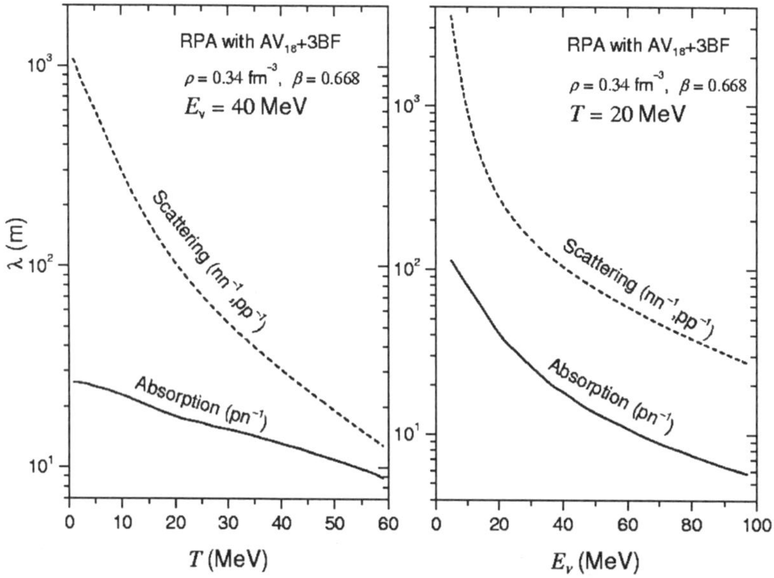

Figure 5. Neutrino mean free path vs. temperature (left) and vs. neutrino energy (right). From Ref. [7].

large values for the mass in the case of hybrid quark-hadron stars (see below).

Another difficulty in the theoretical prediction of the structure of nucleonic star is due to the poor knowledge we still have concerning the behavior of the so-called symmetry energy as a function of the density. The group of Di Toro and collaborators has proposed the existence of a new isovector - scalar interaction, which would increase the symmetry energy only at large density [10,11]. It is interesting to notice that, also in this case, the effect of the new interaction is to increase the maximum value of the mass of a purely nucleonic CS [12], as it can be seen from Figs. 7 and 8. Here the parameter set A corresponds to a phenomenological EOS which is in good agreement with microscopic T-matrix calculations in the high density region, while set B is consistent with microscopic calculations at low densities.

2.4. *Equation of State of β-stable matter including quarks*

One of the most important novelty of the decade is certainly the (re)-discovery of diquark condensation, many years after the seminal works of Bailin and Love [13], which has opened the possibility for a huge number of

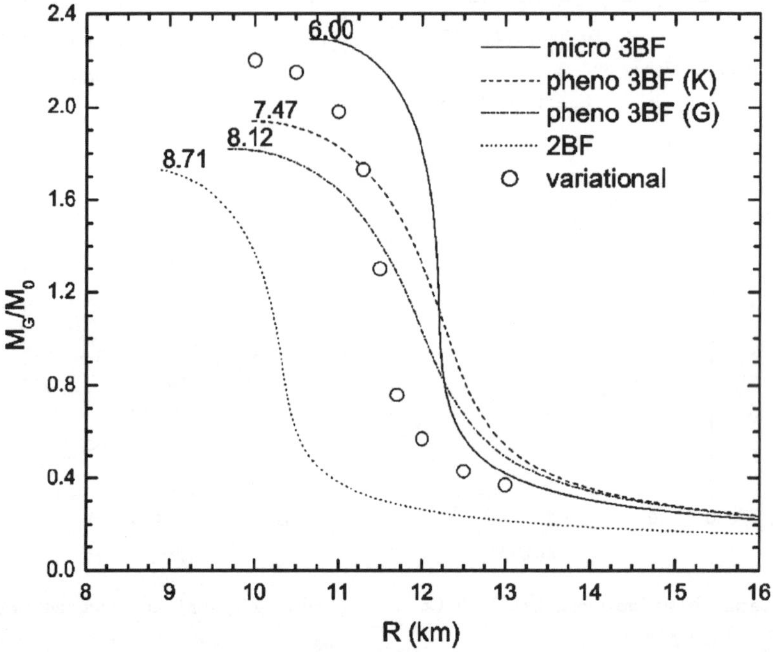

Figure 6. Mass-radius relations of neutron stars evaluated with different equations of state. The numbers denote the central density n_c/n_0 of the maximum mass configuration. From Ref. [9].

different condensates in quark matter (for a review see e.g. Ref. [14]. The effects of the formation of a condensate in the inner part of CSs are various and they concern the mass-radius relation, the cooling rate, the viscosity of the star and its stability.

2.4.1. *Quarks and the mass-radius relation*

We will begin by discussing the effects of the formation of QM on the star mass-radius relation, both in the case in which quarks are decoupled and in the case in which pairing is taken into account. The Catania group has been extremely active in the analysis of the implication of the formation of QM on the EOS and on the value of the maximum stellar mass, in particular [15,16,17,18].

The Catania group has shown that it is rather difficult to reach masses larger than $\sim 1.7\ M_\odot$ in the case of hybrid stars. Their analysis is based on the use of both non-relativistic and also of relativistic EOSs in the

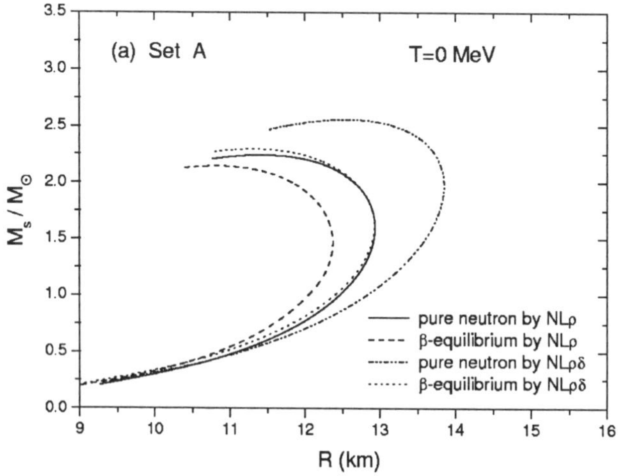

Figure 7. Mass of the neutron star as a function of the radius by set A (see text). From Ref. [12].

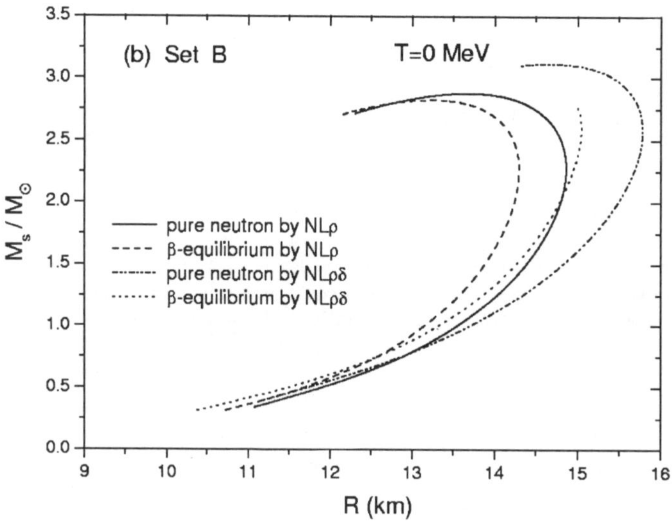

Figure 8. Mass of the neutron star as a function of the radius by set B (see text). From Ref. [12].

hadronic sector (see Fig. 9). The conclusion of their analysis is particularly well established if diquark condensate is not taken into account (although the effect of three-body forces or of the new isospin dependent interaction discussed in Sec.2.3 has still to be discussed in details).

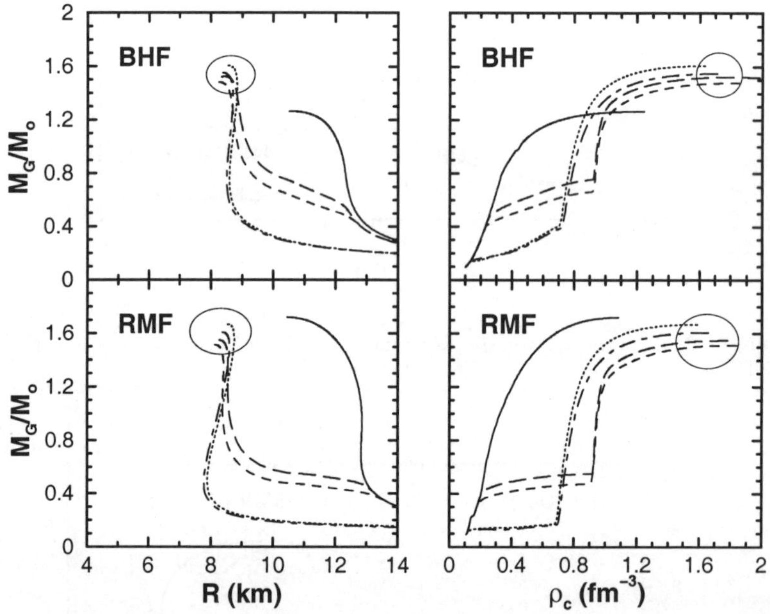

Figure 9. The mass-radius (left panel) and the mass-central density (right panel) relations for a hybrid star are displayed for several parameter sets having different values for the MIT bag constant B, the strange quark mass m_s and the strong coupling constant α_s. The upper panels correspond to a microscopic, non-relativistic EOS, while the lower panels correspond to a relativistic mean-field EOS. From Ref. [16].

The result has been confirmed in a recent calculation in which the so-called Color Dielectric Model [19,20,21,22] has been used [18] (see Fig.10).

When the diquark condensation is considered, the situation seems to be more complicated, since the stability of a hybrid star depends in this case on the existence of a layer of mixed quark-hadron phase inside the star. In the absence of that layer the star becomes unstable as soon as quarks start forming [17], as it is shown in Fig. 11, where all the branches which correspond to the formation of Color Superconducting quarks inside the star (labeled CS in the figure caption) are unstable, since the mass of the hybrid star decreases when the density increases past the critical value.

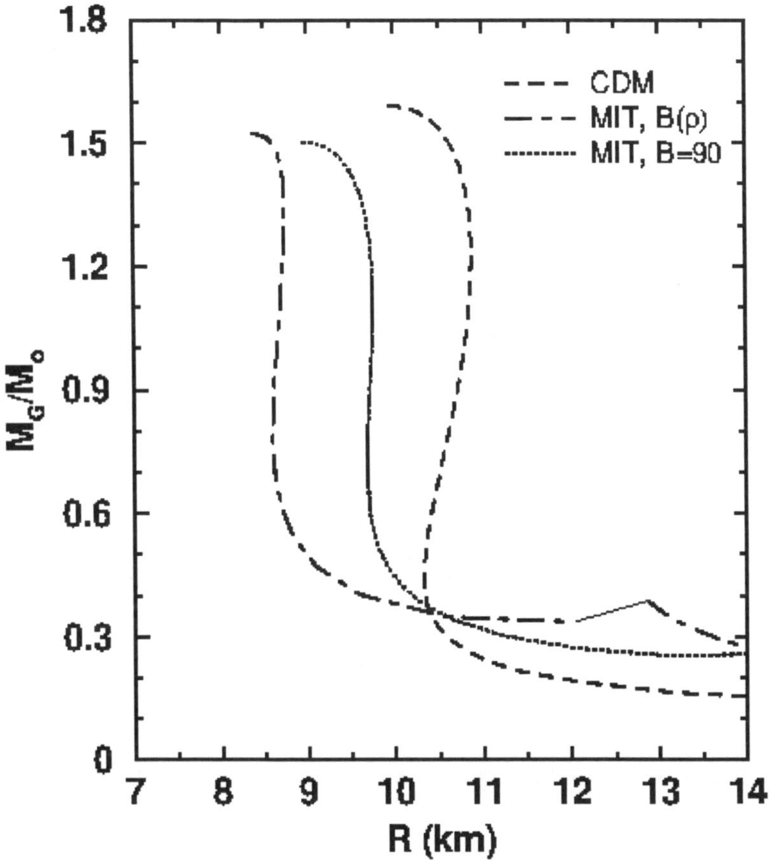

Figure 10. Mass-radius relation, using either the MIT bag model, or a model in which the pressure of the vacuum is density dependent, or the Color Dielectric Model. From Ref. [18].

If, on the other hand, a mixed phase is allowed to exist than hybrid stars are stable and they can reach slightly larger masses, as it has been shown by the Ferrara-Torino collaboration [23] (see Fig. 12, where a Color Flavor Locked (CFL) diquark condensate has been assumed to form).

Recent results seem to confirm that the mass of hybrid stars can indeed reach values up to 2 $M_{\odot}$ if quarks can condensate [24]. The difficulty in describing in a realistic way the transition from the moderate density region, in which hadrons are the relevant degrees of freedom, to the high density region, in which quarks appear, is made even greater by the large variety

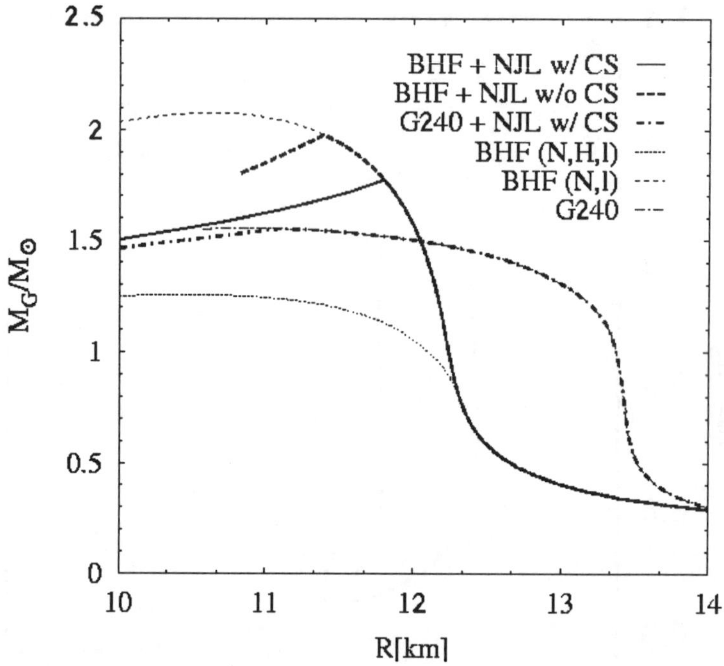

Figure 11. Mass-radius relation, for various EOSs, see text. From Ref. [17].

of different superconducting gaps which can appear in this intermediate region.

2.4.2. *Viscosity of quark matter*

A topic that has received increasing attention during the recent years concerns the stability of a rotating CS. After the discovery of the instability associated with the so-called r-modes [25,26] the search of possible mechanisms damping the instability has concentrated in particular on the evaluation of the bulk viscosity of the matter composing the CS. It has been shown that bulk viscosity, for temperatures of the order of $\sim 10^9$–10^{10} K is large in the case of hyperonic matter and also for uncorrelated QM, so that, for instance, rapidly rotating hot hyperonic stars can exist. On the other hand, bulk viscosity is small for purely nucleonic matter and it is exponentially suppressed if the quarks are completely gapped. The Ferrara-Torino group has investigated the bulk viscosity of a mixed quark-hadron phase, showing that in many cases the viscosity is large enough to stabilize hot hybrid stars [27].

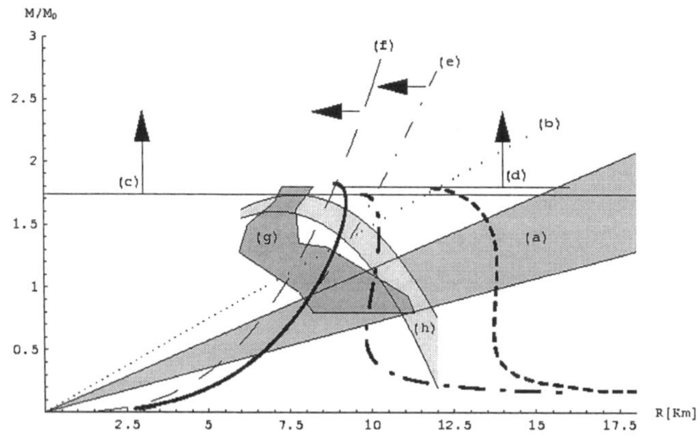

Figure 12. Mass-radius plane with observational limits and representative theoretical curves: thick solid line indicates CFL quark stars, thick dot-dashed line CFL hybrid stars, thick-dashed line hadronic stars. Various observational limits on masses and radii are also displayed. See Ref. [23] for all details.

2.5. *Delayed formation of Quark Matter*

A very important question concerning the possible formation of QM inside a CS concerns the formation scenario. In other terms, is QM forming immediately after the Supernova explosion, i.e. during the cooling of the protoneutron star? Or is it possible that the formation is delayed, so that the CS becomes metastable and only in a later moment the transition from hadronic matter to QM takes place? This question has been investigated in a series of works by the Ferrara-Pisa-Torino groups [28,23,29]. The main idea is that, if a surface tension exists at the interface between hadronic and QM, than it is possible to delay the formation of the first stable drop of QM inside the star. Since a huge amount of energy can be liberated during the formation of QM, the possibility opens that first a Supernova explode and later a second explosion takes place, maybe the one associated with a Gamma Ray Burst. In Fig. 13 a typical scenario is described, in which a purely hadronic star (branch HS) becomes metastable, due for instance to mass accretion. The CS then undergoes a transition to a stable configuration in which QM is present (e.g. a hybrid star, branches HyS). The radius of the star reduces by several kilometers and an energy of order

10^{53} erg is liberated.

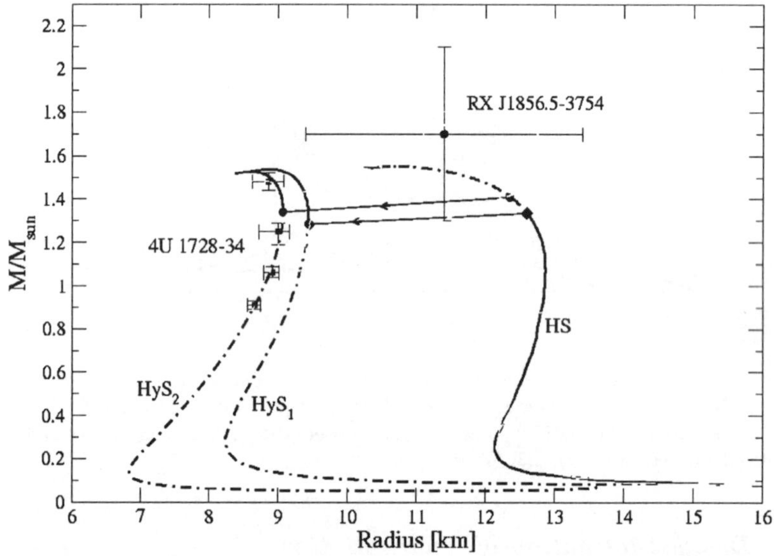

Figure 13. Mass-radius relations for a purely hadronic star (line labeled HS) and for two types of hybrid stars (lines labeled HyS_1 and HyS_2). The two almost horizontal lines indicate the conversion from a metastable hadronic star to a stable HyS (see text). Various observational constraints on masses and radii are also displayed. From Ref. [29].

2.6. *Gravitational wave emission*

During the year 2004 the first data from LIGO gravitational wave detector have been published and VIRGO detector has also started its activity. Moreover, the data obtained by the gravitational bars NAUTILUS and EXPLORER have been analyzed and they suggest the possible first evidence of a burst of gravitational waves. Most of the signals that the gravitational wave detectors will search originate from CSs, either via the merging of two CSs or of a CS and a black hole, or via steady emission of periodic waves due e.g. to r-modes (see Sec.2.4.2), or also via a re-adjustment of the structure of the CS, due for instance to the formation of QM inside the star. This last possibility has been analyzed by the Ferrara group [30] and the results are presented in the contribution of G. Pagliara [31].

2.7. *High velocities of Compact Stars*

A still open question concerns the origin of the high velocities of Supernova remnants. Although various possible explanations have been proposed, in particular associated with the possibility of a strongly asymmetric explosion, up to now none has been able to explain the data, which indicate velocities up to more than 1000 km/s. Moreover, it has recently been proposed that the distribution of the velocities is bimodal, with a first peak for velocities of the order of 100 km/s and a second peak around 700 km/s (see Fig.14). If indeed the distribution of the velocities is bimodal, then it is plausible that two different mechanisms are at work, one responsible for the more modest kicks and a second one more powerful which contributes a large extra velocity in a few cases. This possibility has been discussed by the Pisa group [32] and it has been proposed that the second kick is associated with the (delayed) formation of QM inside the hadronic star.

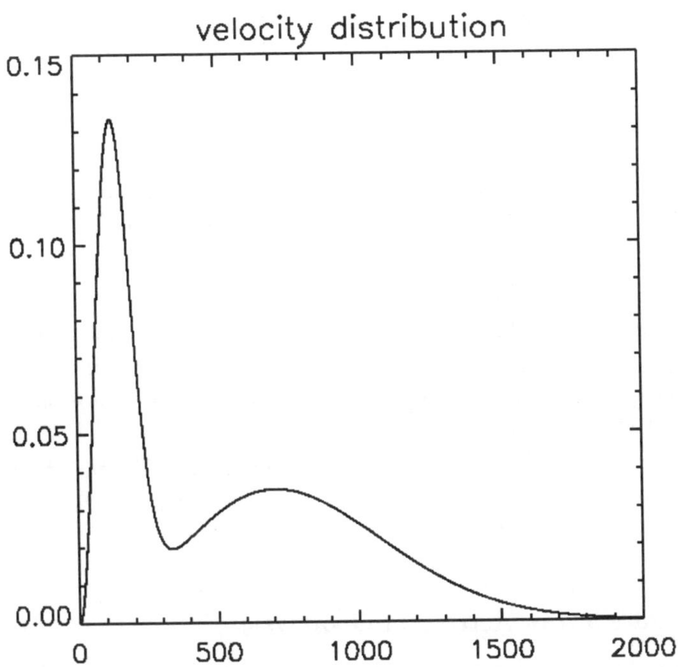

Figure 14. The initial velocity distribution of neutron stars. From Ref. [32].

3. Nuclear Reactions

A traditional topic in Nuclear Astrophysics is the study of nuclear reactions in the stars. On the one side, this is a typical playground for few-body calculations, and the progresses in that research field have been reported by L. Marcucci [33] (an excellent review of ab-initio calculations of low-energy astrophysical reactions can be found in Ref.[34]). It is important to mention a recent work by the Pisa group (Ref. [35]) in which a parameter-free calculation of the threshold S-factors for the solar proton-fusion and *hep* processes has been performed. In their approach, a method is used which combines the high accuracy of standard nuclear physics which the predict power of effective field theories, paving the way to parameter-free predictions for electroweak transitions in light nuclei.

Another research direction is the study of nuclear reactions in connection with the use of non-standard statistics. The idea of this research is to deal with the difficulties of a many-body system, in which long range interactions are present, by introducing a modification to standard distributions. In this way nonextensive statistics like the Tsallis one are introduced. The aim of this research is twofold: from one side it is necessary to provide a better understanding of the microphysics underlying the non-extensive statistics; on the other side, many systems are possible candidates for the use of these techniques, ranging from the nuclear reactions in the Sun, to nuclear reactions in metals, to the study of several plasmas of astrophysical interests. For instance, in Ref. [36] an analysis is made of the *hep* neutrino flux. There it is shown that the use of nonmaxwellian statistics to describe the high-energy tail of the ^{3}He-p momentum distribution results in an increase of *hep* flux which fits the present experimental data. Moreover, other neutrino fluxes remain compatible with the experimental signals and no modification of the temperature and of the density profiles inside the Sun is required.

The research in this field by the Cagliari and Torino groups has been particularly active [37,38,39,40,41,36,42], and it has been summarized in the contribution by A. Lavagno [43].

4. Neutrinos in Astro and Geophysics

A now traditional topic in nuclear astrophysics concerns the physics of neutrino, both as a study of the nuclear reactions in which neutrinos are produced and also as an investigation of neutrino properties which have important consequences for astrophysics. This activity has continued in the recent years due in particular to the Ferrara and Pisa groups (see e.g. Ref. [44]). A new and very promising research line is based on the idea of using neutrinos as a tool to investigate the structure of the Earth [45,46,47,48,49,50]. In particular, neutrinos can help solving the puzzle of the origin of the Earth's heat output. The main question concerns which fraction of the heat originates from natural radioactivity, through the decay chains:

$$
\begin{aligned}
&^{238}\mathrm{U} \rightarrow {}^{206}\mathrm{Pb} + 8\,{}^{4}\mathrm{He} + 6\,e^- + 6\,\bar{\nu}_e + 51.7\,\mathrm{MeV} \\
&^{232}\mathrm{Th} \rightarrow {}^{208}\mathrm{Pb} + 6\,{}^{4}\mathrm{He} + 4\,e^- + 4\,\bar{\nu}_e + 42.8\,\mathrm{MeV} \\
&^{40}\mathrm{K} + e^- \rightarrow {}^{40}\mathrm{Ar} + \nu_e + 1.513\,\mathrm{MeV} \\
&^{40}\mathrm{K} \rightarrow {}^{40}\mathrm{Ca} + \bar{\nu}_e + e^- + 1.321\,\mathrm{MeV}\,.
\end{aligned}
$$

While the neutrinos produced by the Sun completely swamp those emitted by the Earth, it is not so with antineutrinos. These can be detected via the inverse β-decay reaction: $\bar{\nu} + \mathrm{p} \rightarrow \mathrm{n} + e^+ - 1.804$ MeV, which is possible with the antineutrinos from the uranium and thorium chains, but not with the antineutrinos from potassium. The detection of these antineutrinos is within the sensitivity of several detectors, as shown in Fig. 15.

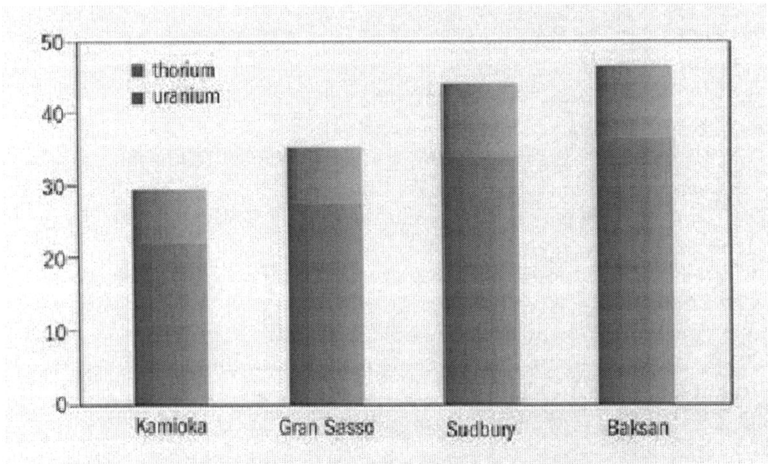

Figure 15. Predicted geoneutrino events in different laboratories per year. From Ref. [51].

5. Conclusions

In this review I have tried to present the main activity of the Italian nuclear physicist community concerning Astrophysics. A rather strict collaboration exists among the various groups. In particular, the experts of hadronic equations of state are more and more interested in exploring the possibility that quark degrees of freedom can be present at the densities reached in the center of compact stars. The researches about the structure of the crust of a compact star are deeply linked to the study of the cooling mechanism which, on the other hand, is also strictly related with the investigation of the inner composition of a compact star. New directions of research have appeared, in particular concerning the connection between the structure of a compact star and the emission of gravitational waves and also between the formation of quark matter and the release of energy powering a Gamma Ray Burst. New ideas about the possible applications of neutrino detectors in geophysics have been proposed and the relevance of non-standard statistics in a variety of physical phenomena have been discussed.

We can expect that the new data obtainable from X-ray satellites, neutrino detectors and, maybe, gravitational wave detectors will provide a further boost to the activities of this already extremely lively community.

References

1. P. Pizzochero, F. Barranco, E. Vigezzi, and R. Broglia, Astrophys.J. **569**, 381 (2002).
2. G. Gori *et al.*, Nucl. Phys. **A731**, 401 (2004).
3. G. Gori *et al.*, sent for publication (2004).
4. W. Zuo *et al.*, Phys. Lett. **B595**, 44 (2004), nucl-th/0403026.
5. A. Lejeune, U. Lombardo, and W. Zuo, Phys. Lett. **B477**, 45 (2000).
6. W. Zuo, A. Lejeune, U. Lombardo, and J. F. Mathiot, Nucl. Phys. **A706**, 418 (2002).
7. C. Shen, U. Lombardo, N. Van Giai, and W. Zuo, Phys. Rev. **C68**, 055802 (2003).
8. J. Margueron, I. Vidana, and I. Bombaci, Phys. Rev. **C68**, 055806 (2003).
9. X. R. Zhou, G. F. Burgio, U. Lombardo, H. J. Schulze, and W. Zuo, Phys. Rev. **C69**, 018801 (2004).
10. B. Liu, V. Greco, V. Baran, M. Colonna, and M. Di Toro, Phys. Rev. **C65**, 045201 (2002).
11. V. Greco, M. Colonna, M. Di Toro, and F. Matera, Phys. Rev. **C67**, 015203 (2003).
12. B. Liu, H. Guo, M. Di Toro, and V. Greco, (2004), nucl-th/0409014.
13. D. Bailin and A. Love, Nucl. Phys. **B205**, 119 (1982).
14. K. Rajagopal, Prepared for Cargese Summer School on QCD Perspectives

on Hot and Dense Matter, Cargese, France, 6-18 Aug 2001.
15. G. F. Burgio, M. Baldo, P. K. Sahu, A. B. Santra, and H. J. Schulze, Phys. Lett. **B526**, 19 (2002).
16. G. F. Burgio, M. Baldo, P. K. Sahu, and H. J. Schulze, Phys. Rev. **C66**, 025802 (2002).
17. M. Baldo *et al.*, Phys. Lett. **B562**, 153 (2003).
18. C. Maieron, M. Baldo, G. F. Burgio, and H. J. Schulze, Phys. Rev. **D70**, 043010 (2004).
19. M. C. Birse, Prog. Part. Nucl. Phys. **25**, 1 (1990).
20. H.-J. Pirner, Prog. Part. Nucl. Phys. **29**, 33 (1992).
21. M. K. Banerjee, Prog. Part. Nucl. Phys. **31**, 77 (1993).
22. A. Drago, U. Tambini, and M. Hjorth-Jensen, Phys. Lett. **B380**, 13 (1996).
23. A. Drago, A. Lavagno, and G. Pagliara, Phys. Rev. **D69**, 057505 (2004).
24. M. Alford, M. Braby, M. Paris, and S. Reddy, (2004), nucl-th/0411016.
25. N. Andersson, Astrophys. J. **502**, 708 (1998).
26. J. L. Friedman and S. M. Morsink, Astrophys. J. **502**, 714 (1998).
27. A. Drago, A. Lavagno, and G. Pagliara, (2003), astro-ph/0312009.
28. Z. Berezhiani, I. Bombaci, A. Drago, F. Frontera, and A. Lavagno, Astrophys. J. **586**, 1250 (2003).
29. I. Bombaci, I. Parenti, and I. Vidana, Astrophys. J. **614**, 314 (2004).
30. A. Drago, G. Pagliara, and Z. Berezhiani, (2004), gr-qc/0405145.
31. G. Pagliara, these Proceedings .
32. I. Bombaci and S. B. Popov, Astron. Astrophys. **424**, 627 (2004).
33. L. Marcucci, these Proceedings .
34. L. E. Marcucci, K. M. Nollett, R. Schiavilla, and R. B. Wiringa, (2004), nucl-th/0402078.
35. T. S. Park *et al.*, Phys. Rev. **C67**, 055206 (2003).
36. M. Coraddu, M. Lissia, G. Mezzorani, and P. Quarati, Physica **A326**, 473 (2003).
37. M. Coraddu, G. Mezzorani, Y. V. Petrushevich, P. Quarati, and A. N. Starostin, Physica **A340**, 496 (2004).
38. M. Coraddu *et al.*, Physica **A340**, 490 (2004).
39. A. Drago, A. Lavagno, and P. Quarati, Physica **A344**, 472 (2004).
40. F. Ferro, A. Lavagno, and P. Quarati, Physica **A340**, 477 (2004).
41. F. Ferro, A. Lavagno, and P. Quarati, Eur. Phys. J. **A21**, 529 (2004).
42. A. Lavagno, Phys. Lett. **A301**, 13 (2002).
43. A. Lavagno, these Proceedings .
44. S. Degl'Innocenti, G. Fiorentini, B. Ricci, and F. L. Villante, Phys. Lett. **B590**, 13 (2004).
45. G. Fiorentini, F. Mantovani, and B. Ricci, Phys. Lett. **B557**, 139 (2003).
46. G. Fiorentini, T. Lasserre, M. Lissia, B. Ricci, and S. Schonert, Phys. Lett. **B558**, 15 (2003).
47. G. Fiorentini, M. Lissia, F. Mantovani, and B. Ricci, (2003), physics/0305075.
48. F. Mantovani, L. Carmignani, G. Fiorentini, and M. Lissia, Phys. Rev. **D69**, 013001 (2004).

49. G. Fiorentini, M. Lissia, F. Mantovani, and R. Vannucci, AHEP **035** (2003).
50. G. Fiorentini, M. Lissia, F. Mantovani, and R. Vannucci, (2004), hep-ph/0409152.
51. G. Fiorentini and A. Pascolini, CERN Cour. **43N8**, 20 (2003).

GRAVITATIONAL WAVES FROM HYBRID STARS

G. PAGLIARA
Dipartimento di Fisica, Università di Ferrara and INFN, Sezione di Ferrara, 44100 Ferrara, Italy

We calculate the Gravitational Wave emission, induced by r-mode instability, from a rotating Hybrid Star. We explore, moreover, a model in which an Hybrid Star can become a Gravitational Waves burster. The GW bursts are produced by sudden phase transitions, around the quark core of the star, induced by r-mode spinning down. The possible connection between the predictions of this model and the bursts signal found by EXPLORER and NAUTILUS detectors during 2001 is also investigated.

1. Introduction

The proof of the existence of the Gravitational Waves (GWs) is yet a challenge for the experimentalists. The first GWs detectors, resonant bars as EXPLORER and NAUTILUS, have not given a strong evidence of the detection of a GWs signal [1] but in the near future new detectors based on Laser Interferometry as LIGO (which is already operating) [2], will have higher sensitivity and probably they will detect the GWs. The first scope of the GW detectors is to prove the existence of GWs, next step will be to use the GWs signals to study the physics of the emitting sources. Neutron Stars or in general Compact Stars (CSs), if the presence in the star of "exotic" particles as hyperons or quarks is considered, are among the most promising sources of GWs. In particular in the case of isolated CSs, from the characteristics (amplitude and frequency) of the emitted signal it will be possible to extract information on the internal structure of the star and therefore on the Equation of State (EOS) of matter at very high density. GW from CSs are generated by the nonradial oscillation modes of the star. In this article, we will not study the emission of GWs through the p-f-g modes of CSs which is described in many papers ([3,4,5]and references therein). We will concentrate on the GW emission from Hybrid Stars (HybSs),i.e. stars containing a core of Quark Matter, induced by r-mode instability. Moreover we will explore a model in which the spinning-down induced by r-modes in an HybS can trigger sudden phase transitions in the inner region of the star which are able to excite nonradial oscillation modes with the subsequent emission of GWs bursts. The results of this model of HybS as a GW burster will be compared with the data taken by NAUTILUS and EXPLORER in 2001.

2. Gravitational Waves induced by r-modes from an Hybrid Star

Recently increasing attention has been focused on the r-modes in rotating CSs. The main characteristic of r-modes is that they are unstable respect to the emission of GWs for all values of the angular velocities of the star [6,7]. This instability is called Chandrasekhar-Friedman-Schutz (CFS) instability and it is the most efficient mechanism by which an isolated CS emits GWs. In a hot-fast rotating Neutron Star, r-mode instability can lead to a strong GWs emission during the first year of the life of the star, reducing its angular velocity to a small fraction of its Keplerian angular velocity [8]. The time evolution of the r-mode instability depends strongly on the composition of the matter of the star: if the value of the viscosity is large, the r-mode can be damped and no GWs are emitted. For CSs containing strangeness in their composition as Hyperonic Stars, Quark Stars or Hybrid Stars the bulk viscosity is large for temperatures of order $\sim 10^9$K or higher and the instability window is very small. It turns out that r-modes, in these type of CSs, are damped till the temperature drops below $\sim 10^9$K [9,10,11]. At lower temperature, it has been shown that Strange Stars, Hyperonic Stars or as we will demonstrate, HybSs, enter a new r-mode instability window, they lose gradually their angular momentum and become steady sources of GWs [12,13,14]. Two scenarios exist in which r-modes instability in this type of CSs produces GW. One is based on a the emission of GWs from a hot and rapidly spinning compact stellar object, which has not lost its angular momentum in the very first part of its existence after the supernova explosion. The second scenario involves older stars which are reaccelerating due to mass accretion from a companion. If mass accretion is present, a sort of "cycle" can develop [15], in which the star periodically goes through the following steps: 1) mass accretion with increase of angular velocity; 2) instability due to r-modes excitations with reheating due to bulk viscosity 3) loss of angular momentum with emission of GWs. In both cases it is crucial to have a large bulk viscosity, because this can split the instability region into two separated temperature window. Also the reheating due to bulk viscosity is so efficient that if the instability region is reached on the low-temperature side the star can be reheated in a small period of time and reach the high-temperature side of the instability region [14]. The evolution of an HybS under r-mode instability is displayed in Fig.2 and it is computed by solving the equations regulating the process of GW emission in the two cases we discussed:

$$\frac{\dot{\alpha}}{\alpha} = \frac{-1}{t_g} - \left(1 - \frac{3\alpha^2 \tilde{J}}{2\tilde{I}}\right)\left(\frac{1}{t_v}\right) - \frac{\dot{M}}{2\tilde{I}\Omega}\left(\frac{G}{MR^3}\right)^{\frac{1}{2}} \tag{1}$$

$$\dot{\Omega} = \frac{\dot{M}}{\tilde{I}}\left(\frac{G}{MR^3}\right)^{1/2} - \frac{\dot{M}\Omega}{M} - 3\Omega\alpha^2\frac{\tilde{J}}{\tilde{I}}\left(\frac{1}{t_v}\right) \tag{2}$$

$$\dot{E}_{\text{thermal}} = \dot{E}_{\text{accretion}} + \dot{E}_{\text{viscosity}} - \dot{E}_{\text{neutrino}}\,. \tag{3}$$

Here α is the dimensionless amplitude of the r-mode, $1/t_v = 1/t_s + 1/t_b$, t_g, t_s and t_b are time scales associated with GW emission, to shear and to bulk viscosity damping, respectively. $\tilde{I}$ and $\tilde{J}$ are dimensionless values of the moment of inertia and of the angular momentum (for all details see e.g. Ref.[12]). Eq. (1) describes the damping of r-modes due to viscosity, Eq. (2) describes angular momentum conservation and, finally, Eq. (3) describes the thermal evolution given by the contributions of the reheating due to mass accretion and shear and bulk viscous dissipation of the r-modes and cooling due to neutrino emission. Obviously, in the first scenario depicted above, mass accretion is not present. To compute the time scale associated with bulk viscosity for HybSs we use the results of Ref.[11] in which the viscosity of MP has been computed. In that paper it has been shown that the viscosity of MP is of the same order of magnitude of the viscosity of pure quark matter if superconducting gaps are not present or it is reduced by a factor $\sim$ 10 if a color superconducting 2SC gap is taken into account [9]. Concerning the value of shear viscosity, we have taken into account not only the contribution associated with pure quark matter [12], but also the contribution associated with the viscous boundary layer which is present in a star having a crust made of nucleonic matter [16]. In Fig. 2 the dimensionless amplitude h and the frequency f of the GWs emitted by a a star in the first scenario are plotted as functions of time. It is even possible that the same CS enters first the instability window associated with the first scenario when the star is relatively young and, after some time, it becomes again unstable due to mass accretion as described in the second scenario. The GW in the two cases is quite similar and we did not display in this

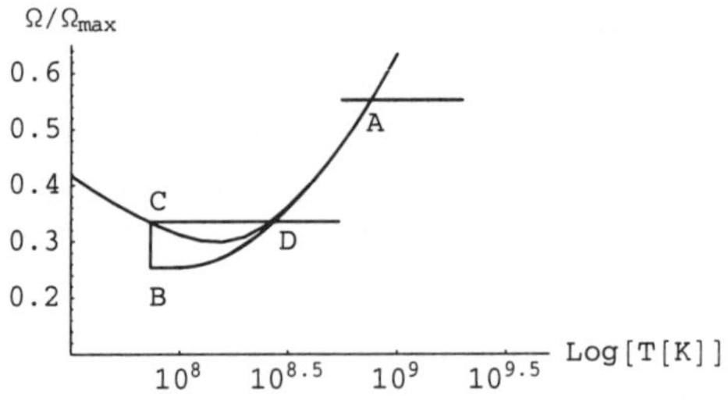

Figure 1. r-modes instability window. Coming from high temperatures (A) the star loses its angular momentum until it exits the instability region (B) (first scenario). By mass accretion the star can be re-accelerated until it reaches the instability window from the low temperature side (C). The excitation of r-modes instability produces a fast reheating of the star due to bulk viscosity dissipation (D) (second scenario). Here we used a mass accretion rate of $10^{-8} M_\odot$ per year.

paper the signal emitted in the second scenario.

As it can be seen in Fig. 2, the initial part of the signal decomposes into bursts lasting few minutes and separated by periods of few days of quiescence. During this initial phase of the emission, the star follows a trajectory in the temperature – angular velocity plane, oscillating around the instability line displayed in Fig. 2 (see also Ref. [12]). After this phase, which can last months or years, the angular momentum is dragged almost continuously and the signal becomes steady for hundreds of years until the star finally exits the instability region.

3. Hybrid Stars as Gravitational Waves bursters

Let us discuss now a model for the emission of GW burst from an HybS. The r-modes are by far the most efficient way of dragging angular momentum from a rotating CS (magnetic dipole radiation can be also an efficient mechanism if the magnetic field is very large). As shown in Refs.[17,18], when a HybS having a MP core slows down, its central density increases and the fraction of the star occupied by MP increases too , till the moment in which the central density reaches the critical value at which pure quark matter starts being produced. During all the spin-down process the radius of the star decreases by several kilometers. This gradual modification of the structure of the star can take place discontinuously, in several (small) steps, if the effect of a non-vanishing surface tension is taken into account. The effect of the surface tension is to delay the phase transitions until a critical

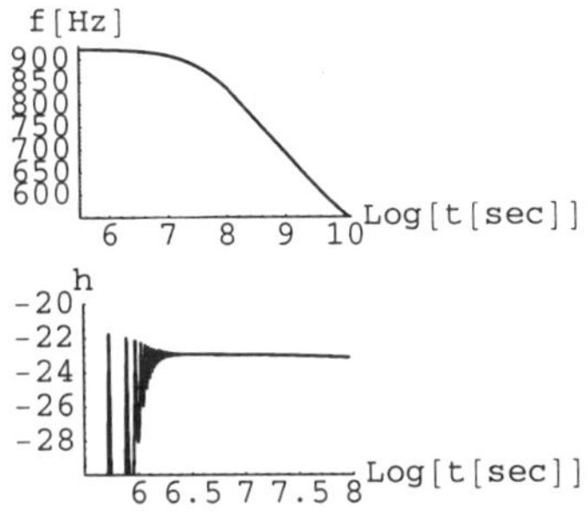

Figure 2. The frequency (upper panel) and amplitude (lower panel) of the GW signal emitted by r-modes instability as functions of time. Here the CS is approaching the instability window from the high temperature side (first scenario).

value of the overpressure is reached. The formation of new germs of stable MP proceeds through quantum tunneling with a probability which depends strongly on the value of overpressure. As shown in Refs.[19,20] the nucleation time needed to form new structures of quark matter can be very long if the overpressure is not large. When the overpressure reaches the critical value, in one randomly chosen site of the star a new drop of quark matter forms. The process of conversion of hadronic into quark matter propagates with finite velocity v_c inside the star and a sudden modification of the composition and of the structure (a sort of mini-collapse) of the star occurs during a timescale $\sim R/v_c$. During this period nonradial modes develops and a few bursts of GWs can be emitted until a new equilibrium configuration is reached.

In order to give a qualitative estimate of the magnitude of the oscillations and therefore of the amplitude of the GWs emitted we resort to the toy model proposed in Ref.[21]. We model the HybS as a spheroid containing a MP core with uniform density ρ_2 and a crust of nuclear matter with uniform density ρ_1. A Newtonian hydrostatic equation (eq. (3) of Ref. [21]) is then solved and in this way the pressure inside the star is analytically determined for a given angular velocity. It is important to calculate which is the variation of the angular velocity $\Delta\Omega/\Omega$ large enough to trigger the formation of a critical drop of quark matter, in a time scale of order days or years. The crucial ingredient in this calculation is the relation between the overpressure $\Delta P/P$ and $\Delta\Omega/\Omega$. The overpressure is determined computing, for a same element of fluid, the difference between the value of the pressure after and before the slow down (lagrangian perturbation). In particular, we are interested in the value of the overpressure in the region immediately surrounding the core of already formed MP. It turns out from the calculation that for a value $\Delta\Omega/\Omega = 0.05$, $\Delta P/P \sim \alpha\ \Delta\Omega/\Omega$, with $\alpha \sim 0.3 \div 1.4$, larger values corresponding to the fastest rotating stars. The computation of the nucleation time for the obtained value of the overpressure and a given value of σ can be done following the formalism developed in Refs.[19,20] and based on quantum tunneling. Within the toy model nucleation time of order days can be obtained, for overpressures corresponding to variation of the angular velocity $\Delta\Omega/\Omega = 0.05$ using values of $\sigma \sim$ a few MeV/fm^2. We can now describe more precisely the GW emission in the mechanism here introduced. As it has been seen in Fig. 2, it is possible to reduce the angular velocity of the star by some 10–20% in $\sim$ 10 years via emission of periodic GWs induced by r-mode instabilities. This reduction corresponds to an increase of the inner pressure by roughly the same amount. As we have seen, when an overpressure of order a few percent is reached, the star will reassess forming a new region of MP. Therefore, we can expect a few periods of GW bursts activity in 10 years, a time structure similar to that of a starquake activity. Also the randomness of the size of the collapsing region is typical of quake phenomena. Once a fraction of the metastable layer has collapsed, the other parts will presumably follow the same fate in a timescale much shorter than the time needed to reach the critical value of the overpressure. In other terms, we expect to have a few bursts taking place in a relatively short period, while a much longer delay (of order years) separates the phases of quake activity. This time structure is similar to the temporal distribution observed in soft-gamma repeaters and interpreted as due to starquake activity [22].

Concerning the energy of the GW bursts, in our model, this quantity can be estimated from the equation:

$$E_{GW} = M\,(\Delta R/R)^2\ , \tag{4}$$

where M is mass in quadrupole motion and ΔR is the amplitude of the oscillation. M is the total mass of the CS and ΔR is of the order of the shrinking of the radius of the star due to formation of a new layer of MP. Using the toy model results, we can estimate that for each mini-collapse the variation of the radius of the star is of order 20–30 m, which corresponds to $\Delta R/R \sim 2 \div 3 \times 10^{-3}$. The energy released in GWs is therefore of order $E_{GW} \sim (0.5 \div 1) \times 10^{-5} M_\odot$. A more realistic model should give larger variations of the radius for a same value of $\Delta\Omega/\Omega$ (see Refs. [17,18]). We can expect in fact that the energy released in GWs can be up to an order of magnitude larger, approaching $E_{GW} \sim 10^{-4} M_\odot$. Concerning the value of the Fourier transformed of the amplitude of the GW, $\tilde{h}(f)$, this can be estimated using the relation:

$$\tilde{h} = 2 \times 10^{-19} \frac{\Delta R}{R} \left(\frac{1\mathrm{kHz}}{f}\right)^{3/2} \left(\frac{1\mathrm{kpc}}{d}\right) \left(\frac{M_\odot}{M}\right)^{1/2} \mathrm{s}\,.$$

f is the frequency of the GW signal and d is the distance from the source.

Let us show now how this model can help to interpret the experimental data taken in 2001 by EXPLORER and NAUTILUS. The analysis of the data shows the existence of coincidences between the signal detected by the two resonant bars[1,23]. In particular, an excess of coincidences respect to the background is concentrated around sidereal hour four, which corresponds to the orientation for which the sensitivity of the bars is maximal for a signal coming from the direction of the galactic center. Although the statistical significance of this signal is debated [24,25,26], it is interesting to investigate the possible origin of the inferred signal using existing models of GW emission. The Fourier transformed of the amplitude of the "detected" signal was $\tilde{h} = 2 \times 10^{-21} Hz^{-1}$. From a time-evolution analysis it turns out that events are clustered with a few events detected in a few days and long periods of quiescence. The time structure of these data seems to be quite similar to the one of the GW burster model we are discussing. Concerning the energy of the GW bursts, using Eq.(5) and taking $\Delta R/R \sim 10^{-2}$ and a distance $d \sim 1$ kpc, the resulting amplitude is of the same order of the one measured in EXPLORER and NAUTILUS experiments. We want to stress that this model, as presented here, is qualitative. A realistic calculation in which we consider a more realistic EOS is in progress [27].

Let us now discuss the stellar objects which can be possible candidates for our model. If we assume that all neutron stars are born with a large value of angular velocity, they will all enter the instability window as described in the first scenario. Taking a neutron star production rate of order 0.02 per year in our galaxy, and assuming the possibility to detect GW bursts up to a distance of order 1 kpc, the probability of finding an active burster in this region if of order percent, if the total duration of the emission phase is of order 50 years. If the possibility of "recycling" described in the second scenario is taken into account, the probability can be larger. A precise estimate of the probability would require a precise knowledge of the number of millisecond pulsars, what is not known at the moment.

In conclusion, we studied two mechanisms by which an HybS emit GWs. The first mechanism produces a periodic GW signal having a duration of many millions of years. The second mechanism, produces burst of GW and have interesting similarities with the data detected by NAUTILUS and EXPLORER on 2001. Both these theoretical predictions will be surely tested by the next generation GW detectors which will be enough sensible to detect these type of signals.

References

1. P. Astone et al., Class. Quant. Grav. **19**, 5449 (2002).
2. B. Abbot et al., Phys. Rev. D. **69**, 102001 (2004).
3. O. Benhar, V. Ferrari, L. Gualtieri, (2004), `astro-ph/0407529`.
4. G. Miniutti, J. A. Pons, E. Berti, L. Gualtieri, and V. Ferrari, Mon. Not. Roy. Astron. Soc. **338**, 389 (2003).
5. K. D. Kokkotas, T. A. Apostolatos, N. Andersson, Mon. Not. Roy. Astron. Soc. **320**, 307 (2001).
6. N. Andersson, Astrophys. J. **502**, 708 (1998).
7. J. L. Friedman and S. M. Morsink, Astrophys. J. **502**, 714 (1998).
8. L. Lindblom, B. J. Owen, S. M. Morsink, Phys. Rev. Lett. **80**, 4843 (1998).
9. J. Madsen, Phys. Rev. Lett. **85**, 10 (2000).
10. L. Lindblom and B. J. Owen, Phys. Rev. **D65**, 063006 (2002).
11. A. Drago, A. Lavagno, and G. Pagliara (2003), `astro-ph/0312009`.
12. N. Andersson, D. I. Jones, and K. D. Kokkotas, Mon. Not. Roy. Astron. Soc. **337**, 1224 (2002).
13. R. V. Wagoner, Astrophys. J. **578**, L63 (2002).
14. A. Reisenegger and A. A. Bonacic, Phys. Rev. Lett. **91**, 201103 (2003).
15. N. Andersson, D. I. Jones, K. D. Kokkotas, and N. Stergioulas, Astrophys. J. **534**, L75 (2000).
16. L. Bildsten and G. Ushomirsky (1999).
17. N. K. Glendenning, S. Pei, and F. Weber, Phys. Rev. Lett. **79**, 1603 (1997).
18. E. Chubarian, H. Grigorian, G. S. Poghosyan, and D. Blaschke, Astron. Astrophys. **357**, 968 (2000).
19. K. Iida and K. Sato, Phys. Rev. **C58**, 2538 (1998).
20. Z. Berezhiani, I. Bombaci, A. Drago, F. Frontera, and A. Lavagno, Astrophys. J. **586**, 1250 (2003).
21. H. Heiselberg and M. Hjorth-Jensen (1998), `astro-ph/9801187`.
22. J. A. de Freitas Pacheco (1998), `astro-ph/9805321`.
23. E. Coccia, F. Dubath, and M. Maggiore (2004), `gr-qc/0405047`.
24. L. S. Finn, Class. Quant. Grav. **20**, L37 (2003).
25. P. Astone et al., Class. Quant. Grav. **20**, S785 (2003a).
26. P. Astone, G. D'Agostini, and S. D'Antonio, Class. Quant. Grav. **20**, S769 (2003b).
27. A. Drago, V. Ferrari, L. Gualtieri, G. Pagliara, work in progress.

NONEXTENSIVE STATISTICAL EFFECTS ON NUCLEAR ASTROPHYSICS AND MANY-BODY PROBLEMS

A. LAVAGNO AND P. QUARATI

Dipartimento di Fisica, Politecnico di Torino and INFN - Sezione di Torino e di Cagliari, Italy

Density and temperature conditions in many stellar core (like the solar core) imply the presence of nonideal plasma effects with memory and long-range interactions between particles. This aspect suggests the possibility that the stellar core could not be in a global thermodynamical equilibrium but satisfies the conditions of a metastable state with a stationary (nonextensive) power law distribution function among ions. The order of magnitude of the deviation from the standard Maxwell-Boltzmann distribution can be derived microscopically by considering the presence of random electrical microfields in the stellar plasma. We show that such a nonextensive statistical effect can be very relevant in many nuclear astrophysical problems.

1. Introduction

The solar core is a neutral system of electron, protons, alpha particles and other heavier nuclei, usually assumed as an ideal plasma in thermodynamical equilibrium described by a Maxwellian ion velocity distribution. Because the nuclear rates of the most important reactions in stellar core are strongly affected by the high-energy tail of the ion velocity distribution, let us start by remanding the meaning of ideal and non-ideal plasma. A plasma is characterized by the value of the plasma parameter Γ

$$\Gamma = \frac{\langle U \rangle_{\rm Coulomb}}{\langle T \rangle_{\rm thermal}} \,, \tag{1}$$

where $\langle U \rangle_{\rm Coulomb}$ is the mean Coulomb potential energy and $\langle T \rangle_{\rm thermal}$ is the mean kinetic thermal energy. Depending on the value of the plasma parameter we can distinguish three regimes:

- $\Gamma \ll 1$ - Diluite weakly interacting gas, the Debye screening length R_D is much greater than the average interparticle distance $r_0 \approx n^{1/3}$, there is a large number of particles in the Debye sphere.

- $\Gamma \approx 0.1 \div 1$ - $R_D \approx r_0$, it is not possible to clearly separate individual and

collective degree of freedom and the plasma is a weakly non-ideal plasma.
- $\Gamma \geq 1$ - High-density/low-temperature plasma, Coulomb interaction and quantum effects dominate and determine the structure of the system.

In the solar interior the plasma parameter $\Gamma_{\odot} \simeq 0.1$ and the solar core can be considered as a weakly nonideal plasma. Similar behavior occurs in other astrophysical systems with $0.1 < \Gamma < 1$, among the others we quote brown dwarfs, the Jupiter core, stellar atmospheres. Weakly nonideal conditions can influence how the stationary equilibrium can be reached within the plasma. In fact, in weakly nonideal astrophysical plasmas we have that the collision time is of the same order of magnitute of the mean time between collisions, therefore, several collisions are necessary before the particle loses memory of the initial state; collisions between quasi-particles (ion plus screening cloud) are inelastic and **long-range** interactions are present.

In the next section we will see how the presence of memory and long-range forces can influence the thermodynamical stability and the stationary distribution function inside the stellar core.

2. Metastable states of stellar electron-nuclear plasma

We can distinguish two kind of thermodynamical equilibrium state [1]:

- global thermodynamical equilibrium: the free energy density is minimized globally
- local thermodynamical equilibrium: free energy density is minimized only in a restricted space, not globally. In this case the system is in a metastable state.

Metastable states are always characterized by **long-range interactions and/or fluctuations** of intensive quantities (like inverse temperature β, density, chemical potential) and the stationary distribution function can be different from the Maxwellian one. In fact, in many-body long-range-interacting systems, it has been recently observed the emergence of long-standing *quasi stationary (metastable) states* characterized by non-Gaussian velocity distributions, before the Boltzmann-Gibbs equilibrium is attained [2,3].

Considering the corrections to an ideal gas due to identity of particles and to inter-nuclear interaction and the black-body radiation emitted, by minimizing the free energy density of the electron-nuclear plasma, we have obtained the following values [4]

$$n_* \simeq 2.74 \cdot 10^{-14}\,\mathrm{fm}^{-3} \quad , \quad \mathrm{k_B}T_* \simeq 5 \cdot \mathrm{keV} \;\text{ and }\; R_* \approx 0.2 R_\odot \,,$$

with a typical stellar chemical composition $\bar{Z} = 1.25$. States with different values of $\mathrm{k_B}T$ (lower) and n (higher) are **metastable** states that can be featured by temperature fluctuations or density fluctuations, by quasi-particle models or by the presence of self-generated magnetic fields or random microfields distributions.

The values obtained above are more than three times higher than the actual temperature of the solar interior and an electron density about half the actual value in the solar core. Therefore, the core of a star like the Sun can not exactly be considered in a global thermodynamical equilibrium state but can be better described as a metastable state and the stationary distribution function could be slightly different form the Maxwellian distribution.

In this context, it has been shown that when many-body long-range interactions are present, in many cases the system exhibits stationary metastable properties with power law distribution well described within the Tsallis nonextensive thermostatistics [5,6,7,8].

3. Microscopic interpretation: random electrical microfield

In this section we want to investigate about a microscopic justification of a metastable power-law stationary distribution inside a stellar core. At this scope, let us start by observing that the time-spatial fluctuations in the particles positions produce specific fluctuations of the microscopic electric field (with energy density of the order of 10^{-16} MeV/fm^3) in a given point of the plasma. These microfields have in general long-time and long-range correlations and can generate anomalous diffusion. The presence of the electric microfield average energy density, $\langle E^2 \rangle$, modifies the stationary solution of the Fokker-Planck equation and the ion equilibrium distribution can be written as

$$f(v) = C \exp\left\{ -\int_0^v \frac{mv dv}{kT(1+\frac{\langle E^2\rangle}{E_c^2})} \right\} , \tag{2}$$

where $E_c = \nu\sqrt{3xmkT/2e^2}$. In the solar core being E not too larger than E_c, the distribution differs slightly from the Maxwellian. Crucial quantity is the elastic collision cross section is the enforced elastic Coulomb cross section $\sigma_0 = 2\pi(\alpha r_0)^2$ where r_0 is the inter-particle distance, α is related to the pair-correlation function $g(R,t)$. The stationary (metastable) distribution (2) for the solar interior can be written as a function of the kinetic energy ϵ_p

$$f(\epsilon_p) = N \exp\left[-\frac{\epsilon_p}{kT} - \delta\left(\frac{\epsilon_p}{kT}\right)^2\right], \tag{3}$$

where the deformation parameter $\delta = (1-q)/2$ can be written as [9]

$$|\delta| \approx \frac{\sigma_0^2}{3\langle\sigma_d^2\rangle} = 12\,\alpha^4\,\Gamma^2 \ll 1\,. \tag{4}$$

A reasonable evaluation of α gives: $\alpha = 0.55$, with $\Gamma \sim 0.1$ and we obtain $q = 0.990$ ($\delta = 0.005$). In the next section we will see as such a small deviation of the MB distribution can be very relevant in several nuclear astrophysical applications.

4. Signals in astrophysical problems

Let us illustrate few problems where we can find signals of the presence of deviations from the MB distribution. Their solutions can be achieved by means of modified (or generalized) rates calculated by means of deformed distributions. Among the others, we quote A) Solar neutrino fluxes; B) Jupiter energy production; C) Atomic radiative processes in electron nuclear plasmas; D) Abundance of Lithium; E) Temperature dependence of modified CNO nuclear reaction rates and resonant fusion reactions. For brevity we will discuss here the last point only. A detailed discussion of the other problems can be found in Ref.s [9,10].

4.1. *Temperature dependence of modified CNO nuclear reaction rates*

The temperature dependence of CNO cycles nuclear rates is strongly affected by the presence of nonextensive effects in Sun like stars evolving towards white dwarfs ($10^7 \div 10^8$ K). Small deviations ($q = 0.991$) from MB distribution strongly increase the rates and can explain the presence of heavier elements (Fe, Mg) in final composition of white dwarfs, consistently with recent limit of the fraction of energy the Sun produces via the CNO fusion cycle (neutrino constraints). We obtain that [11] i) the luminosity yield of the *pp* chain is slightly affected by the deformed statistics, with respect to the luminosity yield of the CNO cycle; ii) the nonextensive CNO correction ranges from 37% to more than 53%; iii) above $T \approx 2 \cdot 10^7$ K, the luminosity is mainly due to the CNO cycle only, thus confirming that CNO cycle always plays a crucial role in the stellar evolution, when the star grows hotter toward the white dwarf stage. Our results are reported in Fig. 1 and Fig. 2. In Fig. 1, we plot the dimensionless luminosity over temperature, for the *pp* chain and the CNO cycle. In Fig. 2, we report the dimensionless equilibrium concentrations of CNO nuclei over temperature.

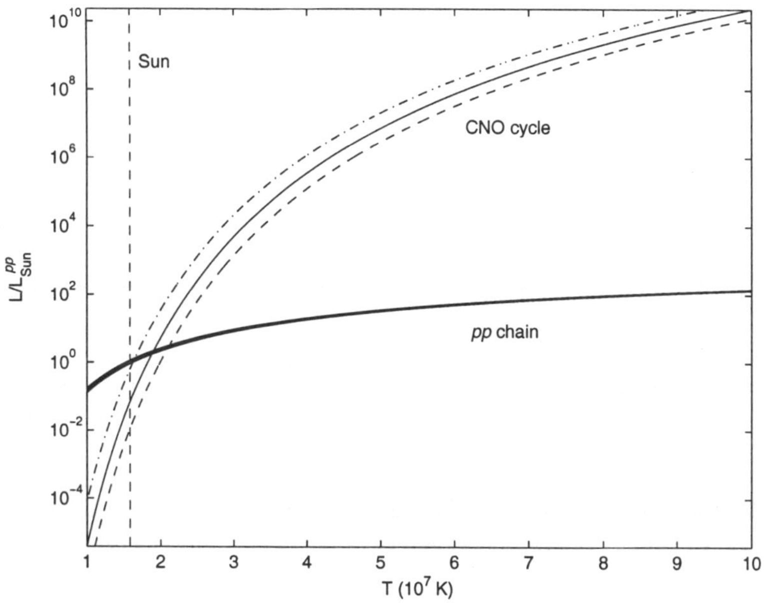

Figure 1. Log-linear plot of dimensionless luminosity over temperature, for the *pp* chain and the CNO cycle. Dashed line, $\delta = +0.0045$, $q = 0.991$; dash-dotted line, $\delta = -0.0045$, $q = 1.009$. The vertical line shows the Sun's temperature. All curves are normalized with respect to the *pp* luminosity inside the Sun.

4.2. *Resonant reaction rates in astrophysical plasma*

Cussons, Langanke and Liolios [12] proposed, on the basis of experimental measurements at energy $E \sim 2.4\,\text{MeV}$, that the resonant behavior of the stellar $^{12}C + ^{12}C$ fusion cross section could continue down to the astrophysical energy range.

The reduction of the resonant rate due to resonant screening correction amounts to 11 orders of magnitude at the resonant energy of 400 keV, with important implications for hydrostatic burning in carbon white dwarfs.

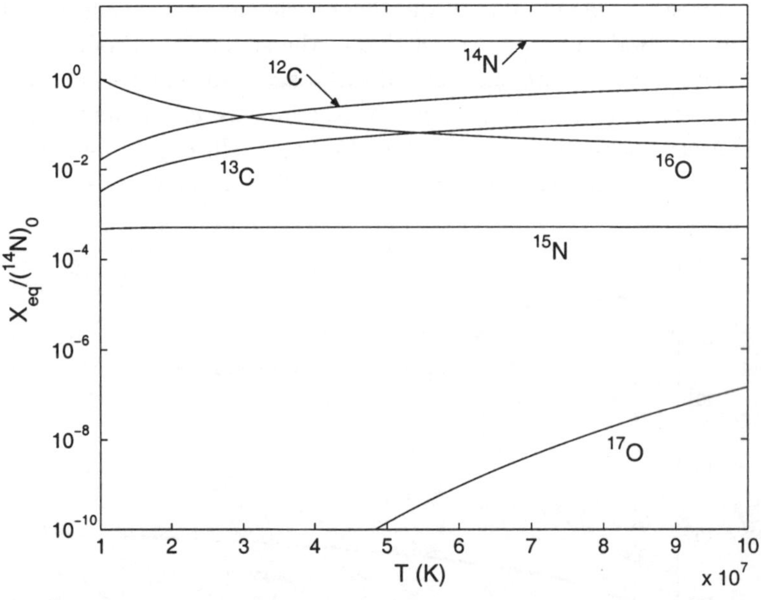

Figure 2. Log-linear plot of dimensionless equilibrium concentrations of CNO nuclei over temperature. Classical statistics has been used. All curves are normalized with respect to the initial density $(^{14}N)_0$ inside the Sun.

We have analytically derived two first-order formulae that can be used to express the non-extensive reaction rate as a product of the classical reaction rate times a suitable corrective factor for both narrow and wide resonances.

Concerning the fusion reactions between two medium-weighted nuclei, for example the $^{12}\mathrm{C} + {}^{12}\mathrm{C}$ reaction, our non-extensive factor, which can be formally defined as follows [13]

$$f_{NE} = 1 + \frac{15}{4}\delta - \left(\frac{E_R}{\mathrm{k_B}T}\right)^2 \delta \, ,$$

gives rise to further correction beside the screening and the potential resonant screening

$$F = f_{NE} \cdot f_S \cdot f_{RS} \, , \tag{5}$$

where f_S and f_{RS} account for the Debye-Hückel screening and the resonant screening effect respectively.

We have applied our results to a physical model describing a carbon white dwarf's plasma, with a temperature of $T = 8 \cdot 10^8\,\mathrm{K}$ and a mass density of $\rho = 2 \cdot 10^9\,\mathrm{g/cm^3}$ (the plasma parameter is, correspondingly,

$\Gamma \simeq 5.6$). Furthermore, we have set a deformation parameter $|\delta| = 10^{-3}$, regardless of its sign, and we have kept the energy of the possible resonance, E_R, as a free parameter. In Fig. 3 we plot our estimation of the effective total factor F as a function of the resonance energy E_R.

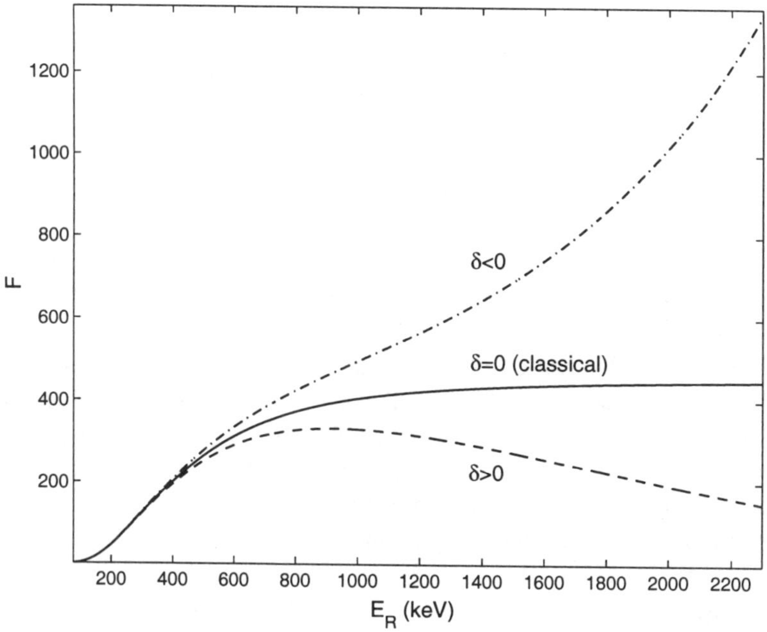

Figure 3. Linear plot of the effective factor F, defined in Eq.(5), against the resonance energy E_R. The dash-dotted (upper) line refers to super-extensivity, the dashed (lower) line to sub-extensivity, while the solid (middle) line describes the classical (MB) result.

All the plasma enhancements due to the presence of long-range many-body nuclear correlations and memory effects are in the direction of still more increasing the effective factor F of nuclear rates of hydrostatic burning and white dwarfs environment.

5. Signals in high-energy nuclear collisions

In this section we want briefly to remark as nonextensive statistical effects can be also very relevant also in the phenomenological interpretation of the high-energy nuclear collisions data. In fact, the quark-gluon plasma close to the critical temperature is a strongly interacting system. For such a system, the color-Coulomb coupling parameter of the QGP can be defined in analogy as

$$\Gamma \approx C \frac{g^2}{r_0 T} > 1 \tag{6}$$

where $C = 4/3$ or 3 is the Casimir invariant for the quarks or gluons, respectively, and $\alpha_s = g^2/(4\pi) = 0.2 \div 0.5$, $r_0 \simeq n^{1/3} \simeq 0.5$ fm.

Near the phase transition, the interaction range is much larger than the Debye screening length (small number of partons in the Debey sphere). In fact, $\lambda_D = 1/\mu \leq 0.2$ fm (using the non-perturbative estimate: $\mu = 6T$)v. The Coulomb radius for a thermal parton with energy $3T$ is given by $< r >= Cg^2/3T = 1 \div 6$ fm. Therefore one obtain $< r > /\lambda_D = 5 \div 30$. Memory effects and long–range color interactions give rise to the presence of non–Markovian processes in the kinetic equation affecting the thermalization process toward equilibrium as well as the standard equilibrium distribution.

A complete description of the applicability of nonextensive statistical effects to high-energy heavy ion collisions lies out the scope of this contribution. However, we want to outline that this aspect has been recently studied by us in connection to a phenomenological interpretation of the SPS data [14,15] and an analysis of the transverse pion momentum spectra and the net proton rapitity distribution measured at RHIC is under investigation.

References

1. G.L. Sewell, Phys. Rep. **57**, 307 (1980).
2. V. Latora, A. Rapisarda, C. Tsallis, Phys. Rev. **E64**, 056134 (2001).
3. M. Montemurro, F. Tamarit, C. Anteneodo, Phys. Rev. **E67**, 031106 (2003).
4. F. Ferro, A. Lavagno, P. Quarati, *Metastable and stable equilibrium states of stellar electron-nuclear plasmas*, submitted Jannuary 2004.
5. M. Gell-Mann, C. Tsallis Eds., *Nonextensive entropy-Interdisciplinary applications*, Oxford University Press, Oxford 2004.
6. E. Borges *et al.*, Phys. Rev. Lett. **89**, 254103 (2002).
7. Y. S. Weinstein, S. Lloyd, C. Tsallis, Phys. Rev. Lett. **89**, 214101 (2002).
8. G. Ananos, C. Tsallis, Phys. Rev. Lett. **93**, 020601 (2004).
9. A. Lavagno, P. Quarati, Phys. Lett. **B498**, 291 (2001).
10. M. Corraddu *et al.*, Physica **A305**, 282 (2002).
11. F. Ferro, A. Lavagno, P. Quarati, Physica **A340**, 477 (2004).
12. R. Cussons, K. Langanke, T. Liolios, Eur. Phys. J. **A15**, 291 (2002).
13. F. Ferro, A. Lavagno, P. Quarati, Eur. Phys. J. **A21**, 529 (2004).
14. W.M. Alberico, A. Lavagno, P. Quarati, Eur. Phys. J. **C12**, 499 (2000); Nucl. Phys. **A680**, 94c (2001).
15. A. Lavagno, Phys. Lett. **A301**, 13 (2002); Physica **A305**, 238 (2002).

FEW-NUCLEON SYSTEMS

A. KIEVSKY

Istituto Nazionale di Fisica Nucleare
and
Dipartimento di Fisica, Universita' di Pisa,
Via Buonarroti 2, 56100 Pisa, Italy

Recent advances in the theoretical description of few-nucleon systems are reported. This research activity has been performed under the Italian project FISICA TEORICA DEL NUCLEO E DEI SISTEMI A MOLTI CORPI. Bound and scattering states as well as specific reactions are analyzed in connection with the current experimental activity.

1. Introduction

One main interest in studying few-nucleon systems is to examine our knowledge of the nuclear interaction. The new generation of nucleon–nucleon (NN) potentials can be used to calculate bound and scattering states and, from a comparison to experimental data, important conclusions about the capability of those interactions to reproduce the dynamics can be extracted. In the framework of the non-relativistic dynamics, it is widely accepted that the potential energy consists in a sum of the pairwise NN interaction plus a term including a pure three-nucleon interaction (TNI). This term is not very well known and, in general, its strength is fixed in order to reproduce the experimental $A = 3$ binding energy. With the recent advances in the solution of the $3N$ and $4N$ continuum, the possibility of using scattering data to improve our knowledge of the TNI is at present feasible.

Parallel to the description of bound and scattering states is the study of electroweak reactions and pion production in few-nucleon systems. To this aim the initial and final state correlations of the nuclear system have to be taken into account. On the other hand, the simple picture in which the nuclear electromagnetic and weak current operators are expressed in terms of individual nucleons is certainly incomplete. The nuclear interaction is mediated by meson exchange mechanisms which lead to many-body current operators. Different terms of the current are related to the Hamiltonian of the system through the continuity equation in such a way that these two subjects, the nuclear interaction and the nuclear current are not independent.

Theoretical studies on few-nucleon systems have a long tradition in Italy. In the last years important results have been achieved and strong efforts have been done in order to support the related experimental activity. The main topics addressed by Italian groups that are currently under way are the following:

(1) *Firenze.* The weak axial nuclear heavy meson exchange currents.[1]

(2) *Padova.* One-pion three–nucleon force effects in $N-d$ scattering,[2] pion electroproduction and proton pionic capture on the deuteron.[3,4]

(3) *Perugia.* Electron disintegration of ^{2}H and ^{3}He at intermediate energies using a generalized Glauber approach.[5]

(4) *Pisa.* Bound and scattering states in nuclei with $A = 3, 4$ using the hyperspherical harmonic expansion,[6–9] radiative $p-d$ and $n-d$ capture reactions and the two-body electrodisintegration of ^{3}He.[10,11]

(5) *Roma.* The polarized response functions of ^{3}He including the final state interaction in the two-body breakup-up channel.[12]

(6) *Trento.* Bound states in nuclei with $A = 3, 4$ using the effective interaction hyperspherical harmonic method,[13] the longitudinal response functions of ^{3}H and ^{3}He and the two-body photodisintegration of ^{4}He using the Lorentz integral transform technique.[14,15]

2. Bound States in $A \leq 4$

In this section the advances in the description of the ^{3}H, ^{3}He and ^{4}He bound states are described. The main efforts in this subject are directed to consider a realistic Hamiltonian which includes two- and three-body interaction terms.

2.1. *The NN potential*

In the last years great efforts have been made to improve the description of the NN interaction. A generation of potentials including explicitly charge independence breaking (CIB) terms appeared. These interactions describe the NN scattering data below $T_{lab} = 300$ MeV with a nearly perfect χ^2/datum≈ 1. The CD-Bonn[16] and Argonne v_{18} (AV18)[17] interactions also allow for charge symmetry breaking (CSB) by providing a neutron-neutron (nn) force, which has been adjusted to the experimental

nn scattering length, whereas the Nijmegen interactions[18] are fitted only to proton-proton and proton-neutron data. Recently, the CD-Bonn potential has been updated.[19] These interactions are quite different from each other in their functional form, but their description of the NN data is almost equally accurate. Therefore a comparison of their predictions in the $A = 3, 4$ systems will give insights into the model dependence.

Very recently NN potentials based on two different approaches has been developed. Although different, these two approaches provide low momentum potentials. From one side we have the potentials based on chiral perturbation theory up to fourth order.[20] Essentially they reproduce the NN date base with χ^2/datum$\approx$ 1 and for this reason they are quantitatively comparable to the phenomenological potentials mentioned above. In the second approach the high momentum components of the realistic interactions are integrated out.[21] The physical condition is that the effective low-momentum interaction reproduces the deuteron pole and the NN phase-shifts below a certain cutoff Λ. The cutoff could be fixed, for example, to reproduce the $A = 3$ binding energy (BE).[22] Applications of this interaction to nuclear systems already started.[23]

2.2. *The $A = 3$ bound state*

All the NN potentials mentioned above can be put in the general form

$$v(NN) = v^{EM}(NN) + v^{\pi}(NN) + v^{R}(NN) \ . \tag{1}$$

The short range part $v^R(NN)$ of all of these interactions includes a certain number of parameters which are determined by a fitting procedure to the NN scattering data and the deuteron binding energy, whereas the long range part is represented by the one-pion-exchange potential $v^{\pi}(NN)$ and the electromagnetic (EM) part $v^{EM}(NN)$. The $v^{EM}(pp)$ term, as used for example in AV18, consists of the one- and two-photon Coulomb terms plus the Darwin-Foldy term, vacuum polarization and magnetic moment interactions. The $v^{EM}(np)$ interaction includes a Coulomb term due to the neutron charge distribution in addition to the magnetic moment interaction. Finally, $v^{EM}(nn)$ is given by the magnetic moment interaction only.

As it is well known, when these interactions are used to describe the $3N$ bound state, an underbinding of about 0.5 MeV to 0.9 MeV depending on the model is obtained. The local potentials lead to less binding than the non-local ones, a characteristic related to the bigger D-state probability predicted for the deuteron. Hence, it seems to be not possible to describe the $A > 2$ systems without the inclusion of TNI terms in the nuclear Hamiltonian. Several TNI models have been studied in the literature mostly based on the exchange of two pions with an intermediate Δ excitation. These interactions include a certain number of parameters not completely determined by theory, therefore some of them can be used to reproduce, for example, the triton binding energy.

In Ref. 6 the Pisa group in collaboration with the Bochum group presented a detailed calculation of the $A = 3$ system including total isospin states $T = 1/2$ and $3/2$. Two different interaction models have been considered, the AV18 and the AV18 plus the TNI of Urbana IX[24] (UR). The results are collected in Table 1.

Table 1. The ^{3}H and ^{3}He binding energies using the AV18 and the AV18+UR potential models.

Hamiltonian	^{3}H	^{3}He
	B(MeV)	B(MeV)
AV18 ($T = 1/2$)	7.618	6.917
AV18 ($T = 1/2, 3/2$)	7.624	6.925
AV18+UR ($T = 1/2$)	8.474	7.742
AV18+UR ($T = 1/2, 3/2$)	8.479	7.750
Expt.	8.482	7.718

A particular attention was given to the difference $D = B(^3\text{H}) - B(^3\text{He})$ as a test of the CSB terms present in the interaction. The experimental value of this quantity is 764 keV, from which only 85% correspond to the standard Coulomb potential. The remaining 15% should come from other CSB terms. Different contributions to D are reported in Table 2.

Table 2. Contributions of the different terms to the $A = 3$ mass difference.

Interaction term	$B(^3\text{H})$-$B(^3\text{He})$
Nuclear CSB	65 keV
Point Coulomb	677 keV
Full Coulomb	648 keV
Magnetic moment	17 keV
Orbit-orbit force	7 keV
n-p mass difference	14 keV
Total (theory)	751 keV
Expt.	764 keV

The Trento group has computed the $A = 3$ binding energies using the effective interaction hyperspherical harmonics (EIHH) technique.[13] This technique can be applied to an A–body Hamiltonian of the form $H^{[A]} = H_0 + V$. Then, one divides the Hilbert space of $H^{[A]}$ into a model space P and a residual space Q. The Hamiltonian $H^{[A]}$ is then replaced by an effective model–space Hamiltonian

$$H_{eff}^{[A]} = PH_0P + PV_{eff}^{[A]}P \qquad (2)$$

that by construction has the same energy levels as the low–lying states of $H^{[A]}$. To find $V_{eff}^{[A]}$, however, is as difficult as seeking the full–space solutions. In the EIHH method, one approximates $V_{eff}^{[A]}$ in such a way that it coincides with V for P$\longrightarrow$ 1, so that an enlargement of P leads to a convergence of the eigenenergies to the *true* values. In Ref. 13 it has been shown how a TNI is incorporated in the EIHH formalism. The results for ^{3}H ($T = 1/2$ component) using the AV18+UR potential model are given in Table 3 and compare to those of Ref. 6.

Table 3. ^{3}H binding energy and L-state probabilities obtained using the EIHH technique and compared to the results of Ref. 6.

	EIHH	Bochum	Pisa
BE [MeV]	8.468	8.470	8.474
S-wave	89.516	89.512	89.509
S′-wave	1.059	1.051	1.055
P-wave	0.135	0.135	0.135
D-wave	9.291	9.302	9.301

From the table we can see that the EIHH results are in close agreement to those obtained by the Pisa and Bochum groups.

2.3. *The $A = 4$ bound state*

Rapid progress has been made during the last few years in the quantitative study of the $A = 4$ nuclear systems. Significant refinements of well-established techniques have allowed the solution of the four-nucleon bound state problem with a control of the numerical error at the level of 10-20 keV, at least for Hamiltonians including only NN interaction models.[25] However, when the Hamiltonian includes two- and three-body forces as for example the AV18+UR interaction model, very few calculations can be found in the literature.[26,27] Moreover these calculations are not in strict agreement. Motivated by this fact a new calculation on the α-particle using the hyperspherical harmonic (HH) formalism has been recently performed.[7] In this formalism the four-nucleon wave function having total angular momentum $J = 0$ and positive parity can be written as

$$\Psi_4 = \sum_{KLST} \sum_{\mu} \frac{u_{KLST,\mu}(\rho)}{\rho^4} \Psi_{\mu}^{KLST}, \tag{3}$$

where K is the grand angular quantum number and Ψ_{μ}^{KLST} is a completely antisymmetric HH function having total angular momentum L, spin S and isospin T. The index μ labels possible choices of the hyperangular, spin and isospin quantum number. The expansion coefficients are the hyperradial functions $u_{KLST,\mu}(\rho)$ depending on the symmetrical variable ρ (the hyperradius) and are determined by the Rayleigh–Ritz variational principle.

The main difficulty in applying the HH technique is the large degeneracy of the basis. In the case of realistic potentials basis elements up to $K \approx 60$ are necessary to obtain a good convergence for the BE. On the other hand, already for values of $K > 20$ it is very difficult to find completely antisymmetric HH states via the Gram-Schmidt procedure due to the loss of precision in the orthogonalization procedure. However, it is possible to separate the HH functions in classes having particular properties and take into account advantageously the fact that the convergent rates of the various classes is rather different.

In Table 4 the HH results for the AV18 and Nijmegen (Nijm II) interactions are given as well as those for the AV18+UR. The convergence of the HH expansion has been studied and estimated to be of the order of 10 keV even when the TNI is included. The results are compare to those obtained solving the Faddeev-Yakubovsky (FY) equations in momentum space,[27] in configuration space[28] and those from the Green Function Monte Carlo (GFMC) method.[26] In the AV18+UR case the HH and FY results are very close to each other. The differences with the GFMC result are however below 1%.

From the results presented in this section, together with others from the literature, it is possible to verify the linear relation between the $A = 3$ and $A = 4$ binding energies. This is shown in Fig. 1 where the AV18, CD-Bonn and Nijmegen potentials have been considered together with the combination with the Urbana IX and Tucson-Melbourne[29] (TM) TNI's. It is possible to observe the nearly linear relation, however the nuclear model interactions consisting a two-body plus a three-body term fixed to reproduce the triton BE produce too much binding in the α particle.

Table 4. The α–particle binding energies B (MeV), the expectation values of the kinetic energy operator $\langle K \rangle$ (MeV), and the P and D probabilities (%) for various realistic interaction models.

Interaction	Method	B	$\langle K \rangle$	P_P	P_D
AV18	HH	24.22	97.84	0.35	13.74
	FY[27]	24.25	97.80	0.35	13.78
	FY[28]	24.22	97.77		
Nijm II	HH	24.43	100.27	0.33	13.37
	FY[27]	24.56	100.31		
AV18+UR	HH	28.47	113.30	0.73	16.03
	FY[27]	28.50	113.21	0.75	16.03
	GFMC[26]	28.34(4)	110.7(7)		

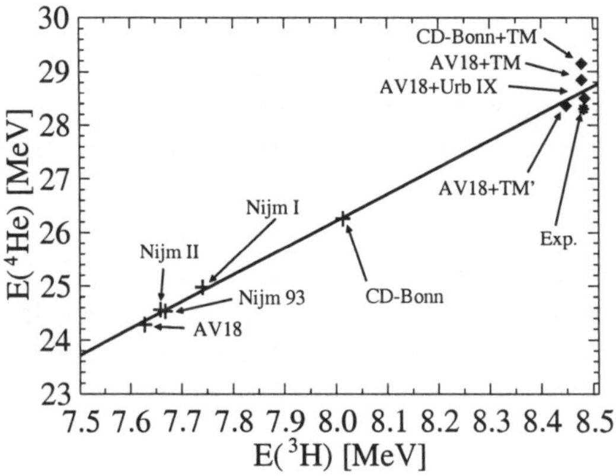

Figure 1. The $A = 4$ binding energy as a function of the $A = 3$ binding energy for different potential models.

3. Continuum States in $A \leq 4$

In this section recent advances in the description of scattering states in few-nucleon systems is reviewed. This research is mainly devoted to improve the theoretical description of observables which are in disagreement with the experimental data. Here two different improvements are reported, the inclusion of the long range electromagnetic force which has been systematically disregarded in $N - d$ scattering and the inclusion of a new type of one-pion exchange TNI which has a new spin structure not considered before.

3.1. $N - d$ *scattering including electromagnetic forces*

The study of the magnetic moment interaction (MM) in the NN system has been subject of many investigations.[30,31] Although the intensity of the MM interaction is very small compared to the nuclear interaction, its long range behavior produces significant effects in NN scattering. As mentioned before, almost all modern NN potentials have been constructed considering the EM interaction used in the Nijmegen partial-wave analysis. However, in the description of the $3N$ continuum the MM interaction and corrections to the Coulomb potential has been systematically disregarded.

In Ref. 8 the Pisa group studied $N - d$ elastic scattering including Coulomb plus MM interactions. A partial-wave decomposition of the scattering process has been performed and for states with low values of the relative orbital angular momentum L of the projectile and the target, the process has been studied by solving the complete $3N$ problem with the Hamiltonian of the system containing nuclear plus Coulomb plus MM interactions. For states with L values sufficiently high, the collision can be considered peripheral and treated as a two-body process. Furthermore, in those states only the EM interaction gives appreciable effects and the corresponding scattering amplitudes can be calculated in Born approximation.

The main results of Ref. 8 are shown if Figs. 2,3. In Fig. 2 the $n - d$ analyzing power A_y calculated using AV18 and AV18+MM are shown together with a calculation in which the MM interaction has been retained in states with low L values (dashed line). The results are compared to a recent experiment.[32] To be noticed the forward-angle dip structure produced by the MM interaction.

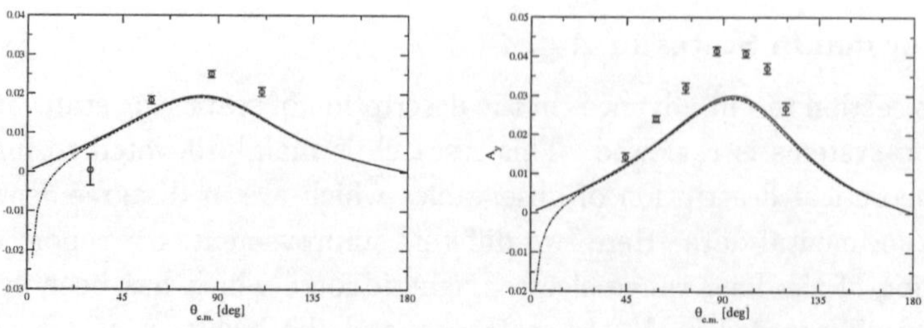

Figure 2. The $n - d$ A_y calculated using AV18 (solid line) and AV18+MM (dotted-dashed line). For the dashed line see text. Experimental points are from Ref. 32.

In order to evaluate the effects of the MM interaction on the vector analyzing powers in presence of the Coulomb field, in Fig. 3 the results of the calculations at $E_p = 1$ and 3 MeV are shown. The solid (dashed) line corresponds to the AV18 (AV18+MM) whereas the dashed-dotted line corresponds to the complete calculation including two more terms of the scattering amplitude as explained in Ref. 8. We see that the major effect of the MM interaction is obtained around the peak and is appreciable at both energies. There is also an improvement in the description of the observable at forward angles, in particular for iT_{11} at $E_p = 3$ MeV.

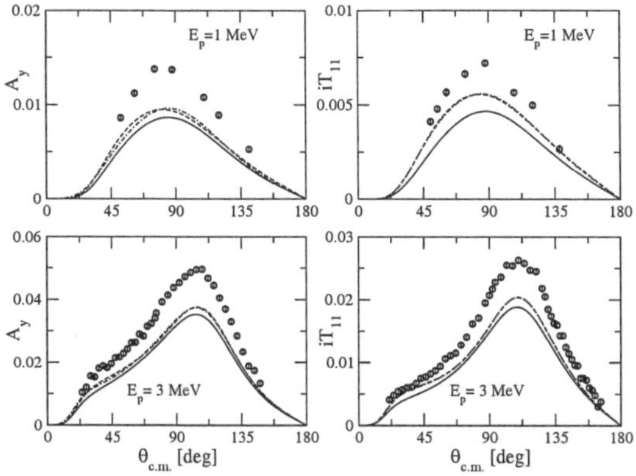

Figure 3. The $p-d$ A_y and iT_{11} calculated using AV18 (solid line) and AV18+MM (dotted-dashed line). For the dashed line see text. Experimental points are from Ref. 33 (1 MeV) and Ref. 34 (3 MeV).

3.2. *Irreducible pionic effects in nucleon-deuteron scattering below 20 MeV*

Recently, irreducible pionic effects in low energy $N-d$ scattering have been studied.[35] This effects appear in the form of a TNI generated by the one-pion exchange diagram when one of the two nucleons involved in the exchange process rescatters with a third one while the pion is in flight.[36] In Ref. 35 the $N-d$ analyzing powers has been studied with this new TNI and encouraging results have been obtained indicating this kind of force as a candidate to solve the long standing problem called "A_y puzzle". In Ref. 2 the study has been extended by considering such an effect with three different NN potentials. In all cases the authors demonstrated that this irreducible pionic effect improves the low-energy vector analyzing powers with negligible effects for the differential cross sections and minor effects for the other spin observables. They have also extended the comparison with the experimental data studying differential cross-sections, vector and tensor analyzing powers, and spin-transfer coefficients at various energies below 20 MeV. As an example, in Fig. 4 A_y is shown at different energies for the Bonn-B (left panel) and CD-Bonn (right panel) potentials.

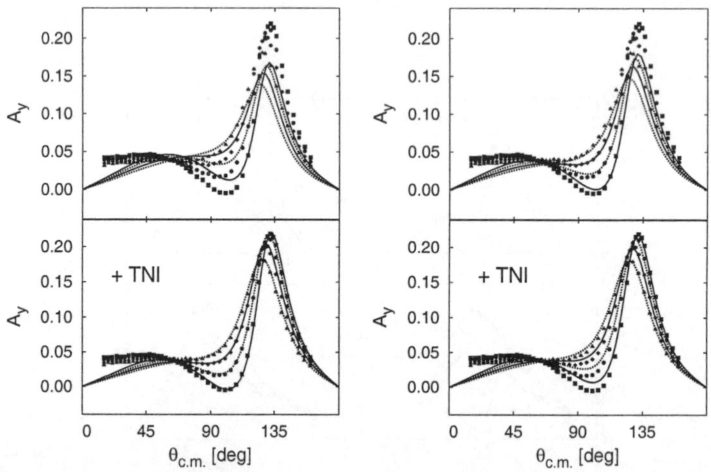

Figure 4. The analyzing power A_y in the range 12-18 MeV. Squares, circles, diamonds and triangles represents *pd* data at 18, 16, 14 and 12 MeV from Ref. 37. The lines in the left (right) panel refer to calculations with the Bonn-B (CD-Bonn) potential. The TNI is included in the lower panels.

To be noticed the correct description of the A_y maximum when the TNI is included.

4. Few-Nucleon Systems using Electroweak probes

In this section different reactions that have been studied experimentally very recently are analyzed. From a theoretical point of view, the main efforts are devoted to include the final state interaction (FSI) and to develop realistic models for the nuclear current. In cases in which the energy of the process is sufficiently high the limits of a description based only on nucleonic degrees of freedom and the use of a non relativistic dynamics can be examined.

4.1. *Longitudinal response functions of ^3H and ^3He*

Inclusive electron scattering can provide detailed information on the transition charge and current densities in nuclei. In the one photon exchange approximation the cross section for this process is given by

$$\frac{d^2\sigma}{d\Omega\, d\omega} = \sigma_M \left[\frac{q_\mu^4}{q^4} R_L(q,\omega) + \left(\frac{q_\mu^2}{2q^2} + \tan^2\frac{\theta}{2} \right) R_T(q,\omega) \right] \tag{4}$$

where R_L and R_T are the longitudinal and transverse response functions respectively, ω is the electron energy loss, q is the magnitude of the electron momentum transfer, θ is the electron scattering angle and $q_\mu^2 = q^2 - \omega^2$. Experimental data for both R_L and R_T are available for a variety of energy and momentum transfers. In Ref. 14 the Trento group calculated trinucleon longitudinal response functions $R_L(q,\omega)$ for q values up to 500 MeV/c. These are the first calculations beyond the threshold region in which both TNIs and Coulomb forces are fully included. Two realistic NN potentials (configuration space BonnA and AV18) and two TNIs (Urbana IX and Tucson-Melborne) have been considered. Complete final state interactions are taken into account via the Lorentz integral transform (LIT) technique.[38,39] Relativistic corrections arising from first order corrections to the nuclear charge operator as well as the reference frame dependence due to the non-relativistic framework employed have been investigated.

A comparison of the ^{3}H and ^{3}He theoretical longitudinal response functions with experimental data[40,41] is shown in Fig. 5 at $q = 250$, 300, and 350 MeV/c. In the peak region one does not find a clear picture, since there is a better agreement once with the TNI (^{3}He) and once without the TNI (^{3}H).

In Fig. 6 (left panel) the R_L of ^{3}H is compared to the experimental data of Ref. 43 at various q in the low energy region. For the lower two momentum transfers there is a rather good agreement of experiment and theory. At $q = 487$ MeV/c the theoretical response functions are larger than the experimental one, in particular very close to threshold. The right panel shows a similar comparison with experimental data but for the R_L of ^{3}He.

In Fig. 6 the theoretical results from the Pisa group are also given.[42] In order to have a clean comparison of the two different calculations, both have been done with the same one body current and the same interaction. For the two higher momentum transfers there is a rather good agreement

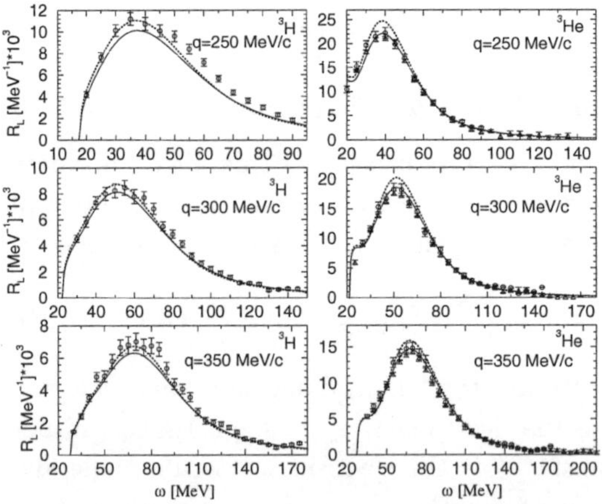

Figure 5. Comparison of theoretical and experimental $R_L^{LAB(B)}(q_{LAB}, \omega_{LAB})$ at q_{LAB} as indicated in figure for ^{3}H (left) and ^{3}He (right) (charge operator: non-relativistic plus DF term): AV18+UR potentials (solid) and AV18 potential (dotted).

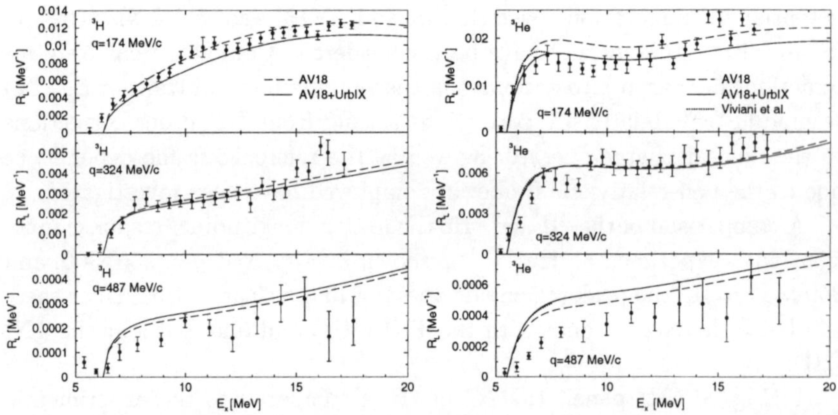

Figure 6. Comparison of theoretical and experimental $R_L^{LAB}(q_{LAB}, E_x)$ for ^{3}H (left panel) and ^{3}He (right panel) at q_{LAB} as indicated in figure. AV18+UrbIX potential (solid) and AV18 potential (dashed); experimental data from Ref. 43 and theoretical result from Ref. 42 (dotted) with AV18+UrbIX potential.

between both calculations. Some differences are visible at $q = 174$ MeV/c. The rather large discrepancy between theory and experiment of the low–energy R_L at $q = 487$ MeV/c requires further theoretical and experimental investigations. For example the inclusion of the relativistic two–body charge operators, although not sufficient to give agreement with experiment, diminishes the discrepancy by about a factor of two.[42]

Extension of the LIT method to describe the two-body photodisintegration of ^{4}He is given in Refs. 15, 44

4.2. *Radiative $p-d$ capture*

Using the ^{3}He bound state wave function and the scattering $A = 3$ wave functions calculated using the pair HH approach,[45,46] the Pisa group have calculated polarization observables in $p-d$ capture. The main ingredients in the description of this process are the the nuclear Hamiltonian H from which the nuclear wave functions are obtained, and the model used to describe the nuclear currents. The nuclear EM current is related to H through current conservation:

$$\mathbf{q} \cdot \mathbf{j}(\mathbf{q}) = [H, \rho(\mathbf{q})] \ , \tag{5}$$

where the nuclear Hamiltonian H includes two- and three-body interactions. To lowest order in $1/m$, Eq. (5) separates into

$$\mathbf{q} \cdot \mathbf{j}_i(\mathbf{q}) = \left[\frac{\mathbf{p}_i^2}{2m}, \rho_i(\mathbf{q})\right] \ , \tag{6}$$

$$\mathbf{q} \cdot \mathbf{j}_{ij}(\mathbf{q}) = [v_{ij}, \rho_i(\mathbf{q}) + \rho_j(\mathbf{q})] \ , \tag{7}$$

and similarly for the three-body current $\mathbf{j}_{ijk}(\mathbf{q})$. To construct the two-body current, it is useful to separate the current $\mathbf{j}_{ij}(\mathbf{q})$ into a model independent and a model dependent part. The former has a longitudinal component and is constructed so as to satisfy Eq. (7), while the later is purely transverse and therefore is unconstrained by the conservation relation. The discussion of this topic has been given in Refs. 10,47. As an illustration, in Fig. 7, the deuteron tensor polarization observables T_{20} and T_{21} calculated using AV18 are shown at $E_{cm} = 2$ MeV. These observables have selected due to their particular sensitivity to the nuclear current. The dotted curves are obtained including only the one-body current contributions, the dashed curves are obtained with the model for the nuclear current operator of Ref. 42, and the dotted-dashed curves are obtained in the long-wavelength-approximation (LWA), applying the Siegert theorem. Finally, the solid curves are obtained including the one- and two-body contributions necessary to satisfy the conservation relation with the AV18 nuclear Hamiltonian.

By inspection of the figure we can observe the good agreement between the LWA and the calculations obtained satisfying the conservation relation.

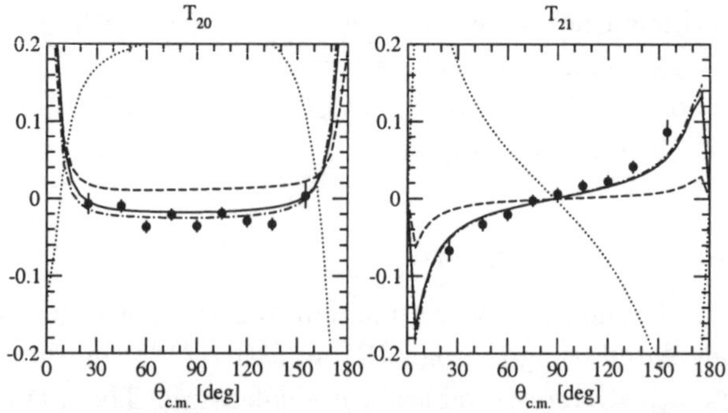

Figure 7. Deuteron tensor polarization observables T_{20} and T_{21} for pd radiative capture at $E_{c.m.} = 2$ MeV, obtained with the AV18 Hamiltonian models. See text for explanations of the different curves. The experimental data are from Ref. 48.

Conversely the results of Ref. 42 are in disagreement since in those calculations the conservation relation was only partially fulfilled. It is encouraging to observe that there is a rather good description of the data when the conserved current is used.

4.3. *Exclusive processes in ^2H and ^3He within a Generalized Glauber Approach*

Exclusive lepton scattering could be very useful to investigate the limits of validity of the description of nuclei in terms of the solution of the non relativistic Schrödinger equation containing realistic NN interactions, since it might yield relevant information on the nuclear wave function. To this end, the theoretical description of the initial and final states involved in the scattering process should be described within a consistent, reliable approach. In the case of few-nucleon systems, a consistent treatment of initial and final states is nowadays possible at low energies, as described in the previous sections, but at higher energies, when the number of partial waves sharply increases and the NN interaction becomes highly inelastic, the Schrödinger approach becomes impractical and other methods have to be employed.

In Ref. 5 a systematic theoretical investigation of the exclusive process $A(e,e'p)B$ off ^{2}H and ^{3}He has been performed based on a reliable description of initial state correlations (ISC), treated by the use the state-of-the-art few-body wave functions[45] corresponding to the AV18 interaction; and FSI, treated within the Generalized Glauber approach (GGA) which is a generalization of the standard Glauber approach.[49–52] The application of the GA to the treatment of $A(e,e'p)B$ processes requires the following approximations: *i*) the NN scattering amplitude is obtained within the eikonal approximation; *ii*) the nucleons of the spectator system $A-1$ are stationary during the multiple scattering with the struck nucleon (the *frozen approximation*), and *iii*) only perpendicular momentum transfer components in the NN scattering amplitude are considered. In the GGA the *frozen approximation* is partly removed by taking into account the excitation energy of the $A-1$ system, which results in a correction term to the standard profile function of GA, leading to an additional contribution to the longitudinal component of the missing momentum.

In this formalism the cross section of the process $A(e,e'p)(A-1)$ can be factorized in the following expression

$$\frac{d^6\sigma}{dE'd\Omega'd\mathbf{p}_m} = K(\bar{Q}^2, x, \mathbf{p}_m)\,\sigma^{eN}(\bar{Q}^2, \mathbf{p}_m)P_A^{FSI}(\mathbf{p}_m, E_m) \tag{8}$$

where K is a kinematical factor, σ^{eN} the cross section describing electron scattering by an off-shell nucleon, $x = Q^2/2M_N q_0$ the Bjorken scaling variable and $P_A^{FSI}(\mathbf{p}_m, E_m)$ is the *Distorted Spectral Function*. In order to illustrate some of the results presented in Ref. 5, in Fig. 8 a comparison to experimental results are given for the processes ^{2}H$(e,e'p)n$ and ^{3}He$(e,e'p)^2$H. For the first process the effective momentum distribution $N_{eff}(|\mathbf{p}_m|)$ (or reduced cross section) is compare to the data of Ref. 53 whereas for the two-body breakup of ^{3}He the cross section is compared to a recent set of data from the E89044 Jlab Collaboration.[54] In the figure the dashed line represents the calculation in plane wave impulse approximation (PWIA) whereas the solid line includes the final state rescattering. There is a noticeable improvement in the description of the processes when the FSI is taken into account.

4.4. *Transverse asymmetry of $^3\vec{\mathrm{He}}$ and the magnetic form factor of the neutron*

In the last few years, an impressive amount of experimental work has been devoted to an accurate investigation of nucleon EM form factors, obtaining unexpected results like the puzzling ratio G_E/G_M for the proton.[55] In order to have a complete knowledge of the nucleon form factors, one has to face with the particularly difficult problem represented by the extraction of the neutron EM form factors, since free neutron targets do not exist in nature. To this end the use of polarized ^{3}He targets has been relevant and a huge amount of experimental and theoretical work has been devoted to this issue. However it is not a trivial task to disentangle the neutron information

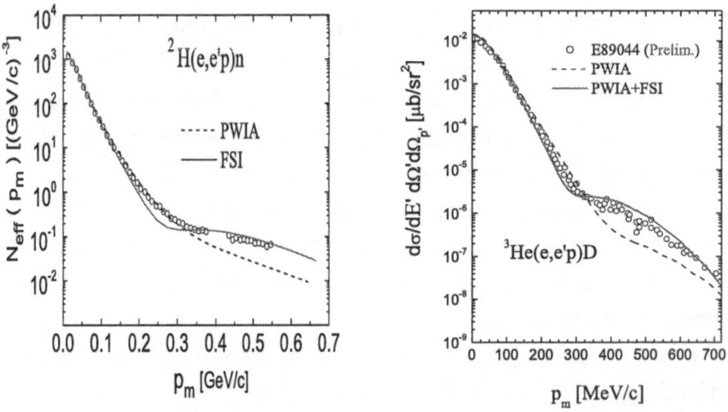

Figure 8. Left panel: reduced cross section as a function of the missing momentum. Right panel: total cross section as a function of the missing momentum at $Q^2 = 1.55$ $(\mathrm{GeV}/c)^2$ and $x = 1$.

from the nuclear-structure effects, given the many effects playing relevant roles, like: *i*) the "small" components of the bound-state wave function; *ii*) the Δ excitation; *iii*) the inclusion of the final state interaction, between the knocked-out nucleon and the interacting spectator pair; *iv*) the meson exchange currents (MEC) and *v*) the relativistic dynamics. As a first step, the analysis of inclusive responses of polarized ^{3}He was carried out within the PWIA,[56–58] where realistic NN interactions were adopted and in the final state the interaction between the spectator pair and the knocked-out nucleon was disregarded. As a step further, the Rome group recently presented an analysis of the inclusive responses of a polarized ^{3}He target including FSI in the two-body breakup channel.[12]

The inclusive scattering of polarized electrons (with helicity h) by a polarized ^{3}He target ($\vec{e}\ +\ ^3\vec{\mathrm{He}} \rightarrow e' + X$) is given by

$$\frac{d^2\sigma(h)}{d\Omega d\omega} = \Sigma + h\,\Delta \tag{9}$$

with Σ depending on the unpolarized responses R_L and R_T and Δ depending on the polarized responses $R_{T'}$ and $R_{TL'}$. The asymmetry A is defined as A$=\Delta/\Sigma$. The nuclear structure effects are included in four response functions. In PWIA the responses can be described in terms of the spectral function of ^{3}He. However, using for example the pair HH $p-d$ wave function,[46] it is possible to include the rescattering between the interacting pair and the asymptotically free particle, as explained in Ref. 12. To illustrate the main results of that work, in Fig. 9 the asymmetry A calculated in PWIA and including FSI in the two-body breakup channel is compared to the experimental data.[59] As can be seen from the figure the inclusion of the FSI produces a strong effect and a net improvement in the description of the data.

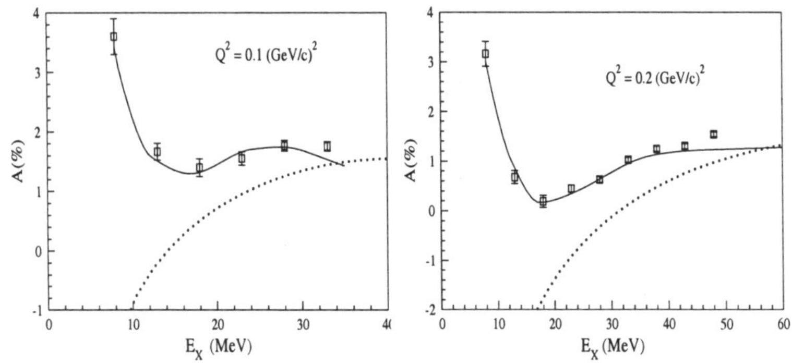

Figure 9. The asymmetry A for $Q^2 = 0.1\ (GeV/c)^2$ (left) and $Q^2 = 0.2\ (GeV/c)^2$ (right) vs the missing energy, E_X, at low energy transfer. Solid line: results with FSI; dotted line: PWIA. Experimental data from Ref. 59.

4.5. *Nuclear effects in positive pion electroproduction on the deuteron*

Valuable information on the structure of nucleons and nuclei can be obtained from high precision experiments on pion electroproduction, including measurements of polarization observables and the separation of the structure functions (SF's). These possibilities have already triggered a number of novel theoretical approaches developed for the treatment of pion photo- and electro-production off the nucleon.[60] Furthermore, the data[61] on neutral pion production on the proton near threshold also renewed the interest in this reaction.

More understanding of mechanisms of pion photo- and electro-production can be gained in studies of the reaction on bound systems of nucleons. Apart from offering means to extract the amplitude of pion production off the neutron, the processes like

$$\gamma\ (\gamma^*) + {}^2\mathrm{H} \to \pi + NN \tag{10}$$

could shed light on the details of the nuclear structure.

For the description of photomeson processes on nuclei, the impulse approximation (IA) is commonly used. However, in general, there are not enough grounds to use this approximation when high-momentum components of the nuclear wave function are probed. Following Ref. 3 some applications of an alternative formalism for the description of many-body currents and off-shell effects in the theory of photomeson processes on nuclei are given.[62,63] The formalism is based on the unitary transformation method in combination with a gauge independence calculations through an extension of the Siegert theorem.

The differential cross section of the $^2\mathrm{H}(\mathrm{e},\mathrm{e}'\pi^+)nn$ reaction with unpolarized particles in the one-photon exchange approximation can be expressed as

$$\frac{\mathrm{d}^4\sigma}{\mathrm{d}\Omega'\ \mathrm{d}E'\ \mathrm{d}\Omega_\pi\ \mathrm{d}E_\pi} = \sigma_\mathrm{M} E_\pi \mid \boldsymbol{p}_\pi \mid \ (\ \xi^2 W_\mathrm{C}\ +\ (-\frac{\xi}{2} + \eta) W_\mathrm{T}\ +$$

$$+\ \xi(-\xi + \eta)^{\frac{1}{2}} \cos\varphi W_\mathrm{I}\ -\ \frac{\xi}{2} \cos 2\varphi\ W_\mathrm{S}\ )\ ,$$

with $\xi = q^2/\boldsymbol{q}^2$, $\eta = \tan^2\theta/2$, σ_M is the Mott cross section and W_α ($\alpha =$C, T, I and S) are the four SFs containing the information on the nuclear dynamics. The operator of pion production on a bound nucleon has been constructed using the generalized Born approximation based on pseudovector (PV) πNN coupling with the EM vertices taken for the on-mass-shell hadrons. Using an explicit gauge independent expression for the amplitude the differential cross sections of the $^1\mathrm{H}(\mathrm{e},\mathrm{e}'\pi^+)n$ and $^2\mathrm{H}(\mathrm{e},\mathrm{e}'\pi^+)nn$ reactions have been calculated for the kinematics of Ref. 64.

In Fig. 10, the plots of the $^2\mathrm{H}(\mathrm{e},\mathrm{e}'\pi^+)nn$ cross section are displayed. The cross section is presented as a function of the "missing" mass M_x (invariant mass of the nn pair). The shown results suggest that, the nn FSI leads to some increase of the π^+ production rate within the pion energy range covered in the experiment. It is seen also that the result obtained using the unitary transformation method in conjunction with the extension of the Siegert theorem does not much differ from the one provided by the IA indicating a weak sensitivity of the $^2\mathrm{H}(\mathrm{e},\mathrm{e}'\pi^+)nn$ differential cross section to the two-body contributions for the kinematics of Ref. 64.

The description of the polarized proton pionic capture in deuterium can be find in Refs. 4, 65.

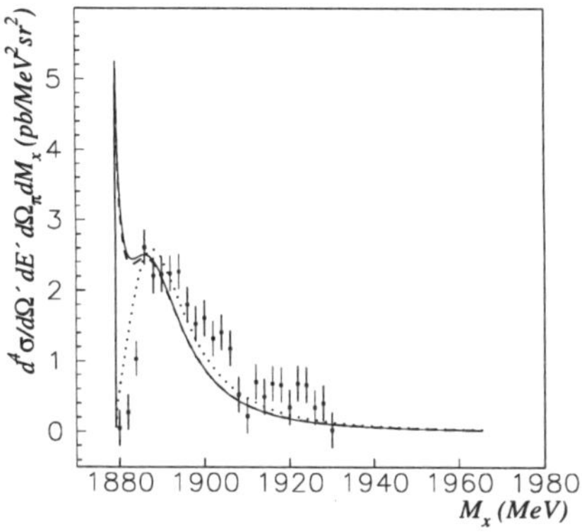

Figure 10. Differential cross sections of the ^{2}H(e, e$'\pi^+$)nn reaction for the kinematics of experiment[64] (solid line). The dashed (dotted) curve refers to the IA calculation with (without) the $^1S_0$ state nn FSI included.

4.6. *Weak axial nuclear heavy meson exchange currents*

In Ref. 1 the construction of the weak axial nuclear exchange currents (WANECs) of the heavy meson range is described. The aim is to produce currents suitable for calculations in the standard nuclear physics with nuclear wave functions generated from the Schrödinger equation. For the construction of the WANECs, the nucleon Born terms have been added to the weak axial two–nucleon relativistic amplitudes derived in Ref. 67. The WANECs are then defined in analogy with the electromagnetic MECs [68] as the difference between these relativistic amplitudes and the first Born iteration of the weak axial one–nucleon current contribution to the two–nucleon scattering amplitude satisfying the Lippmann–Schwinger equation. The WANECs defined in this way satisfy the nuclear PCAC equation.

The relative strength of various parts of the space component of the WANECs has been studied in the transition $^3S_1 -^3 D_1 \rightarrow {}^1S_0$ for the weak deuteron disintegration by low energy neutrinos in the neutral current channel

$$\nu_x + d \rightarrow \nu_x' + p + n\,, \tag{11}$$

where ν_x refers to any active flavor of the neutrino. This reaction is important for studying the solar neutrino oscillations ($x = e$). The results for the cross section are presented in Table 5 and can be compared to previous ones from Ref. 66, however not compatible with chiral invariance.

Table 5. Cumulative contributions to the cross section $\sigma_{\nu d}$ ($\times 10^{-42}$ cm^2) for various neutrino energies are displayed. The number in the bracket is the ratio of the n-th cross section to the cross section in the row above.

E_ν [MeV]	5	10	15	20
IA	0.0929 (-)	1.082(-)	3.269 (-)	6.648 (-)
$+\delta$IA	0.0926 (0.996)	1.077 (0.996)	3.254 (0.995)	6.616 (0.995)
$+\Delta(\pi)$	0.0977 (1.056)	1.145 (1.063)	3.472 (1.067)	7.083 (1.071)
$+\Delta(\rho)$	0.0964 (0.986)	1.127 (0.984)	3.414 (0.983)	6.958 (0.982)
$+$p(π)	0.0956 (0.992)	1.117 (0.991)	3.382 (0.991)	6.888 (0.990)
$+$p(ρ)	0.0959 (1.003)	1.120 (1.003)	3.392 (1.003)	6.911 (1.003)
$+\rho$-π	0.0967 (1.008)	1.131 (1.010)	3.427 (1.010)	6.987 (1.011)
$+$p(ω)	0.0966 (0.999)	1.130 (0.999)	3.422 (0.998)	6.975 (0.998)
$+a_1$-ρ	0.0964 (0.999)	1.128 (0.999)	3.417 (0.999)	6.965 (0.999)

5. Summary

The main achievements obtained in recent years by Italian groups in the theoretical description of few-nucleon systems have been reported. This research has been done in close connection to the current experimental activities. The comparison between data and theoretical results helped to to clarify different points. For example, in order to improve the description of bound state properties and several observables in scattering processes in $A = 3$, the long range potential $v^{EM}(NN)$ has been considered as well as a new TNI term with a particular spin structure. A complete study of bound and scattering states in the $A = 4$ system, which shows a greater sensitivity to the interaction models, has been started.

In the study of few-nucleon systems with electroweak probes two main lines are intensively pursued. From one side big efforts have been done to go beyond the PWIA. This could be problematic at high energies and different strategies has been discussed to include the final or initial state correlations. A second problem is the study of the mechanism of the reaction. To this aim, polarization observables could be extremely useful since they are sensitive to many-body current terms. Some of them can be measured with high accuracy allowing for a detailed study of the nuclear current.

Finally I would like to stress the fact that the intense research activity here described has stimulated a fruitful exchange with several theoretical groups and the main experimental laboratories in the world.

References

1. B. Mosconi, P. Ricci and E. Truhlík, nucl-th/0212042
2. L. Canton, W. Schadow, and J. Haidenbauer, *Eur. Phys. J.* **A14**, 225 (2002)
3. L. G. Levchuk, L. Canton, and A. Shebeko *Eur. Phys. J.* **A21**, 29 (2004)

4. L. Canton, L. G. Levchuk, nucl-th/0411030
5. M.A. Braun, C. Ciofi degli Atti, and L.P. Kaptari, *Eur. Phys. J.* **A19**, 143 (2004); C. Ciofi degli Atti and L.P. Kaptari, *Phys. Rev.* **C**, in press, nucl-th/0407024
6. A. Nogga, A. Kievsky, H. Kamada, W. Glöckle, L.E. Marcucci, S. Rosati and M. Viviani, *Phys. Rev.* **C67**, 034004 (2003)
7. M. Viviani, A. Kievsky and S. Rosati, *Phys. Rev.* **C**, in press, nucl-th/0408019
8. A. Kievsky, M. Viviani and L.E. Marcucci, *Phys. Rev.* **C69**, 014002 (2004)
9. P. Barletta and A. Kievsky, *this proceedings*
10. M. Viviani, L.E. Marcucci, A. Kievsky, S. Rosati, and R. Schiavilla, *Eur. Phys. J.* **A17**, 483 (2003)
11. L.E. Marcucci, M. Viviani, R. Schiavilla, A. Kievsky, and S. Rosati, *Eur. Phys. J.* **A**, in press, nucl-th/0411082
12. A. Kievsky, E. Pace, and G. Salme', *Eur. Phys. J.* **A19**, 87 (2004)
13. N. Barnea, V.D. Efros, W. Leidemann, G. Orlandini, *Few-Body Syst.*, in press, nucl-th/0404086
14. V.D. Efros, W. Leidemann, G. Orlandini, and E.L. Tomusiak, *Phys. Rev.* **C69**, 044001 (2004)
15. S. Quaglioni, W. Leidemann, G. Orlandini, N. Barnea, and V.D. Efros, *Phys. Rev.* **C69**, 044002 (2004)
16. R. Machleidt, F. Sammarruca, and Y. Song, *Phys. Rev.* **C53**, R1483 (1996)
17. R.B. Wiringa, V.G.J. Stoks, and R. Schiavilla, *Phys. Rev.* **C51**, 38 (1995)
18. V.G.J. Stoks *et al.*, *Phys. Rev.* **C49**, 2950 (1994)
19. R. Machleidt, *Phys. Rev.* **C63**, 024001 (2001)
20. Entem and R. Machleidt, *Phys.Rev.* **C68**, 041001 (2003); E. Epelbaum *et al.*, nucl-th/0405048
21. S.K. Bogner *et al.*, *Phys. Rep.* **386**, 1 (2003)
22. A. Nogga*et al.*, nucl-th/0405016
23. A. Gargano *et al.*, *this proceedings*
24. B.S. Pudliner *et al.*, *Phys. Rev.* **C56**, 1720 (1997)
25. H. Kamada *et al*, *Phys. Rev.* **C64**, 044001 (2001)
26. R. B. Wiringa, S. C. Pieper, J. Carlson, and V. R. Pandharipande, *Phys. Rev.* **C62**, 014001 (2000).
27. A. Nogga *et al.*, *Phys. Rev.* **C65**, 054003 (2002).
28. R. Lazauskas and J. Carbonell, *Phys. Rev.* **C70**, 044002 (2004)
29. S.A. Coon *et al.*, *Nucl. Phys.* **A317**, 242 (1979); S.A. Coon and W. Glöckle, *Phys. Rev.* **C23**, 1790 (1981)
30. L.D. Knutson and D. Chiang, *Phys. Rev.* **C18**, 1958 (1978)
31. V.G.J. Stoks and J.J. de Swart, *Phys. Rev.* **C42**, 1235 (1990)
32. E.M. Neidel et al., *Phys. Lett.* **B552**, 29 (2003)
33. M.H. Wood *et al.*, *Phys. Rev.* **C65**, 034002 (2002)
34. S. Shimizu et al., *Phys. Rev.* **C52**, 1193 (1995)
35. L. Canton and W. Schadow, *Phys. Rev.* **C64**, 031001(R) (2001).
36. L. Canton and W. Schadow, *Phys. Rev.* **C62**, 044005 (2000).
37. K. Sagara *et al.*, *Phys. Rev.* **C50**, 577 (1994).
38. V.D. Efros, W. Leidemann and G. Orlandini, *Phys. Lett.* **B338**, 130 (1994).

39. V.D. Efros, W. Leidemann and G. Orlandini, *Phys. Rev. Lett.* **78**, 4015 (1977)
40. K. Dow *et al*, *Phys. Rev. Lett.* **61**, 1706 (1988).
41. C. Marchand *et al*, *Phys. Lett.* **B153**, 29 (1985).
42. M. Viviani, A. Kievsky, L.E. Marcucci, S. Rosati, and R. Schiavilla, *Phys. Rev.* **C61**, 064001 (2000).
43. G.A. Retzlaff *et al*, *Phys. Rev.* **C49**, 1263 (1994).
44. S. Quaglioni *et al.*, *this proceedings.*
45. A. Kievsky, M. Viviani, and S. Rosati, *Nucl. Phys.* **A577** , 511 (1994); A. Kievsky, M. Viviani, and S. Rosati, *Nucl. Phys.* **A551**, 241 (1993)
46. A. Kievsky, M. Viviani, and S. Rosati, *Phys. Rev.* **C64**, 024002 (2001).
47. L. E. Marcucci *et al.*, *this proceedings.*
48. M.K. Smith and L.D. Knutson, *Phys. Rev. Lett.* **82**, 4591 (1999).
49. L.L. Frankfurt, W.R. Greenberg, G.A. Miller, M.M. Sargsian, M.I. Strikman, *Z. Phys.* **A352**, 97 (1995).
50. L.L. Frankfurt, M.M. Sargsian, M.I. Strikman, *Phys. Rev.* **C56**, 1124 (1997).
51. M.M. Sargsian, *Int. J. Mod. Phys.* **E10**, 405 (2001).
52. R. J. Glauber, *Lectures in Theoretical Physics*, W. E. Brittin *et al* Editors, New York (1959).
53. P.E. Ulmer *et al.*, *Phys. Rev. Lett.* **89** 062301 (2002).
54. D. W. Higinbotham, *for the Jefferson Lab Hall A Collaboration*, *Eur. Phys. Jou.* **A19**, 171 (2004).
55. O. Gayou *et al.*, *Phys. Rev. Lett.* **88**, 092301 (2003).
56. C. Ciofi degli Atti, E. Pace, and G. Salmè, *Phys. Rev.* **C46**, R1591 (1992); C. Ciofi degli Atti, E. Pace, and G. Salmè, *Phys. Rev.* **C51**, 1108 (1995).
57. A. Kievsky, E. Pace, G. Salmè, and M. Viviani, *Phys. Rev.* **C56**, 64 (1997).
58. R. W. Schültze and P.U. Sauer, *Phys. Rev.* **C48**, 38 (1993).
59. F. Xiong *et al.*, *Phys. Rev. Lett.* **87**, 242501 (2001).
60. Y. Surya and F. Gross, *Phys. Rev.* **C53**, 2422 (1996); C. H. M. van Antwerpen and I. R. Afnan, *Phys. Rev.* **C52**, 554 (1995); S. Nozawa, T.-S. H. Lee, and B. Blankleider, *Phys. Rev.* **C41**, 213 (1990).
61. E. Mazzucato *et al.*, *Phys. Rev. Lett.* **57**, 3144 (1986); R. Beck *et al.*, *Phys. Rev. Lett.* **65**, 1841 (1990); J. C. Bergstrom *et al.*, *Phys. Rev.* **C53**, R1052 (1996); H. B. van den Brink *et al.*, *Phys. Rev. Lett.* **74**, 3561 (1995).
62. L. G. Levchuk and A. V. Shebeko, *Phys. At. Nuclei* **58**, 923 (1995).
63. L. G. Levchuk and A. V. Shebeko, *Sov. J. Nucl. Phys.* **50**, 607 (1989).
64. R. Gilman *et al.*, *Phys. Rev. Lett.* **64**, 622 (1990).
65. L. Canton *et al.*, *this proceedings.*
66. S. Nakamura *et al.*, *Nucl. Phys.* **A707**, 561 (2002).
67. F.C. Khanna and E. Truhlík, *Nucl. Phys.* **A673**, 455 (2000).
68. J. Adam, E. Truhlík, and D. Adamová, *Nucl. Phys.* **A494**, 556 (1989).

ELECTROMAGNETIC STRUCTURE OF FEW-BODY NUCLEAR SYSTEMS

L. E. MARCUCCI, M. VIVIANI, A. KIEVSKY AND S. ROSATI
Department of Physics, University of Pisa, I-56100 Pisa, Italy
INFN, Sezione di Pisa, I-56100 Pisa, Italy

R. SCHIAVILLA
Department of Physics, Old Dominion University, Norfolk, Virginia 23529, USA
Jefferson Lab, Newport News, Virginia 23606, USA

Recent advances in the study of *pd* and *nd* radiative capture reactions in a wide range of center-of-mass energy below and above deuteron breakup threshold are presented and discussed.

1. Introduction

The electromagnetic transitions for few-body nuclear systems have been extensively studied by several research groups (see Ref. [1] for a review). The advantage of investigating these processes is that nowadays several methods can provide accurate bound- and scattering-state wave functions using realistic Hamiltonian models. Therefore, different models for the nuclear electromagnetic current operator can be tested with the large body of available experimental data. In the present study, we concentrate our attention on the *pd* radiative capture reaction in a wide range of center-of-mass energy ($E_{c.m.}$), and on the *nd* radiative capture at thermal energies and at $E_{c.m.}$ = 6.00 and 7.20 MeV. These processes are studied using pair-correlated hyperspherical harmonics (PHH) wave functions obtained from a realistic Hamiltonian model consisting of the Argonne v_{18} two-nucleon [2] and Urbana IX three-nucleon [3] interactions (AV18/UIX). The nuclear electromagnetic current operator includes, in addition to the one-body term, also two- and three-body contributions, constructed with the goal of satisfying the current conservation relation (CCR) with the AV18/UIX Hamiltonian model. This aspect of the calculation is reviewed in the following section. In Sec. 3, the results for the *pd* and *nd* radiative captures are presented and discussed. Finally, some concluding remarks are given in Sec. 4.

2. The nuclear electromagnetic current operator

The nuclear electromagnetic current operator can be written as sum of one- and many-body terms that operate on the nucleon degrees of freedom. The one-body operator is derived from the non-relativistic reduction of the covariant single-nucleon current, by expanding in powers of $1/m$, m being the nucleon mass [1]. To construct the two-body current operator, it is useful to adopt the classification scheme of Ref. [4], and separate the current into model-independent (MI) and model-dependent (MD) parts. The MD two-body current is purely transverse and therefore is un-constrained by the CCR. It is taken to consist of the isoscalar $\rho\pi\gamma$ and isovector $\omega\pi\gamma$ transition currents, as well as the isovector current associated with excitation of intermediate Δ resonances, as in Ref. [5]. The MI two-body currents have longitudinal components and have to satisfy the CCR with the two-nucleon interaction. The MI terms arising from the momentum-independent terms of the AV18 two-nucleon interaction have been constructed following the standard procedure of Ref. [6], hereafter quoted as meson-exchange (ME) scheme. It consists in fitting the isospin-dependent spin-spin and tensor components of the interaction with the exchange of phenomenological pseudoscalar and vector mesons. The corresponding current operators are then obtained by minimal substitution in the effective coupling Lagrangians. It can be shown that these two-body current operators satisfy *exactly* the CCR with the first six operators of the AV18. The two-body currents arising from the spin-orbit components of the AV18 could be constructed using again ME mechanisms [7], but the resulting currents turn out to be not strictly conserved. The same can be said of those currents arising from the quadratic momentum-dependent components of the AV18, if obtained, as for example in Ref. [5], by gauging only the momentum operators, but ignoring the implicit momentum dependence which comes through the isospin exchange operator (see below). It was already suggested in Ref. [5] and it has been clearly shown in Refs. [8,9] that the deuteron tensor observables T_{20} and T_{21} in *pd* radiative capture cannot be described by the theory if the two-body MI current operators do not satisfy the CCR with the complete AV18 two-nucleon interaction. Therefore, the currents arising from the momentum-dependent terms of the AV18 interaction have been obtained following the procedure of Ref. [10], which will be quoted as minimal-substitution (MS) scheme. The main idea of this procedure, reviewed and extended in Ref. [9], is that the isospin-dependence of the isospin-conserving part of all realistic two-nucleon interactions is given by the isospin operator $\tau_i \cdot \tau_j$, which is formally equivalent to an implicit momentum dependence [10]. In fact, $\tau_i \cdot \tau_j$ can be expressed in terms of the space-exchange operator (P_{ij}) using the relation $\tau_i \cdot \tau_j = -1 - (1 + \sigma_i \cdot \sigma_j)P_{ij}$, valid when operating on antisymmetric wave functions. The space-exchange operator is defined as $P_{ij}\, f(\mathbf{r}_i, \mathbf{r}_j) \equiv \mathrm{e}^{\mathbf{r}_{ji}\cdot\nabla_i + \mathbf{r}_{ij}\cdot\nabla_j} f(\mathbf{r}_i, \mathbf{r}_j) = f(\mathbf{r}_j, \mathbf{r}_i)$, where the ∇-operators act only on the generic function $f(\mathbf{r}_i, \mathbf{r}_j)$ and not on the vectors $\mathbf{r}_{ji} = \mathbf{r}_i - \mathbf{r}_j = -\mathbf{r}_{ij}$ in the exponential. In the presence of an electromagnetic field, minimal substitution is performed both in the momentum dependent terms of the two-nucleon interaction and in the space-exchange operator P_{ij}. The resulting current operators are then obtained with standard procedures [9,10]. Explicit formulas can be found in Ref. [9].

Both the ME and the MS schemes can be generalized to calculate the three-body current operators induced by the three-nucleon interaction (TNI). Here, these three-body currents have been constructed within the ME scheme to satisfy the CCR with the Urbana-IX TNI [3]. Details of the calculation can be found in Ref. [9].

In summary, the present model for the many-body current operators retains the two-body terms obtained within the ME scheme from the momentum-independent part of the AV18, those ones obtained within the MS scheme from the momentum-dependent part of the AV18, the MD terms quoted above, and the three-body terms obtained within the ME scheme from the UIX TNI. Thus, the full current operator satisfies *exactly* the CCR with the AV18/UIX nuclear Hamiltonian.

3. Results

The theoretical predictions of the *pd* radiative capture observables at $E_{c.m.}$ =5.83–18.66 MeV obtained with the AV18/UIX Hamiltonian model are compared with the available experimental data of Refs.[11,12,13,14,15,16] in Figs. 1– 3. In all the figures, the dashed, dotted-dashed and solid lines correspond to the calculation with one-body only, with one- and two-body, and with one-, two- and three-body currents. From inspection of the figures, we can conclude that: (i) the present "full" model for the nuclear electromagnetic current operator provides an overall nice description for all the observables, with the only exception of the differential cross section at the two highest values for $E_{c.m.}$, and of the deuteron vector analyzing power $A_y(d)$ at small center-of-mass angles at $E_{c.m.}$=5.83 and 9.66 MeV.

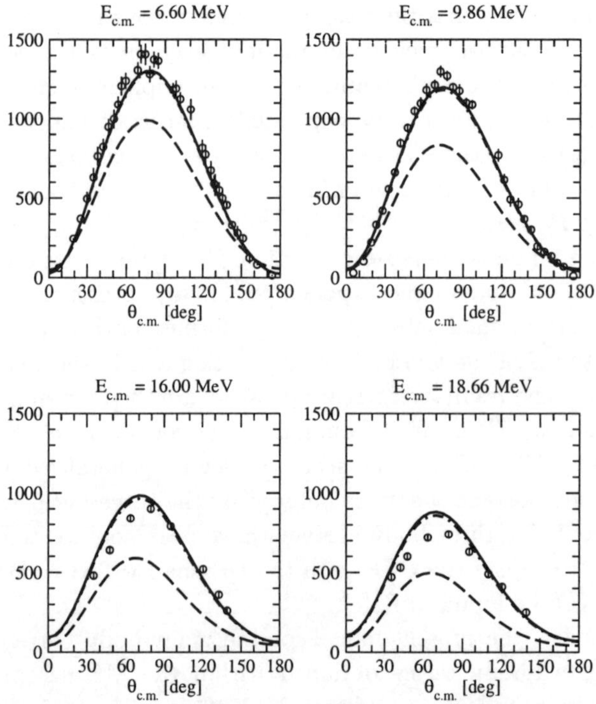

Figure 1. Differential cross section in μb/sr for pd radiative capture up to $E_{c.m.}$ =18.66 MeV as function of the center-of-mass γ-p scattering angle, obtained with the AV18/UIX Hamiltonian model. See text for explanation of the different lines. The experimental data are from Ref. [11] for $E_{c.m.}$ =6.60 and 9.86 MeV, and from Ref. [12] for the two other cases.

The origin of these discrepancies is currently under investigation. In the case of the differential cross section, it should be pointed out that the experimental data are few and old. In the case of the $A_y(d)$ observable, a possible explanation could be a deficiency in the used model for the TNI, in particular in the absence of three-nucleon spin-orbit terms [17]. (ii) Some small three-body current effects are noticeable, especially in the deuteron tensor polarization observable A_{yy}. This is an indication of the fact that if a Hamiltonian model with two- and three-nucleon interactions is used, then the model for the nuclear current operator should include the corresponding two- and three-body contributions.

The calculated total cross section for nd radiative capture at thermal energies obtained with the AV18/UIX Hamiltonian model is 0.566±0.013 μb, when the "full" model for the nuclear current operator is used. The

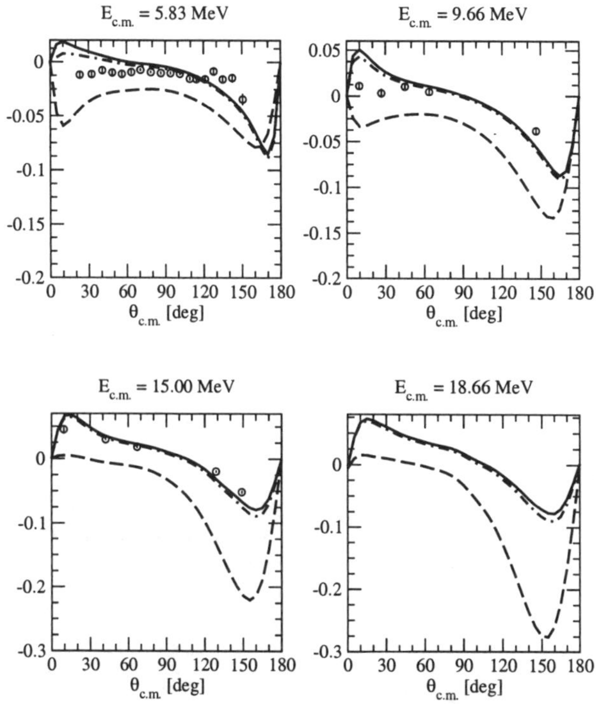

Figure 2. Deuteron vector analyzing power for *pd* radiative capture up to $E_{c.m.}$ =18.66 MeV as function of the center-of-mass γ-p scattering angle, obtained with the AV18/UIX Hamiltonian model. See text for explanation of the different lines. The experimental data are from Ref. [13] for $E_{c.m.}$ =5.83 MeV and from Ref. [14] for $E_{c.m.}$ =9.66 and 15.00 MeV.

quoted error is the statistical uncertainty due to the Monte Carlo integration procedure. This results is to be compared with the experimental value of 0.508±0.015 μb [18]. The origin of the 10 % overprediction of the experimental value is still under investigation. Note that when only the one-body and the one- and two-body current contributions are retained, the calculated total cross section is 0.226±0.006 μb and 0.531±0.013 μb, respectively. Finally, we would like to recall the previous calculation of Ref. [19], where the *nd* radiative capture was studied using PHH wave functions obtained with the AV18/UIX Hamiltonian model, and the nuclear current operator retained one- and two-body contributions, the latter calculated within the ME scheme, and therefore not strictly conserved. The value for the total cross section listed in that work is 0.578, about 14 % larger that the experimental value and 9 % larger than the value here obtained retaining one-

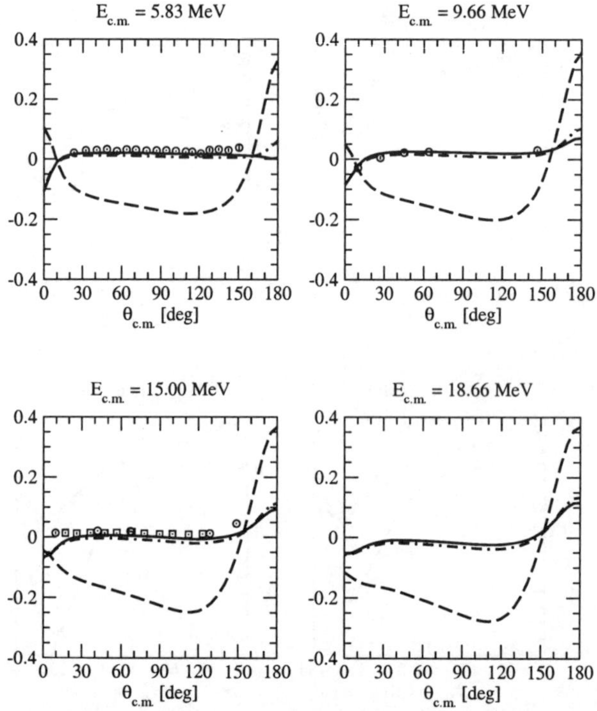

Figure 3. Deuteron tensor analyzing power A_{yy} for pd radiative capture up to $E_{c.m.}$ =18.66 MeV as function of the center-of-mass γ-p scattering angle, obtained with the AV18/UIX Hamiltonian model. See text for explanation of the different lines. The experimental data are from Ref. [13] for $E_{c.m.}$ =5.83 MeV, from Ref. [14] for $E_{c.m.}$ =9.66 MeV, and from Refs. [14,15,16] for $E_{c.m.}$ =15.00 MeV.

and two-body current contributions.

The theoretical prediction of the nd differential cross section at $E_{c.m.}$=6.00 and 7.20 MeV is compared with the experimental data of Ref. [20] in Fig. 4. The calculation is performed using PHH wave functions obtained from a Hamiltonian model consisting of the AV18 two-nucleon interaction only. Therefore, the "full" model for the nuclear current operator (thick-solid lines) retains one- and two-body contributions, but no three-body terms. The results obtained using only one-body current (dashed lines) are also shown. The present results are compared with the ones of Ref. [21] (dotted-dashed lines), obtained using the Bonn-A two-nucleon interaction. From inspection of the figure, we can conclude that: (i) the "full" calculation agrees fairly well with the experimental data, which are however

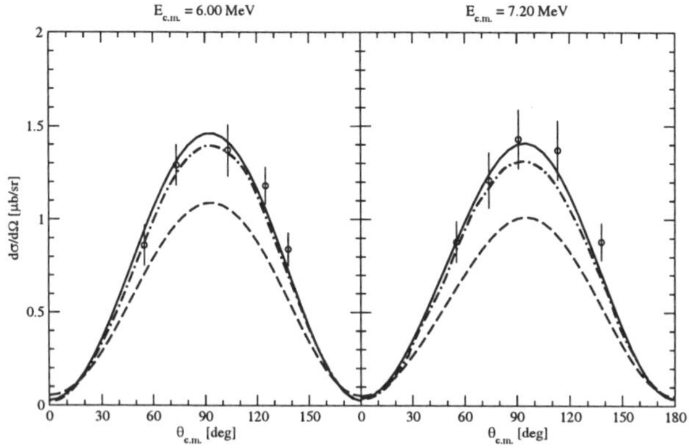

Figure 4. Differential cross section in μb/sr for *nd* radiative capture at $E_{c.m.}$=6.00 and 7.20 MeV as function so the center-of-mass γ-p scattering angle, obtained with the AV18 Hamiltonian model. See text for explanation of the different lines. The experimental data are from Ref. [20].

affected by a large experimental uncertainty; (ii) there is a small difference between the present "full" calculation and the one of Ref. [21]. This is due to the fact that the differential cross section is sensitive to the three-nucleon binding energy, and the Bonn-A two-nucleon interaction binds the triton more than the AV18 one. In fact, our calculation has been compared with a similar one performed by Skibiński and collaborators, using Faddeev wave function calculated with the AV18 potential [22]. Their results coincide with ours. Finally, we consider the so-called fore-aft asymmetry, defined as $a_s = [\sigma(55°) - \sigma(125°)]/[\sigma(55°) + \sigma(125°)]$. Our predicted values are -0.057 and -0.076 at $E_{c.m.}$=6.00 and 7.20 MeV, respectively. The corresponding experimental data [20] are -0.115±0.028 and -0.136±0.032. Therefore, there is a strong discrepancy between theory and experiment, which was already seen in Ref. [21]. The origin of this discrepancy is puzzling and still unknown.

4. Summary and Outlook

We have reported new calculations for *pd* radiative capture observables in a wide range of $E_{c.m.}$ above deuteron breakup threshold, and for *nd* radiative capture total cross section at thermal energies and at $E_{c.m.}$=6.00 and 7.20 MeV. These calculations use accurate bound and scattering state wave functions obtained with the PHH technique from the Argonne v_{18} two-nucleon and, in the case of the *pd* capture and *nd* capture at thermal energies, the Urbana IX three-nucleon interactions. The model for the electromagnetic current operator includes one- and many-body components, constructed so as to satisfy *exactly* the CCR with the given Hamiltonian model. An overall nice description of all the observables has been achieved, with few remarkable exceptions: the differential cross section in *pd* capture at $E_{c.m.}$=16.00 and 18.66 MeV, the $A_y(d)$ observable in *pd* capture at small angles at $E_{c.m.}$=5.83 and 9.66 MeV, the fore-aft asymmetry in *nd* capture at $E_{c.m.}$=6.00 and 7.20 MeV.

The work of R.S. was supported by the U.S. DOE Contract No. DE-AC05-84ER40150, under which the Southeastern Universities Research Association (SURA) operates the Thomas Jefferson National Accelarator Facility.

References

1. Carlson, J., Schiavilla, R., *Rev. Mod. Phys.*, **70**, 743 (1998).
2. Wiringa, R.B., Stoks, V.G.J., Schiavilla, R., *Phys. Rev.*, **C51**, 38 (1995).
3. Pudliner, B.S., Pandharipande, V.R., Carlson, J., Wiringa, R.B., *Phys. Rev. Lett.*, **74**, 4396 (1995).
4. Riska, D.O., *Phys. Rep.*, **181**, 207 (1989).
5. Viviani, M., Kievsky, A., Marcucci, L.E., Rosati, S., Schiavilla, R., *Phys. Rev.*, **C61**, 064001 (2000).
6. Riska, D.O., *Phys. Scr.*, **31**, 107 (1985); *Phys. Scr.*, **31**, 471 (1985).
7. Carlson, J., Riska, D.O., Schiavilla, R., Wiringa, R.B., *Phys. Rev.*, **C42**, 830 (1990).
8. Marcucci, L.E., *et al.*, nucl-th/041182 and nucl-th/041183.
9. Marcucci, L.E., Viviani, M., Schiavilla, R., Kievsky, A., Rosati, S., in preparation.
10. Sachs, R.G., *Phys. Rev.*, **74**, 433 (1948); Nyman, E.M., *Nucl. Phys.*,**B1**, 535 (1967).
11. Belt, B.D., *et al.*, *Phys. Rev. Lett.*, **C24**, 1120 (1970).
12. Anghinolfi, M., Corvisiero, P., Guarnone, M., Ricco, G., Zucchiati, A., *Nucl. Phys.* **A410**, 173 (1983).
13. Akiyoshi, H., *et al.*, *Phys. Rev.*, **C64**, 043001 (2001).
14. Jourdan, J., private communications.
15. Jourdan, J., *et al.*, *Nucl. Phys.* **A453**, 220 (1986).
16. Anklin, H., *et al.*, *Nucl. Phys.* **A636**, 189 (1998).
17. Kievsky, A., *Phys. Rev.*, **C60**, 034001 (1999).
18. Jurney, E.T., Bendt, P.J., Browne, J.C., *Phys. Rev.* **C25**, 2810 (1982).
19. Viviani, M., Schiavilla, R., Kievsky, A., *Phys. Rev.* **C54**, 534 (1996).
20. Mitev, G., Colby, P., Roberson, N.R., Weller, H.R., *Phys. Rev.* **C34**, 389 (1986).
21. Schadow, W., Nohadami, O., Sandhas, W., *Phys. Rev.* **C63**, 044006 (2001).
22. Skibiński, R., private communications.

PION-FEW-NUCLEON PROCESSES FROM A PHENOMENOLOGICAL PERSPECTIVE*

LUCIANO CANTON
Istituto Nazionale di Fisica Nucleare, Padova, via Marzolo, n. 8 Italy

LEONID G. LEVCHUK
NSC Kharkov Institute of Physics and Technology, 61108 Kharkov, Ukraine

We analyze pion production from nucleon-deuteron collisions, with the outgoing three-nucleon system in bound state. Potentially, these reactions could be used to dissection three-nucleon force diagrams, which lead to three-nucleon potential operators. The experimental data, from threshold up to the Δ resonance, are compared with calculations using accurate nuclear wavefunctions. Pion production amplitudes are obtained through matrix elements involving pion-nucleon rescattering mechanisms in S- and P-waves. We assume the hypothesis that S-wave rescattering includes an isoscalar contribution which is generally suppressed for low-energy pion-nucleon scattering, but is enhanced for pion production because the kinematical regime is different, with involvment of high-momenta contributions. P-wave rescattering includes also explicitly the Δ degrees of freedom. Initial-state interactions (ISI) between the proton and the deuteron have sizable effects on the spin-averaged observables. These ISI effects become important for spin observables such as the deuteron tensor analyzing powers T_{20}. For spin observables involving interference terms amongst the various helicity amplitudes, such as for the nucleon vector analyzing power A_y, ISI effects are very important. **Keywords:** Polarization phenomena. Pion production. Few-nucleon systems.

1. Introduction

Pion production from nucleon-deuteron collisions represents an interesting topic of research. It interconnects low-energy few-nucleon physics with intermediate-energy physics, pion dynamics, etc. With this reaction it is possible to study the $NN \to NN\pi$ inelasticities in the most simple (complex) nuclear environment, the three-nucleon system, where rigorous few-body techniques have been developed to describe adequately the nucleon

*TALK PRESENTED BY L.CANTON

dynamics. But these reactions could be also used to probe, in an independent and complementary way, the diagrams that presumably contribute to the three-nucleon forces. Traditionally, 3NF's are constructed phenomenologically in low-energy few-nucleon physics to overcome some deficiencies in the three-nucleon and more-nucleon systems, with parameters adjusted *ad hoc* to reproduce some data that could not be reproduced with a given set of conventional 2N potentials. An important validating feature on the structure of 3NF could come from observations which probe in an independent and complementary way the structure of 3NF diagrams. Pion production reactions could be used, amongst other processes, to shed light on these diagrams.

2. Overview of the model

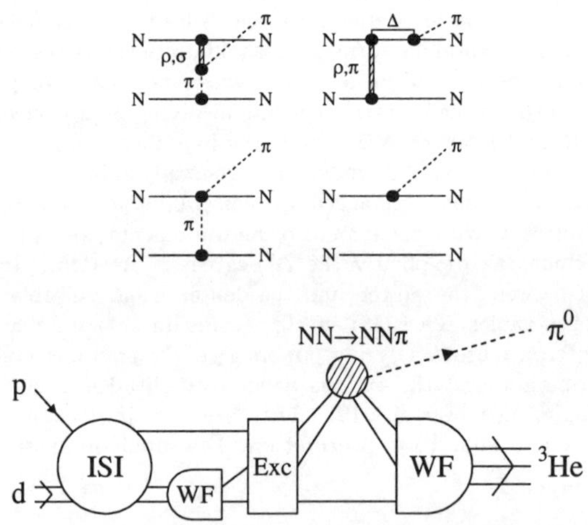

Figure 1. Diagrammatic representation of the calculation required for determining the pion-production amplitude from nucleon-deuteron collisions. *Top*, the elementary production mechanisms. *Bottom*, the overall diagram.

In Fig.1 the base calculation for pion production is illustrated. It involves use of accurate 3N bound-state wavefunctions, calculation of ISI via Faddeev-Alt-Grassberger-Sandhas techniques[1], and an exhorbitant number of multidimensional integrals for the partial-wave evaluation of the elementary pion-production processes. Details and updates for the present calculations can be found in Refs. [2].

The elementary production mechanisms (shown on top of Fig.1) are calculated starting from the phenomenological low-energy interaction Hamiltonian, coupling the pion with the nucleon field. The calculations refer to the following coupling parameters $f^2_{\pi NN}/4\pi = 0.0735$, $f^2_{\pi N\Delta}/4\pi = 0.32$. Here we give only some details on the S-wave rescattering mechanisms, for other details we refer to the References already given.

To construct the S-wave rescattering mechanisms, we adopt the form

$$H_{\text{int}} = 4\pi \frac{\lambda_I}{m_\pi^2} \bar{\Psi}\gamma^\mu \vec{\tau}\Psi \cdot \left[\vec{\Phi} \times \partial_\mu \vec{\Phi}\right] + 4\pi \frac{\lambda_O}{m_\pi} \bar{\Psi}\Psi \left[\vec{\Phi} \cdot \vec{\Phi}\right] . \tag{1}$$

The two parameters have been set to $\lambda_I = 0.045$ and $\lambda_O = 0.006$.

We considered the following off-shell extrapolations of the two λ parameters in the evaluation of the π-production matrix elements (t is the square of the transverse four-momentum of the πN system):

$$\lambda_I^{OFF} = \lambda_I^{ON} \frac{m_\rho^2}{m_\rho^2 - t} \frac{\Lambda_\rho^2}{\Lambda_\rho^2 - t} \qquad \lambda_O^{OFF} = \lambda_O^{ON} \frac{a_{SR} + a_\sigma \frac{m_\sigma^2}{m_\sigma^2 - t}}{a_{SR} + a_\sigma} . \tag{2}$$

The form on the left denotes the isospin-odd contribution in terms of a ρ-exchange model, while on the r.h.s. we describe the isospin-even term as the combined effect of phenomenological short-range (SR) processes and an effective scalar-meson (σ) exchange. The two combined effects act in opposite directions[3]. The form on the right leads to an enhancement of the probability amplitude for pion production in the scalar-isoscalar channel. The $NN \to NN\pi$ inelasticities have been studied extensively in the case of the simpler reactions $pp \to \pi^+ d$; $pp \to pp\pi^o$; and $pp \to \pi^+ pn$. Reference to earlier works can be found in [4]; a more updated review is Ref. [5]. An interesting element of debate concerns the possible mechanisms responsible to the production yield for the process $pp \to pp\pi^o$ at threshold. This yield has been explained by resorting to two different mechanisms, represented in Fig. 2. Studies of pion production from nucleon-nucleon collisions have not been able to exclude neither one or the other of the two mechanisms shown in Fig. 2.

3. Consequences for 3NF diagrams

Clearly, the two mechanisms lead to different structures for the 3NF, as shown in Fig. 3. One option leads to 3NF-structure of short-range/pion-

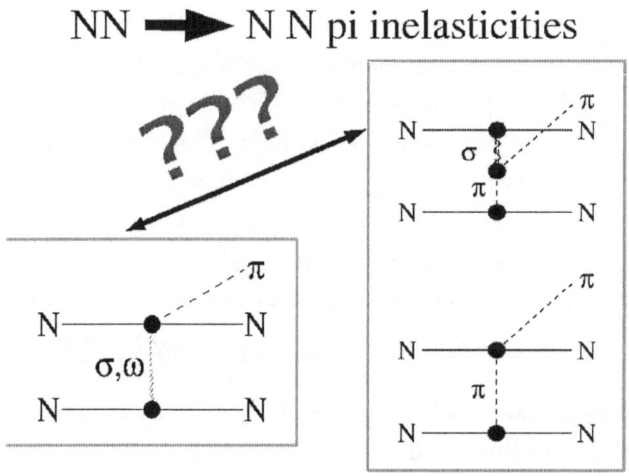

Figure 2. Alternative mechanisms suggested for the $pp \to pp\pi^o$ reaction at threshold. *Left*, pion coupling to short-range two-body exchange currents[6]. *Right*, pion rescattering in the isoscalar channel[7].

range type, the other option does not. Such 3NF mechanisms of short-range/pion-range have been suggested[8] in the past, and the hope was that this 3NF structure could explain the so-called A_y puzzle, in elastic nucleon-deuteron scattering at low energy[9]. Presently, there is no confirmation that this 3NF diagram, alone, could explain the A_y structure; on the other hand it has been shown that the pion dynamics in the 3N system could generate, via meson-retardation effects, a different mechanism that has the potential to dissolve the A_y discrepancy. This situation is illustrated in Fig. 4.

4. Pion capture of nucleons on deuterons

One hopes that the reactions $pd \to \pi^o\ {}^3He$ and $pd \to \pi^+\ {}^3H$ could help to clarify the question of which is, if any, the correct mechanism that describe the process in the isospin even channel. These three-nucleon-type reactions are extremely complicated, and therefore much more difficult to analyze theoretically. However, here the interference effects amongst the various

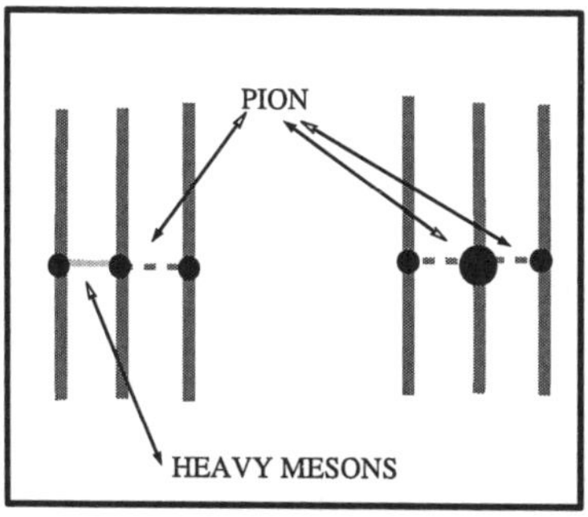

Figure 3. The two alternative mechanisms suggested in the previous figure leads to quite different operators for the 3NF. *Left*, 3NF of short-range/pion range. *Right*, Effects already included in the standard 2-pion-exchange 3NF

mechanisms are much more important than for two-nucleon collisions at threshold and therefore these reactions might represent a more stringent test for the possible mechanisms that describe the process. In the following, we will present results obtained assuming that the production process in the scalar-isoscalar channel is dominated by the rescattering model (the two mechanisms on the l.h.s of Fig. 2); the mechanism depicted on the r.h.s of the same figure will be possibly analyzed in the future.

Fig. 5 shows on the l.h.s. the differential cross-section for the $pd \to \pi^o\ {}^3He$ process. Calculations are for various 2N potentials. It is seen that ISI have a significant effect on the angular dependence of the differential cross-section, in particular at backward angles. The same figure shows on the r.h.s the dramatic effect that ISI's have on the proton analyzing powers A_y. In the region of interference between S-wave and P-wave mechanisms, which corresponds approximately to $\eta \approx 0.5$, the structure of A_y exhibits a rapid variation in sign, with the appearance of an additional "bump" in the angular distribution. This structure is reproduced by our complete model independently of the selected 2N potentials, once the effects of ISI are taken into account.

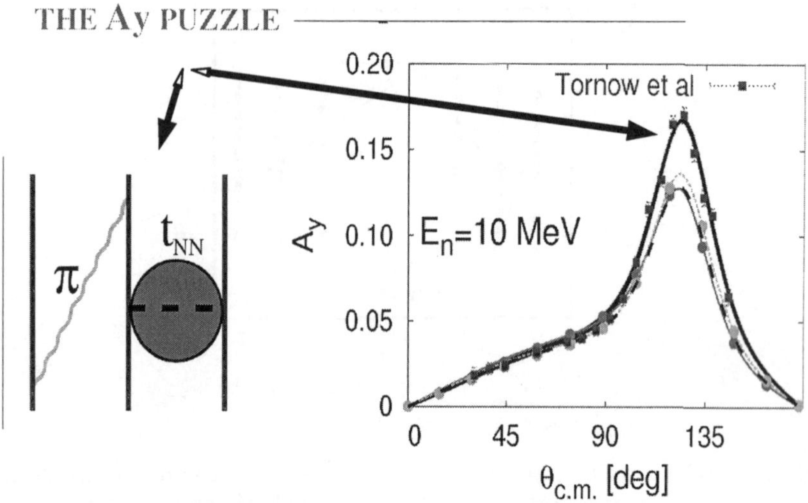

Figure 4. The irreducible one-pion-exchange diagram and its effect on A_y[10]. Experimental data from Ref. [11]. Calculations with modern 2N potentials (AV18, CD-Bonn, Bonn-B).

Another interesting observable that has been analyzed is the deuteron tensor analyzing power T_{20} (Fig. 6). The production reaction acts at threshold as an helicity selector, in that the observed T_{20} is close to its geometrical limit -$\sqrt{2}$. This limit can be obtained in plane-wave calculations with pure isovector πN S-wave rescattering[12]. However, the trend with energy is much better reproduced once the πN rescattering in the scalar-isoscalar channel are also considered. The description of the observable is good once ISI are taken into account[13].

5. Conclusions

Progresses made recently on the pion-production reaction $pd \to \pi^o\ ^3He$ have been reported. The phenomenology of this reaction is quite compli-

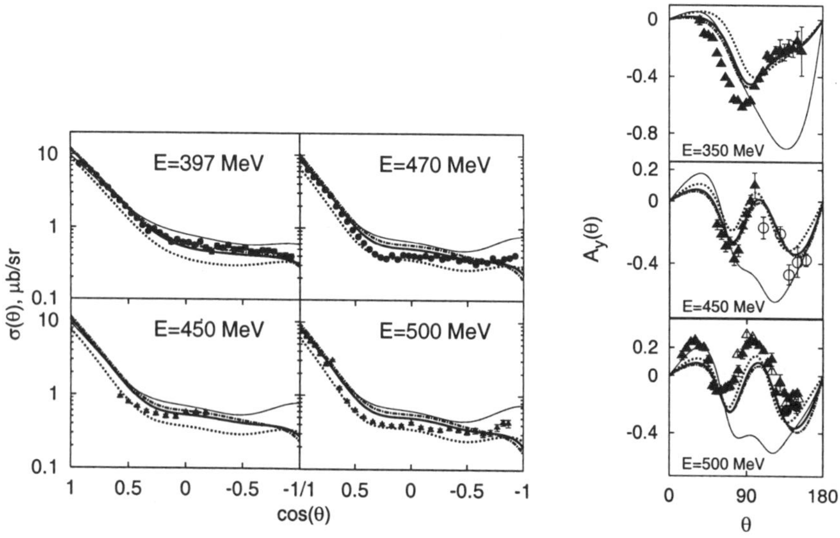

Figure 5. *(Left)* Differential production cross-section. *(Right)* Proton analyzing power in the "interference" region between P-wave and S-wave mechanisms. In all panels, the thin solid line denotes plane-wave calculations with Bonn-B interaction. The thick lines denote ISI calculations for different 2N potentials, dashed (Bonn A), solid (Bonn-B), dotted (Paris). Cosy/Triumf data[14].

cated, especially if one starts to consider the spin-structure of the process. It is shown that a reasonable understanding of the process is possible, provided that the variety of elementary production mechanisms used to describe pion-production from 2N collisions are taken into account, and the nuclear 3-body aspects (bound-states and scattering effects in the initial state) are carefully calculated. Also, the interference effects amongst the various production mechanisms are quite important in this process. We have shown that it is possible, in principle, to dissection the various diagrams that form the postulated 3NF operators in low-energy few-nucleon physics, and analyze their consequences in the phenomenology of pionic-capture processes in nucleon-deuteron collisions.

Acknowledgments We thank T. Melde, G. Pisent, W. Schadow, A. Shebeko, and J.P. Svenne for discussions and collaboration on many topics that have been of relevance to this subject. This work has been co-funded by the Italian MIUR-PRIN project "Fisica Teorica del Nucleo e dei Sistemi

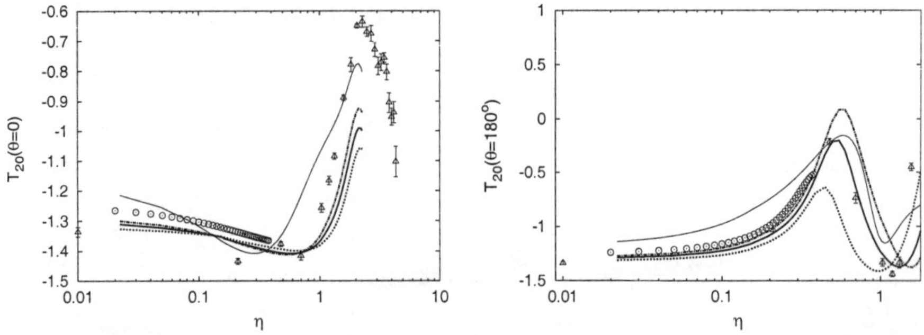

Figure 6. Deuteron tensor analyzing power T_{20} in forward (*left*) and backward (*right*) collinear kinematics. The thin solid line denotes plane-wave calculations with Bonn-B interaction. The thick lines denote ISI calculations for different 2N potentials, dashed (Bonn A), solid (Bonn-B), dotted (Paris). Saclay data[14].

a Molti Corpi".

References

1. E.O. Alt, P. Grassberger, W. Sandhas, Nucl. Phys. **B2**, 167 (1967).
2. L. Canton and W. Schadow, Phys. Rev. C **56**, 1231 (1997); L. Canton *et al.* Phys. Rev. C **57**, 1588 (1998); L. Canton *et al.* Nucl. Phys. **A684**, 417c (2001); L. Canton and W. Schadow, Phys. Rev. C **61**, 064009 (2000); L. Canton and L. Levchuk, submitted for publication.
3. J. Hamilton, *High Energy Physics*, E.H.S. Burhop ed. (Academic, New York, 1967); O.V. Maxwell, W. Weise, and M. Brack, Nucl. Phys. **A348**, 338 (1980).
4. H. Garcilazo and T. Mizutani, *πNN Systems* World Scientific, Singapore (1990).
5. C. Hanart, Phys. Rept. **397**, 155 (2004).
6. T.S-H. Lee and D.O. Riska, Phys. Rev. Lett **70**, 2237 (1993).
7. E. Hernandez and E. Oset, Phys. Lett. **B350** 158, (1995).
8. S.M. Coon, M.T. Peña, and D.O. Riska, Phys. Rev. C 52, 2925 (1995).
9. D. Hüber, J.L. Friar, A. Nogga, H. Witala, and U. van Kolck, Few-Body Syst. **30**, 95 (2001).
10. L. Canton and W. Schadow, Phys. Rev. C **64**, 031001R (2001).
11. W. Tornow *et al.* , Phys. Rev. Lett. **49** 312 (1982).
12. J.F. Germond and C. Wilkin, J. Phys. G **16**, 381 (1990).
13. L. Canton *et al.* in preparation.
14. C. Kerboul *et al.* Phys. Lett. B **181**, 28 (1986); V. Nikulin *et al.* Phys. Rev. **54** 1732 (1996); M. Betigeri *et al.* Nucl. Phys. **A690** 473 (2001); S. Abdel-Samad *et al.* Phys. Lett. **B553**, 31 (2003); J. Cameron, Phys. Lett. **103 B**, 317 (1981).

VARIATIONAL ESTIMATES FOR THREE-BODY SYSTEMS USING THE HYPERSPHERICAL ADIABATIC APPROXIMATION WITHIN THE DISCRETE VARIABLE APPROXIMATION

P. BARLETTA AND A. KIEVSKY

Dip. di Fisica "E.Fermi", Università di Pisa, and INFN, Sezione di Pisa
Via Buonarroti 2
56100 Pisa, Italy
E-mail: barletta@df.unipi.it

The Discrete Variable Representation (DVR) is used in combination with the Hyperspherical Adiabatic Approximation (HAA) to obtain variational estimates for the solution of the Schrödinger equation for a three-body system. This method is particularly appropriate to study weakly bound systems, such as nuclear or van der Waals complexes. It can be applied to bound states as well as to scattering observables. Applications will be shown for a three-nucleon system using a simple scalar interaction, restricted to the lowest adiabatic curve, and to zero partial angular momenta.

1. Introduction

The hyperspherical adiabatic framework[1] consists of two elements: the choice of hyperspherical coordinates to parametrize the internal degrees of freedom of the system under study, and the expansion of its wavefunction into a complete set of hyperangular adiabatic eigenfunctions multiplied by unknown hyperradial functions. In this approach the solution of the Schrödinger equation is divided in two steps: firstly one needs to obtain the hyperangular adiabatic eigenfunctions and eigenvalues, then one has to solve the infinite set of coupled one-dimensional differential equations for the hyperradial functions. The retention of only the lowest adiabatic channel leads to the "strict adiabatic approximation", whereas the inclusion of all the adiabatic channels (until convergence) is indicated as the "coupled-adiabatic-channel method". The strict adiabatic approximation allows the uncoupling between the hyperangular and hyperradial degrees of freedom, reducing the original three-dimensional problem to a "2 + 1" problem. This approximation works well for those states lying deep in the lowest adiabatic potential, and is not very good for higher energy states, where the coupling with the next adiabatic curves is strong.

We have chosen the Discrete Variable Representation (DVR) to solve both the adiabatic equations and the uncoupled hyperradial differential equation. The DVR is widely used in molecular and atomic physics, and its main advantage is that it avoids the necessity of calculating numerically complicated matrix elements. We would like to employ this method to describe few-nucleon systems. Its implementation leads to the solution of an eigenvalue problem for bound states, and to a linear non homogeneous problem for scattering states. The matrices involved are typically very large and sparse. An efficient procedure to obtain solutions to the linear problems associated with the DVR is the diagonalization-truncation method.[2] If one retains only the λ lowest adiabatic basis functions, the diagonalization-truncation method is equivalent to the coupled-channel method with the lowest λ adiabatic channels.

The number of DVR basis functions needed to obtain a good convergence for the eigenfunctions and eigenvectors of the hyperangular adiabatic equations (which depend parametrically on the hyperradius) increases for larger hyperradii.[3] This complication is due in part to the larger physical space spanned by the hyperangular coordinates, in part to the clusterization that the adiabatic basis undergoes.

The properties of the adiabatic eigenfunctions and eigenvalues have been extensively studied. In particular their asymptotic behavior for large hyperradii is well known. Following the work of Nielsen *et al.*,[3] it is possible to solve in a numerically much more efficient way the hyperangular differential equation for large hyperradii by reducing it to an one-dimensional differential equation. This approach reduces the computational effort noticeably.

The main limit of the DVR is its non-variational character, which is a consequence of the approximations introduced in the numerical calculation of the matrix elements. In a previous paper,[4] we have devised a simple procedure to restore the variational character of the DVR, by calculating a very small number of accurate integrals. This approach, besides being helpful in improving the convergence for shallow bound states, showed to be particularly useful in obtaining a stable pattern of convergence for scattering observables.

In this paper we aim to combine the DVR and the HAA to solve the Schröedinger equation for a three-nucleon system in a wide portion of the spectrum, including bound states and dissociative states, below and above break-up. Here our study is limited to the lowest adiabatic channel, and we will consider only zero partial angular momenta. In Section 2, we will provide a description of the method. In Section 3, we will show an application to the three-nucleon system using a simple soft-core scalar potential. Finally, in Section 4, we will briefly comment on the work, and its possible extensions.

2. Theoretical method

Let us consider a system of three identical particles, with spin zero, where the i-th particle has coordinate $\mathbf{r}_i$. For simplicity, we will consider the particles in a state of zero relative angular momenta. Furthermore, we will consider the three particles interacting through a sum of three-pairwise scalar potentials, whose explicit form will be given in the next Section.

2.1. *The Hyperspherical Adiabatic Expansion*

The Jacobi set of coordinates $\{\mathbf{x}_i, \mathbf{y}_i\}$ for this system can be defined as follows:

$$\mathbf{x}_i = \frac{1}{\sqrt{2}}(\mathbf{r}_j - \mathbf{r}_k), \quad \mathbf{y}_i = \sqrt{\frac{2}{3}}[\mathbf{r}_i - (\mathbf{r}_j + \mathbf{r}_k)/2], \tag{1}$$

with $\{i, j, k\}$ cyclic. The associated hyperspherical coordinates ρ, θ_i, μ_i are defined as follows:

$$\rho^2 = x_i^2 + y_i^2, \quad \cos\theta_i = x_i/\rho, \quad \mu_i = \hat{\mathbf{x}}_i \cdot \hat{\mathbf{y}}_i. \tag{2}$$

The volume element is $\rho^5 \cos^2\theta_i \sin^2\theta_i d\rho d\mu_i d\theta_i$. We will refer to the set of hyperangular coordinates $\{\theta_i, \mu_i\}$ as Ω_i, and we will omit the index i when not necessary. The kinetic energy operator can be written as

$$-\frac{\hbar^2}{2m}\left\{\frac{d^2}{d\rho^2} + \frac{5}{\rho}\frac{d}{d\rho} + \frac{L_{\Omega_i}^2}{\rho^2}\right\} = T_\rho + T_\Omega, \tag{3}$$

where $L_{\Omega_i}^2$ is the hyperangular operator. Its eigenfunctions are the Hyperspherical Harmonics (HH) $\{B_k(\Omega)\}$, which, for zero partial angular momenta, take the following simple expression:

$$B_k(\Omega) = \mathcal{N}_k \sum_{i=1}^{3} P_k(\cos(2\theta_i)), \tag{4}$$

where the P_k are normalized Chebyschev of the second kind polynomials, and $\mathcal{N}_k$ is a normalization factor. The associated eigenvalues are $-2k(2k+$

4), for $k = 0, 2, 3, \ldots$. The sum over i in Eq. (4) ensures the symmetrization of the basis.

The adiabatic basis $\{\Phi_\lambda\}$ is constituted by the orthonormalized eigenfunctions (with associated eigenvalues U_λ) of the hyperangular operator T_Ω plus the interaction V:

$$(T_\Omega + V)\Phi_\lambda(\rho;\Omega) = U_\lambda(\rho)\Phi_\lambda(\rho;\Omega), \qquad \int \Phi_\lambda(\rho;\Omega)\Phi_{\lambda'}(\rho;\Omega)d\Omega = \delta_{\lambda,\lambda'}. \quad (5)$$

The functions Φ_λ must possess the desired symmetry under particle permutations: in this work we will require them to be completely symmetric as we deal with spinless bosons.

The wavefunction Ψ_n for the n-th eigenstate of the hamiltonian, with energy E_n, can be expanded in terms of the adiabatic basis:

$$\Psi_n(\rho,\Omega) = \sum_{\lambda=1}^{\infty} u_\lambda^{(n)}(\rho)\Phi_\lambda(\rho;\Omega). \quad (6)$$

As a consequence of the strict adiabatic approximation, we will retain only the first (lowest) term, $\Psi_n = u^{(n)}(\rho)\Phi(\rho;\Omega)$, and drop th eindex λ for simplicity.

The function $u^{(n)}$ is the solution of the lowest uncoupled equation:

$$(T_\rho + U(\rho) + Q(\rho) - E_n)u^{(n)}(\rho) = 0, \quad (7)$$

$$Q(\rho) = \langle\Phi|T_\rho|\Phi\rangle_\Omega. \quad (8)$$

2.2. *Resolution Methods*

By means of the HAA we have reduced the original three-dimensional problem to a $2+1$ problem, with a two-step procedure. Consequently, one must firstly solve Eq. (5), then Eq. (7). There are several methods that could be employed to solve the hyperangular adiabatic equation. Here the function Φ is expanded on HH:

$$\Phi(\rho;\Omega) = \sum_{k=1}^{N} C_k(\rho)B_k(\Omega). \quad (9)$$

The coefficients C_k have been obtained by diagonalization of the matrix $\langle B_k|T_\Omega + V|B_{k'}\rangle$. The kinetic matrix elements where computed analytically, whereas the potential matrix elements are computed by using a finite basis representation in θ, and an exact integration over μ. That is, for a N-dimensional expansion:

$$\langle B_k|V|B_{k'}\rangle_\Omega = \int_{-1}^{1} \sum_{j=1}^{N} B_k(\theta_j,\mu)B_{k'}(\theta_j,\mu)V(\rho,\theta_j,\mu)w_j d\mu, \tag{10}$$

where the variables θ and μ refer to a specific Jacobi permutation, and the integration points θ_j and weights w_j are the Gaussian points related to the Jacobian $\cos^2\theta\sin^2\theta$.

For large values of ρ the HH expansion of Eq. (9) needs a very high number of terms to achieve convergence. It is possible to overcome this difficulty exploiting that the asymptotic behaviour of the adiabatic functions is well known from the literature. In particular, the lowest adiabatic function tends to the dimer wavefunction, with an opportune normalization, and the associated eigenvalue tends to the dimer energy. Both the function and its energy can be obtained by solving a dimer-like one-dimensional equation, as showed in [3]. The short and large-distance solutions are then opportunely combined in an intermediate region.

The function $u^{(n)}$ is obtained by means of a DVR in ρ, following [4]. For bound states, the wavefunction is expanded on Laguerre polynomials:

$$u^{(n)}(\rho) = \sum_i A_i^{(n)} L_i^{(5)}(\beta\rho)\exp(-\beta\rho/2), \tag{11}$$

where β is a non-linear parameter.

For scattering states of energy E, the wavefunction is decomposed into a short-range part u_c, and an asymptotic part:

$$u^{(E)}(\rho) = u_c(\rho) + f(\rho) + \mathcal{L}g(\rho). \tag{12}$$

The function u_c is expanded in terms of Laguerre polynomials as in Eq. (11). The asymptotic functions f and g represent the asymptotic solutions of Eq. (7). Their explicit form depend also on the energy E. For $E = E_d$ (where E_d is the deuteron energy), $\mathcal{L} = a$, where a is the scattering length, so they read

$$f(\rho) = \sqrt{\frac{m}{\hbar^2}}s(\rho)\rho^{-3/2}, \quad g(\rho) = \sqrt{\frac{m}{\hbar^2}}s(\rho)\rho^{-5/2}. \tag{13}$$

For $E = E_d + \hbar^2k^2/2m$, $\mathcal{L} = \tan\delta$, and

$$f(\rho) = \sqrt{\frac{m}{\hbar^2 k}}s(\rho)\frac{\sin(k\rho)}{\rho^{5/2}}, \quad g(\rho) = \sqrt{\frac{m}{\hbar^2 k}}s(\rho)\frac{\cos(k\rho)}{\rho^{5/2}}, \tag{14}$$

where $s(\rho)$ is a factor regularizing f and g at the origin, namely

$$s(\rho) = (1 - \exp(-\rho/\gamma))^p. \tag{15}$$

The scattering matrix $\mathcal{L}$ is obtained by means of the Khon variational principle

$$[\mathcal{L}] = \mathcal{L} - 2\langle\Psi|\mathcal{H} - E|\Psi\rangle, \tag{16}$$

and the two-step procedure outlined in [4]. The reference (scattering) functions can be expressed in hyperspherical coordinates, Eqs. (13) and (14), using the relations in [1, 5].

3. Results

In order to show a practical application of the method outlined in the previous Section, we have considered a system of three-nucleon interacting through a simple scalar potential, $V(r) = A\exp(-(r/r_0)^2)$, with $A = -51.5$ MeV fm^2, and $r_0 = 1.6$ fm. The particle mass is such that $\hbar^2/m = 41.4696$ MeV fm^2. The two-body compound supports only one bound state with zero angular momentum, with energy $E_d = -0.39774$ MeV, whereas the two-body scattering length is 11.36 fm.

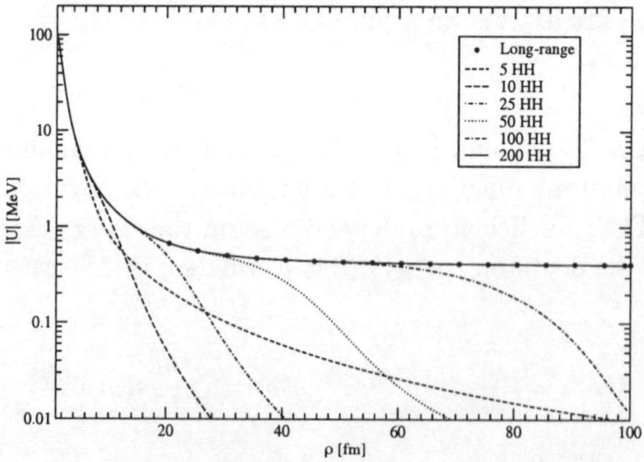

Figure 1. Convergence of the lowest adiabatic curve $U(\rho)$ for different sizes of the hyperangular basis set. The circles represent the long-range eigenvalues of Eq. (5) obtained with the asymptotic approach.

The first step is represented by calculating the hyperangular adiabatic functions. For the integral over μ we have used the Gauss-Legendre formula, with up to 500 points. The eigenfunctions and eigenvalues of Eq. (5) are found by an expansion on up to 200 HH, for the small ρ values. The matching point between the short and long-range approach was fixed at $\rho = 40$ fm. Figure 1 shows the convergence for the lowest adiabatic curve $U(\rho)$, for different numbers of HH included in the expansion. For simplicity, we have plotted $|U|$ in a logarithmic scale, in order to show both the long and short-range regions. The circles represent the asymptotic behaviour of $|U|$, as calculated using the approach showed in [3]. It is possible to see how the area of convergence increases with the number of HH included in the expansion.

Table 1. Convergence patterns for the ground state energy E_0, the excited state energy E_1, the scattering length a, and the phase-shift δ (in degrees) at two energies, namely -0.1 MeV and 0.1 MeV. The convergence is given as a function of the number N_ρ of DVR points included.

	bound		scattering		
N_ρ	E_0 [MeV]	E_1 [MeV]	a [fm]	δ (-0.1 MeV)	δ (0.1 MeV)
10	-9.3048	-0.4589	-32.4998	63.6917	46.7470
20	-9.7419	-0.4780	-30.9813	65.2499	48.6178
30	-9.7555	-0.4782	-30.9655	65.2735	48.6374
40	-9.7559	-0.4783	-30.9642	65.2749	48.6383
50	-9.7559	-0.4783	-30.9642	65.2749	48.6383

Table 1 show the convergence pattern for the DVR solutions of eq. (7). We have chosen to present the convergence for five selected quantities related to five different eigenstates of the Hamiltonian. We have followed the approach of [4], thus all results shown are variational. The non-linear parameters were chosen to be, for all calculations, $\beta = 1$ fm^{-1}, and, in Eq. (15), $p = 4$, and $\gamma = 1$ fm.

Figure 2 shows the phase-shift δ for a range of values of the total energy of three-body system.

4. Conclusions

In this paper we have applied the DVR method to the solution of the three-body problem, using Hyperspherical coordinates to build the DVR. However, the limit of this approach can be seen in Fig. 1, where it is shown

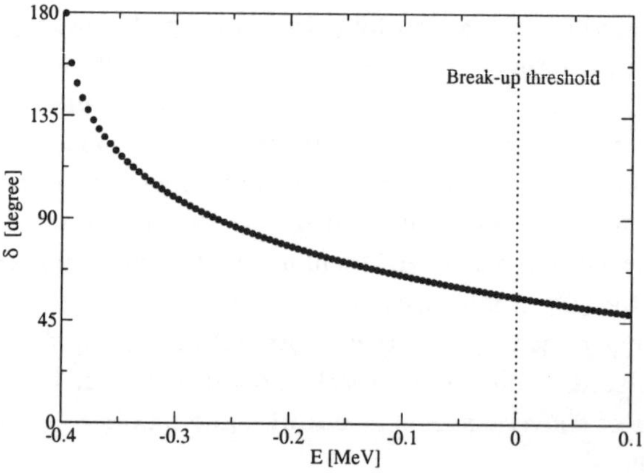

Figure 2. Phase-shift δ for different energies of the three-body system.

the very slow convergence of the hyperangular DVR basis for large values of ρ. This problem is general, not related to the particular choice of coordinates, and is due to the clusterization of the wavefunction at large ρ. Nielsen *et al* [3] have proposed an efficient procedure to solve the hyperangular equations (5) directly for large values of ρ. We have thus combined their method for the long-range part with a DVR in the short-range. The resulting hyperradial equation (7) has been solved using a variational DVR in ρ as in [4]. In this way we have been able to compute bound and scattering states in a wide portion of the spectrum.

Here we have restricted the method to the lowest adiabatic curve and to the lowest (zero) angular momenta. Our aim in future will be to overcome such approximations, in order to describe three-nucleon systems using realistic interactions.

References

1. M. Fabre de la Ripelle, H. Fiedeldey and S. A. Sofianos *Phys. Rev. C* **38**, 449 (1988).
2. R. M. Whitnell and J. C. Light, *J. Chem. Phys.* **90**, 1774 (1989).
3. E. Nielsen, D. V. Fedorov and A. S. Jensen, *J. Phys. B: Mol. Opt. Phys.* **31**, 4085 (1998).
4. M. Lombardi, P. Barletta and A. Kievsky, *Phys. Rev. A* **70**, 032503 (2004).
5. M. Fabre de la Ripelle, *Few-Body Syst.* **14**, 1 (1993).

HIGHLIGHTS ON HEAVY ION REACTIONS AROUND THE FERMI ENERGY

A.BONASERA
Laboratori Nazionali del Sud, INFN, via S.Sofia 44 95123, Catania-Italy
E-mail: bonasera@lns.infn.it

We review some features of heavy ion collisions around the Fermi energy with particular emphasis on results obtained in Italy within a broad theoretical and experimental collaboration. Microscopic vs. macroscopic equation of state (EOS) are discussed and compared to data both in low density and finite temperature regions as well as at higher then normal densities. Fragmentation and a possible liquid gas phase transition is discussed. At higher densities the role of three body forces are discussed and contrasted to microscopic calculations of the EOS and of the binding energy of light nuclei.

1. Introduction

One among the many purposes to collide heavy ions at beam energies below 100 MeV/A is the study of the nuclear matter equation of state (EOS) at finite densities and temperatures. Infact, in such conditions the two nuclei, initially in their ground state, are compressed and heated up. After tens of fm/c, the maximum compression is reached and a compound is formed which then expands and depending on the excitation energy reached might even break into many pieces (multifragmentation). In such a scenario there are many factors at play. In the compression stage it is very important the EOS of the system and the viscosity. Thus data sensitive to the early stages such as energetic protons, neutrons and more complex fragments, as well as photons and pions, will give valuable informations and put constraints on the EOS, as for instance, to the relevance of two body versus three body forces. We have seen infact that the interplay betweeen two and three body forces is very delicate and it turns out that for instance to fit the ground state properties of nuclear matter (NM) in microscopic calculations it is necessary to introduce a three body force[1]. This is demonstrated in fig.(1) for microscopic Bruckner Hartree Fock (BHF) calculations [1]. In the figure, the calculations with two body forces only are given by the dotted line. The

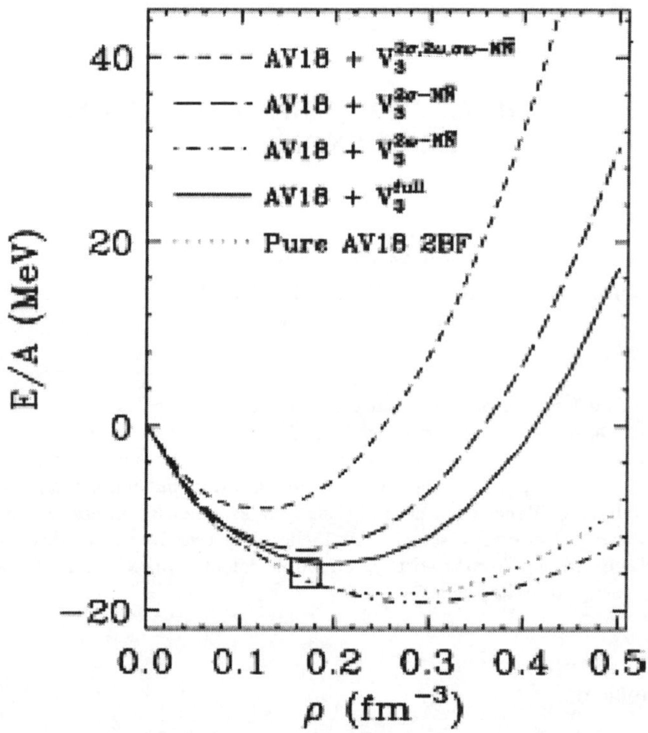

Figure 1. BHF calculations of the nuclear matter EOS including three body forces.

two body force is parametrized to fit the nucleon-nucleon data. A particular form called AV18 is used in this calculation, but any other parametrization of the force gives very similar results. We clearly see that the approach does not work, infact it gives a ground state density of about $0.3fm^{-3}$ and a binding energy of about -20 MeV, recall that the experimental data gives $0.15fm^{-3}$ and -16MeV respectively (square symbol in fig.1). In order to improve the agreement to data, a genuine three body force was included in the calculations. The contribution from different channels are displayed in the figure and the final result is given by the full line. The effect is indeed dramatic. The ground state density is shifted to $0.19fm^{-3}$ and the binding energy to the experimental value. The three body force is obtained trough a fit to the binding energies of light nuclei, t and ^3He essentially [2]. Thus in the NM calculations there are no adjusted free parameters. The fact that the calculations are not yet perfect implies that something is still missing. This is also true when calculating binding energies of other light nuclei where one finds a not so good reproduction of data [2]. One, very naive, idea would be to introduce 4 body forces and so on to reproduce the binding energies of light nuclei. Of course, it would not be too much instructive to add more parameters for each nucleus! Otherwise, it might be as well that the models at hand miss something, for instance a fully relativistic approach and so on. We note on passing, that variational calculations of NM give worse results than BHF[1]. Some light on this problem could be given by experimental data on nucleon production in heavy ion collisions. We will show below that such data does *not* support the need for a strong three body force.

The equation of state discussed so far is at zero temperature. Naturally a complete knowledge of the EOS requires informations at finite temperatures. Microscopic calculations as above but at finite temperatures show, as espected, that the EOS of NM looks like a Van Der Waals(VDW). Infact the nucleon-nucleon force has an attractive tail and a repulsive hard core such as many classical systems. At variance with classical systems the ground state is not a solid but a particular Fermi liquid. However, other properties such as liquid to gas phase transition at finite temperatures and small densities, are of the VDW type. In figure 2 we plot pressure versus density obtained in classical calculation with forces taylored to nuclear properties[3].

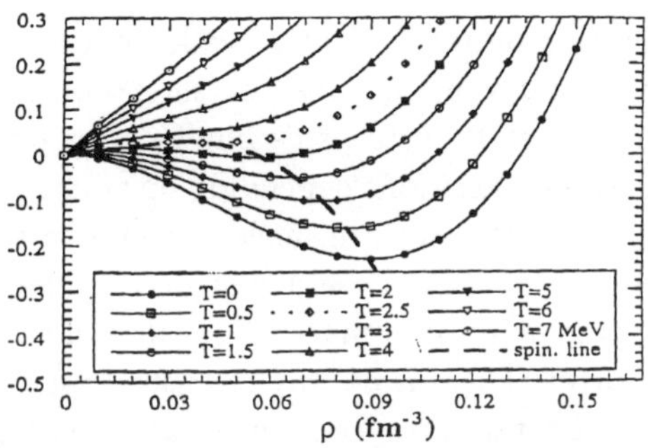

Figure 2. EOS for a classical many body system.

Different temperatures ere explicitly indicated in the figure as well as the instability region for liquid to gas transition. The critical point is at 2.5 MeV temperature and about $0.05 fm^{-3}$ density. Mean field calculations, as well as finite temperature BHF give a critical temperature of about 9 to 15 MeV. Experimental data, which we will discuss in more detail below, give a critical temperature of the order of 5 MeV. Notice however, that microscopic calculations refer to infinite nuclear matter. Finite size effects plus the Coulomb force might decrease somehow the value of the critical temperature. It is somehow surprising that a simple classical molecular dynamics model, whose EOS is given in fig.2, gives many qualitative features in agreement to experimental data on nucleus-nucleus collisions[3]. This is a consequence of the universal features of phase transitions: it is enough a simple scaling of basic quantities like ground state densities of different systems, binding energies etc. and results can be compared to each other. One such comparison between nuclei and metallic cluster is given in [4].

In the following sections we will discuss in some detail some phenomena in systems that have, presumably or to our best selection, reached equilibrium and extract some informations especially on the instability region with a possible first or second order phase transition. Next we will see how we can obtain also some relevant data from non equilibrium situations especially when high densities and excitations energies are reached in the heavy ion collisons.

2. Phase transitions in finite systems: First order, Second order, or...?

From the consideration on the NM EOS we know that at finite temperatures and small densities a phase transition should occur from the nuclear liquid to the gas. Similar to water, this implies that we should have first and second order phase transitions depending on volume, temperature and pressure of the system. However, in nuclear collisions we are not dealing with infinite matter for which we could change somehow the physical conditions. Our system is finite, it is charged and we can only collide two ions in order to excite them. There is no way to control the collisions apart the beam energies and target and projectile mass combinations. Thus we have to change strategy, i.e. we have to opportunely select the data and choose the one that looks more or less equilibrated and for which we can deduce a density where fragmentation occurs (liquid-gas transition) and determine its excitation energy. This is surely an incredible task, infact already the extension of the concept of a phase transition in finite system is a little controversial. However there are proofs in other fields, for instance solid to liquid phase transitions[3], where the transition can be defined with less than 10 particles. In nuclear physics, the possibility of a phase transitions has been demonstrated in microcanonical statistical models where finite sizes and coulomb forces are taken into account[5]. The price to pay in such models is the definition of a freeze-out volume, i.e. a volume where the system is assumed to be in equilibrium and decays in various channels. Such a volume is usually assumed around 6 times the ground state nuclear volume. The agreement of such statistical models to data is sometime very impressive[5]. This of course could be due to the fitting of 2-3 parameters of the models and to the available phase space that is well taken into account. Infact if one adds a nuclear force, which still plays a role at such freeze-out volumes, the good agreement to data is lost[6]. Apart the theoretical considerations discussed above the data shows so many features of phase transitions that in any case it is hard to imagine other effects at play.

One of such features is scaling. We have seen already an example in[4], where a simple scaling let us compare nuclei and clusters.

To see this in more detail, we plot in figure 3 the mass distributions obtained from an expanding classical system initially at temperature T. One can see typical features of evaporation at low T i.e. one big fragment

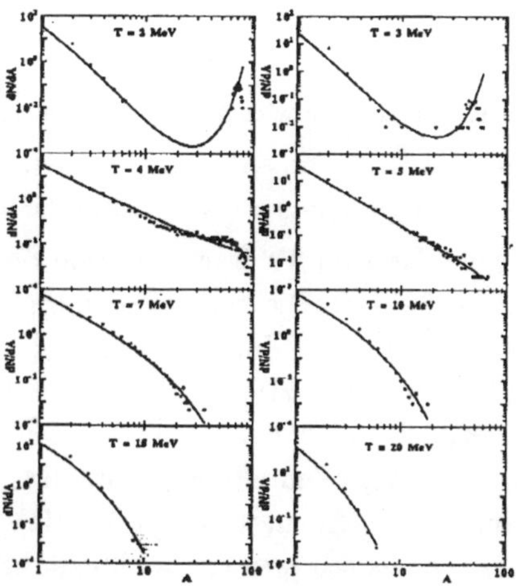

Figure 3. Mass distibution for an expanding system of 100 particles at different initial temperatures. Full lines are the Fisher's fits to the distributions.

and many small ones, and vaporization at high T, i.e. no big fragments but only very small ones, are present. At 5 MeV initial temperature (critical temperature) the mass yield is a power law with exponent -2.31. This is reminescent of Fisher's law for a second order phase transition. This gives a mass yield: $Y = y_0 V^A S^{A^{2/3}} A^{-\tau}$. Where V (volume term) and S (surface term) can be used as free parameter to fit the mass distributions, full lines in figure 3. At the critical point only the power law survive, while above the critical point the power law and the volume terms remain. Thus at the critical point the mass distribution is self-similar. In figure

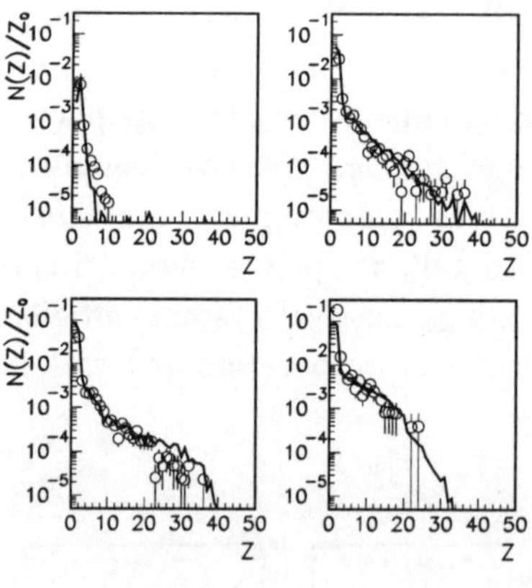

Figure 4. Mass distributions in Au+Au with Z_0 = 85 to 138 (full lines) and Z_0 = 79 at excitation energies 1.5, 3, 4.5 and 7 MeV.

4 we plot the mass yield for Au+Au fragmentation obtained in different experimental conditions, for instance central or peripheral collisions, for excitation energies between 1.5 to 7 MeV[7]. One clearly sees, evaporation in the top right graph (1.5MeV), a typical U shape which indicates that one big fragment is formed and this evaporates many little ones. The almost disappearance of a big residual fragment (3MeV) in the top left plot which gives the onset of fragmentation. At 4.5 MeV excitation energy, the mass distribution becomes a power law with exponent 2.1 typical of a liquid to gas transition(bottom right figure). At higher excitation energy the distribution is exponential i.e. the system has been vaporized. These features are very similar to the classical calculations [3] and in agreement to

the Fisher model.

Two different "critical" analyses have been performed, the first based on the moments of the size distribution[3] and the second on the Fisher droplet model[7]. This last method shows a straight line in the scaled yield of fragments of mass A, $n_A/(q_0 A^{-\tau})$ *vs.* the scaled temperature $\epsilon A^{\sigma}/T$ in a semi-logarithmic presentation (see Fig. 5). This scaling[7] gives "critical"

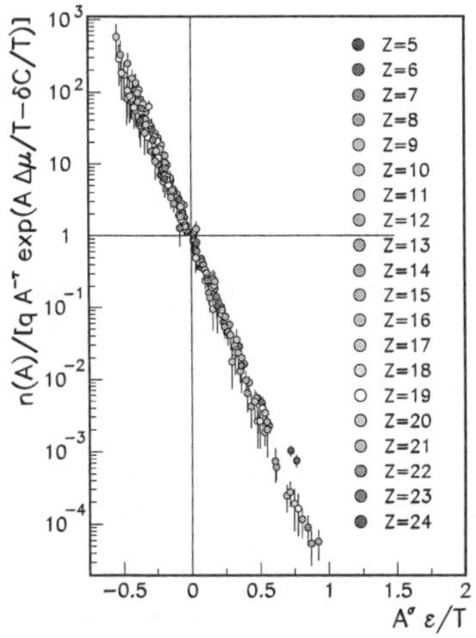

Figure 5. Fisher's scaling for Au fragmentation

parameters σ and τ very similar to the ones for a liquid-gas phase transition at a "critical" excitation energy $\epsilon_c \approx 4.4\ AMeV$, the very same parameters obtained in a moment analysis[7]. The scaling is consistent with similar results obtained by the EOS and ISIS[3] collaborations.

The signals discussed above are necessary conditions for the system undergoing a second order phase transition. The values of the critical parameters, because of indeterminations due to experimental uncertainties and the finite size of the system, are somewhat in between the percolation and the liquid-gas values[3]. Thus nothing definite can be concluded at this time. However, assuming that the signals are for a second order liquid to gas phase transition, we can explore the possibility for a first order transition as well. One way of doing that is by extracting excitation energies and temperature from the data.

The excitation energy can be considered as composed by a configurational part, due to Coulomb and Q-values of the partitions and a kinetic part consisting in the internal energy of the fragments at the freeze-out stage and the translational energy:
$E^* = E_{config} + E_{kin} = E_{coul}(V) + Q_v + E_{int}(T) + T_{tr}(T)$.
The configurational energy depends on the volume and therefore allows to backtrace the density of the system, whereas the kinetic part allows to extract information on the temperature. Everything is under the constraint of energy conservation[7].

The average volume can be obtained in an iterative way by putting the reconstructed primary fragments in a given volume and let them move in the Coulomb field and undergo secondary decay. The volume compatible with experimental data is obtained when the comparison of the obtained observables (for instance the mean kinetic energy of the final fragments) to the measured quantities is satisfactory. The best agreement has been obtained, for excitation energies greater about 4.4 AMeV , with a volume about 2.7 times the normal density one. One weak point in this analysis is if the volumes are so small (relatively) should nuclear forces play a role[6]?

The temperature can be obtained from an energy balance starting from the average kinetic energy[7] $\langle E_{kin}\rangle = \frac{3}{2}\langle m-1\rangle T + \langle \sum aA_{IMF}\rangle T^2$ or from translational energy as $T = \langle \frac{E_{tr}}{\frac{3}{2}m-1}\rangle$. The results obtained are in very good agreement within each other[7] and to the values obtained by independent measurements. A further confirmation of the validity of the temperatures obtained is given by a comparison of the internal excitation energy of the fragments to the experimental results of Indra collaboration[7]. The first result of this procedure is that the system do not follow the liquid-drop line in the caloric curve, similarly to the results obtained by the Aladin collaboration[7]. The thermodynamical analysis allows to extract the microcanonical heat capacity. Indeed one can obtain the heat capacity in terms of kinetic energy fluctuations: $\frac{C_{kin}}{C} = 1 - \frac{\sigma^2_{kin}}{\sigma^2_{can}}$, where $\sigma^2_{can} = T^2 C_{kin} = T^2 \frac{dE_{kin}}{dT}$. The kinetic energy are equal to configurational fluctuations, since in a microcanonical ensemble the total energy is fixed.

If σ^2_{kin} exceed canonical fluctuations one would expect negative values for the heat capacity. The results are shown in Fig. 6 where two divergences and a negative branch of the heat capacity clearly appear, both for peripheral and central collisions. The distance between the two divergences is related to the latent heat.

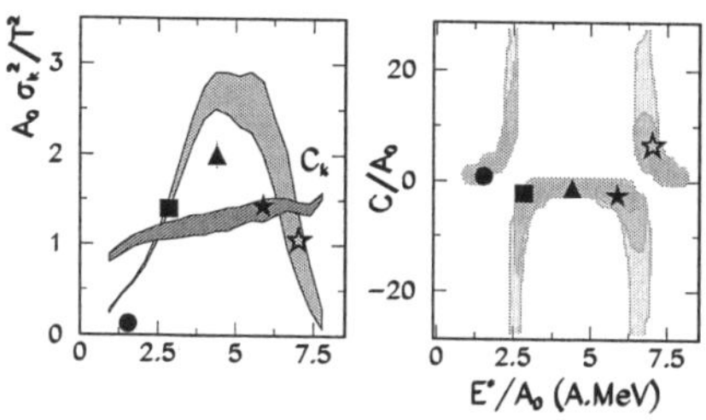

Figure 6. Left panel: normalized partial energy fluctuations for QP events (gray contours) and central Au + C, Au + Cu and Au + Au central events. estimation of the heat capacity per nucleon of the source.

These results are a strong evidence of a first order phase transition and have been confirmed by other collaborations, like Indra, in particular as far as the second divergence is concerned[7].

3. Spinodal Fragmentation

The observation of several intermediate mass fragments (IMF) in heavy ion collisions and the possible connection to the occurrence of liquid-gas phase transitions in nuclear systems are still among the most challenging problems in heavy ion reaction physics.

Recently many efforts have been devoted to indentify relevant observables that may signal the occurrence of phase transitions in finite nuclei [8,9,10,11,12]. These studies are mostly based on the investigation of thermodynamical properties of finite systems at equilibrium and on critical behaviour analyses. In such a context, a description of the dynamics of fragment formation can provide an important complementary piece of information to learn about the path of the fragmentation process. Indeed the study of the dynamical evolution of complex excited nuclear systems may shed some light on the relevant fragmentation mechanisms and on density and temperature conditions where fragments are formed. This opens the possibility to get insight into the behaviour of the system in such regions, far from normal values, encountered along the fragmentation path, and hence on its EOS. Eventually at the freeze-out configuration one may check how the available phase space has been filled and compare results obtained for some relevant degrees of freedom and physical observables to the values expected at thermodynamical equilibrium. At the same time one may try to identify some specific observables that keep the memory of the fragmentation mechanism and are sensitive to the EOS.

We discuss here a study performed within the framework of stochastic mean field approaches [13,14,15,16]. As extensively discussed also in Ref.[17], according to this theory, the fragmentation process is dominated by the growth of volume (spinodal) and surface instabilities encountered during the expansion phase of the considered excited systems. Hence the fragmentation path is driven by the amplification of the unstable collective modes. The correct description of the degree of thermal agitation, that determines the amplitude of the stochastic term incorporated in the treatment, is essential since fluctuations provide the seeds of fragment formation.

As shown by these simulations, the presence of spinodal decomposition is signaled by the formation of nearly equal-sized fragments, due to the dominance of a few collective unstable modes in the linear regime. Therefore, one may search for events that keep the memory of this early configuration, typical of spinodal decomposition, even after the secondary decay. With this aim, one may employ the correlation analysis proposed by Moretto *et al.* [18] which reveals the presence of such events even if they are only relatively few. It basically consists of producing a two-dimensional histogram (occasionally called a LEGO plot) in terms of the extracted mean and dispersion of the IMF charge distribution, $\langle Z \rangle$ and ΔZ.

The Figure 7 presents the results of such an analysis made for the stochastic simulations [14] of central $Xe + Sn$ collisions at $32MeV/A$ [19]. It can be seen that a peak at very small values of ΔZ, *i.e.* approximately equal-size fragments, stands out quite clearly against a rather structureless background, even though it contains only a few per cent of the yield. The background can be well accounted for by statistical considerations (for instance it can be reproduced from artificial events constructed by sampling the fragments randomly from different events. Thus, the emerging peak cannot be of statistical origin but must be regarded as a very specific signal of the spinodal decomposition scenario. This analysis presents a powerful tool for identifying the spinodal disassembly scenario in experiments, see Ref.s [19,12].

4. Neck Fragmentation

Summarizing the main experimental observations, we would like to stress the following peculiarities of a "dynamical" IMF production mechanism in semi-peripheral collisions:

1. An enhanced emission is localized in the mid-rapidity region, inter-

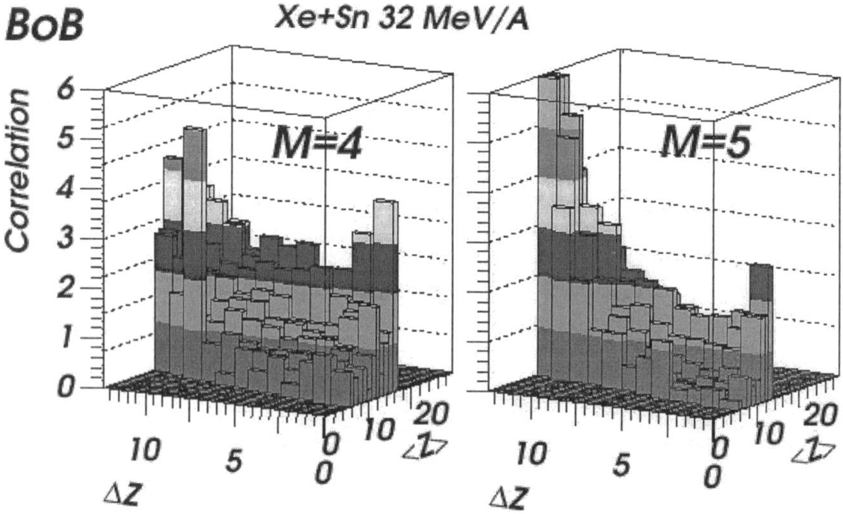

Figure 7. Fragment size correlations.

The IMF size correlations displayed as a two-dimensional histogram where the axes represent the average fragment charge $\langle Z \rangle$ and the associated dispersion ΔZ, as obtained with BOB simulations. The peak at very small ΔZ contains events having approximately equal-size fragments, while the background can be well reproduced by event mixing.

mediate between PLF and TLF sources, especially for IMF's with charge Z from 3 to 15 units.

2. The IMF's relative velocity distributions with respect to PLF (or TLF) cannot be explained in terms of a pure Coulomb repulsion following a statistical decay. A high degree of decoupling from the PLF (TLF) is also invoked.

3. Anisotropic IMF's angular distributions are indicating preferential emission directions and an alignment tendency.

4. For charge asymmetric systems the light particles and IMF emissions keep track of a neutron enrichment process that takes place in the neck region.

A fully consistent physical picture of the processes that can reproduce observed characteristics is still a matter of debate and several physical phenomena can be envisaged, ranging from the formation of a transient neck-like structure that would break-up due to Rayleigh instabilities or through a fission-like process, to the statistical decay of a hot source, triggered by the proximity with PLF and TLF [20,21,22]. Dynamical transport models suggest since long time the possibility of observing neck emission [23,24,25,26], [27,28,29]. We show in Fig.8 the density contour plots of a neck fragmetation event at $b = 6fm$, for the reaction $^{124}Sn+^{64}Ni$ at $35AMeV$, where nice data are appearing from the $CHIMERA$ collaboration at the LNS [30,31], as obtained from the numerical calculations of Ref.[32].

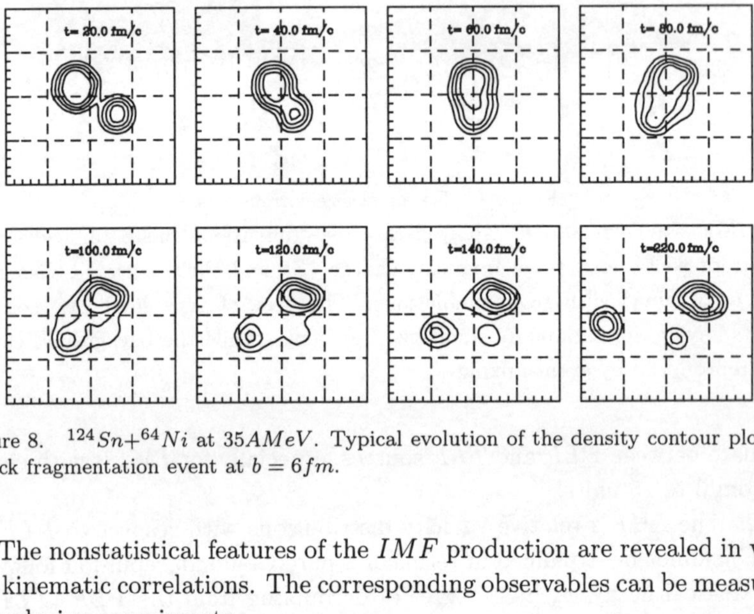

Figure 8. $^{124}Sn+^{64}Ni$ at $35AMeV$. Typical evolution of the density contour plot for a neck fragmentation event at $b = 6fm$.

The nonstatistical features of the IMF production are revealed in various kinematic correlations. The corresponding observables can be measured in exclusive experiments.

An interesting kinematical observable is the asymptotic relative velocitity of the neck-produced $IMFs$ with respect to the PLF (TLF), $v_{rel}(PLF, TLF) \equiv |\mathbf{v_{PLF,TLF}} - \mathbf{v_{IMF}}|$. This is compared with the relative velocity reached in a pure Coulomb-driven separation, signature of a statistical fission process of a compound PLF^* or TLF^* system, as provided by the Viola systematics [33,34]:

$$v_{viola}(1,2) = \sqrt{\frac{2}{M_{red}}(0.755\frac{Z_1 Z_2}{A_1^{1/3} + A_2^{1/3}} + 7.3)} \quad (1)$$

where A_1, A_2, Z_1, Z_2 are the mass and charge numbers of the fission products and M_{red} is the corresponding reduced mass.

For each Neck-IMF we can evaluate the ratios
$r = v_{rel}(PLF)/v_{viola}(PLF)$, $(r1 = v_{rel}(TLF)/v_{viola}(TLF))$.

In Figure 9 we plot $r1$ against r for each IMF. The solid lines represent the loci of the PL-$(r = 1)$ and TL-$(r1 = 1)$ fission events respectively. The values $(r, r1)$ appear simultaneously larger than 1 suggesting a weak IMF correlation with *both* PLF and TLF, in contrast to a statistical fission mechanism. The process has some similarities with the participant-spectator scenario. However the dynamics appear much richer than in the simple sudden abrasion model, where the locus of the $r - r1$ correlation should be on the bisectrix, apart the *Goldhaber* widths, see ref.[21]. Here the wide distributions of Fig.9 reveal a broad range of fragment velocities, typical of the instability evolution in the neck region that will lead to large dynamical fluctuations on IMF properties.

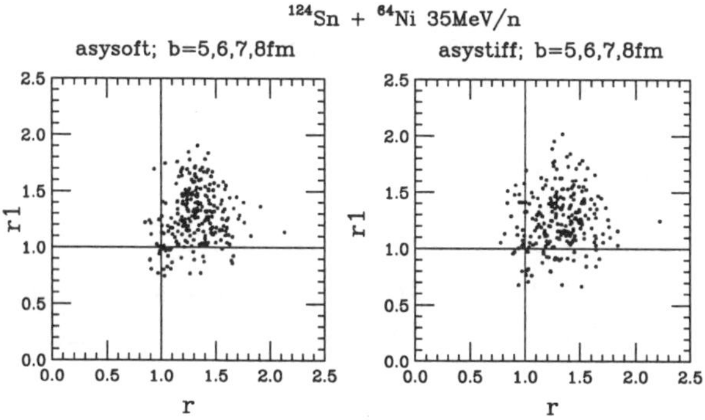

Figure 9. Correlation between deviations from Viola systematics, see text. Results are shown for two parameterizations of the symmetry term in the EOS.

5. The EOS at high densities: compressibility vs. viscosity

Non equilibrium effects play an important role in the first stages of the reaction. This stage has a time duration of the order of hundred fm/c and determines the fate of the reaction i.e. if the system reachs equilibration or not and what kind of temperature if the equilibrium is reached. Thus it is important to have a good grasp on this part of the reaction and this goal has been fulfilled using transport equations[36,37]. The transport approach contains essentially two ingredients, the mean field and a collision term. These are responsible for the highest density the system can reach for a given beam energy and impact parameter and for the production of preequilibrium particles such as photons, pions and energetic nucleons. These features can be experimentally determined looking at different observables. One of such observables is the collective flow. This is the flow of fragments along preferential directions for nuclei colliding at non zero impact parameters. It is determined by plotting the transverse momentum (to the beam direction) of the particle versus its rapidity (along the beam axis). The larger value of collective flow gives a high compressibility of the EOS. Also viscosity plays an important role, i.e. no viscosity will result in nuclei which are transparent that is small flow. Thus it is the interplay between EOS and two body collisions which determines the flow. For sufficiently high energies hydrodynamics scaling laws apply as it can be easily seen by plotting normalized quantities, i.e. transverse momenta divided beam momenta and rapidity divided by beam rapidity. One such a plot is given in figure 10 which shows a good scaling for different colliding nuclei and beam energies [35]. The scaling holds down to low energies around the Fermi energy where, as we have seen, matter becomes unstable due to the occurrence of a phase transition. The absolute value of the maximum transverse momentum depends sensitively on viscosity and compressibility of the EOS, thus we need other quantities that are sensitive to one or the other ingredient. However, the conclusion so far is that flow can be reproduced with an EOS which has a compressibility of about 200MeV and some momentum dependence.

In order to put more constraints on viscosity and EOS one looks at pion or photon production. The result is that very energetic particles need cooperative effects in order to be produced. Thus one has to include three or more body collisions to have energetic particles[36]. This has been nicely shown in a Medea experiment at the LNS in Italy[38].

In the experiment Ni+Ni at 30 MeV/A, protons were detected in coincidence with heavier fragments. The number of participating nucleons were estimated (i.e. the impact parameter) and for each impact parameter the number of protons with given energies were measured. Such a quantity is plotted in figure 11 for various cuts in energy. One sees that for low energies the relation between multiplicities and the number of participant is linear but it starts to deviate for large energies. This implies that cooperative effects (three or higher order collisions) become more and more important the larger is the energy of the detected particle. Microscopic BNV calculations which include a two body collision term are also displayed in the figure. One notices that BNV calculations with a momentum dependent force fit the data better (open circles) for protons energies less than 120MeV, while

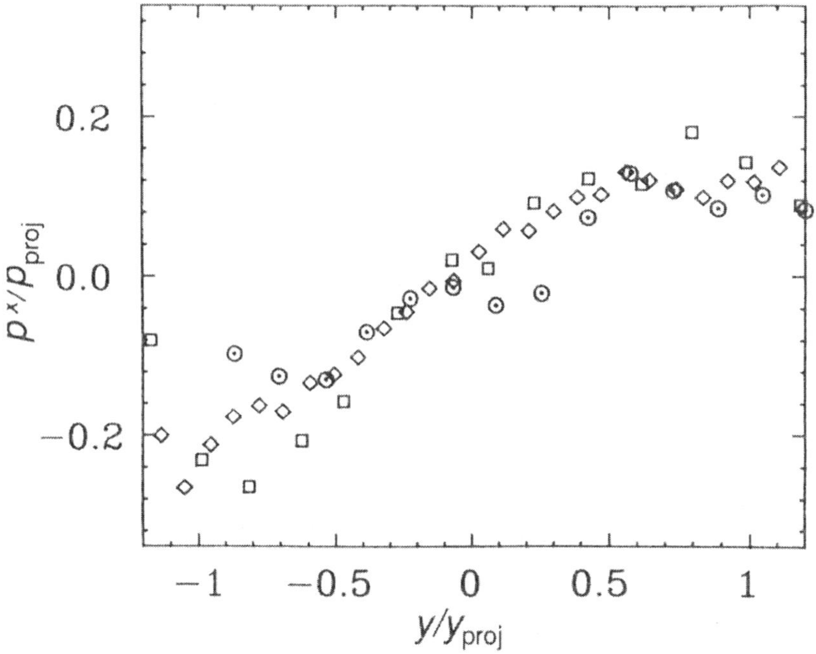

Figure 10. Normalized transverse momenta vs. normalized rapidity for different nuclei and beam energies.

calculations with momentum independent forces fail already at lower energies, which again puts constraints on the EOS. At the highest energies both calculations fail[38]. Thus three body correlations are important for those energetic particles, however we have to stress that the amount of particles of such energy produced are small (two order of magnitude less than the protons below 80 MeV), thus the effect is small. This should be constrasted with microscopic BHF calculations [1] and calculations of light nuclei binding energies [2] where the effect of three body forces seem to be crucial. If these forces are so important should one detected more energetic particles? This is an important point which deserves more different experimental investigation.

6. Summary

In this work we have reviewed some features of heavy ion collisions at intermediate energies. In particular we have explored the possibility of a

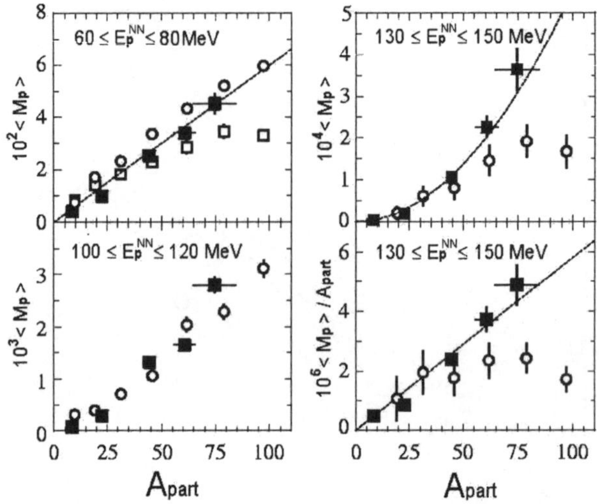

Figure 11. Proton multiplicities vs number of participants for various energy cuts in the proton multiplicities.

'liquid-gas' type phase transition in nuclear systems and derived some relevant features such as critical exponents, critical temperature and density. In this context, both calculations and data seem to converge in the sense that there is some interesting signal of a possible transition but with still many open problems. The first one is if the system really gets in equilibrium at high excitation energies. Infact in such conditions, i.e. when the excitation energy is of the order of the binding energy of nuclei, it is not clear what holds the particles long enough for equilibrium to be reached. Rather, with increasing energies more and more collective flow appears. Neverthless, either because of thermal or dynamical effects the nuclei break somehow and some critical signals might appear in a percolation like mode. If the problem of equilibrium at high excitation energies is not unambigously resolved the occurrence of a first order phase transition and negative specific heat cannot be firmly established. Instabilities occur also for dynamical reasons, i.e. either because the system enters quickly into an instability region (spinodal) or because of hydrodynamical instabilities. We have discussed both theoretical results and data. For the first compression stage, important informations which put severe constraints on the EOS and viscosity, have been somewhat discussed. New data with more and more performing apparata which give informations on the 4π distribution for charges and masses is needed together with production of new particles such as gammas, pions and kaons for instance. An aspect we have not discussed here is the mechanism of production of such particles and in particular if one can have information on incoherent vs. coherent production.

From a purely theoretical point of view, we have seen in the past twenty years or so a sensitive improvement in the quality of the models also in parallel with the raise of the computer speeds and capability. Early simple cascade, Boltzmann type mean field models have been slowly replaced by models which take into account correlations and some quantum physics. We believe that the coming times will be extremely important to get a good picture of the EOS at various densities and temperatures both from a theoretical and experimental point of view. At the same time we will know more about forces in nuclear systems.

Acknowledgments We thank many people for providing most of the material discussed in this paper and for stimulating discussions. In particular we would like to thank M.D'Agostino, M.Bruno, M.Colonna, R.Coniglione, M.Di Toro, U.Lombardo, M.Papa and P.Sapienza.

References

1. Zuo W, Lejeune A, Lombardo U, et al. Eur. Phys. J.**A 14**, 469(2002); Zuo W, Lejeune A, Lombardo U, et al. Nucl. Phys. A706, 418(2002).
2. A. Kievski, these proceedings.
3. A.Bonasera, M.Bruno, C.O.Dorso and P.F. Mastinu, *Rivista Nuovo Cimento* **23,n.2**, 1 (2000).
4. A.Bonasera, Phys. World,, feb.1999,20.
5. D.Gross *Rep.Progr.Phys.* **53**,605(1990).
6. S.K.Samaddar, J.N.De and A.Bonasera, *nucl-th*0402068 and PRC(submitted).
7. M.D'Agostino et al., .Lett. **B473** 219, (2000); M.D'Agostino et al., *Nucl.Phys.* **A734**,512 (2004).
8. R.Botet et al., Phys. Rev. Lett. **86** (2001) 3514; J.D.Frankland et al., contribution to the XLth Int. Wint. Meet. on Nuclear Physics, Bormio (Italy), January 21-25 2002.
9. J.B.Elliott et al., Phys. Rev. Lett. **88** (2002) 042701
10. M.D'Agostino et al., Nucl. Phys. **A699** (2002) 795
11. Ph.Chomaz and F.Gulminelli, Nucl. Phys. **A647** (1999) 153; F.Gulminelli, Ph.Chomaz, Al.H.Raduta, Ad.R.Raduta, Phys.Rev.Lett. **91** (2003) 202701
12. G.Tabacaru et al., Eur. Phys.J. **A18** (2003) 103
13. A.Guarnera, M.Colonna, Ph.Chomaz, Phys. Lett. **B373** (1996) 297
14. Ph.Chomaz et al., Phys. Rev. Lett. **73** (1994) 3512; A.Guarnera et al., Phys. Lett. **B403** (1997) 191
15. M.Colonna et al., Nucl. Phys. **A642** (1998) 449
16. F.Matera, A.Dellafiore, G.Fabbri, Phys. Rev. **C67** (2003) 034608

17. Ph.Chomaz, M.Colonna, J.Randrup, Phys. Rep. **389** (2004) 263
18. L.G.Moretto et al., Phys. Rev. Lett.**77** (1996) 2634
19. B.Borderie et al., Phys. Rev. Lett. **86** (2001) 3252
20. U. Brosa, S. Grossman, A. Muller, Phys. Rep. 197 (1990) 167.
21. J. Lukasik et al., Phys. Lett. B566 (2003) 76.
22. A.S. Botvina et al., Phys.Rev. C 59 (1999) 3444.
23. A.Bonasera, G.F.Bertsch and E.El Sayed, Phys.Lett. B141 (1984) 9.
24. M.Colonna, N.Colonna, A.Bonasera, M.Di Toro, Nucl.Phys. A541 (1992) 295.
25. M.Colonna, M.Di Toro, A.Guarnera, Nucl.Phys. A589 (1995) 160.
26. L.G. Sobotka, Phys.Rev. C 50 (1994) 1272R.
27. J.F.Dempsey et al., Phys.Rev. C54 (1996) 1710.
28. J.Lukasik et al., Phys.Rev. C55 (1997) 1906.
29. L.G.Sobotka et al., Phys.Rev. C55 (1997) 2109.
30. A.Pagano et al., Nucl.Phys. A681 (2001) 331c.
31. A.Pagano et al., Nucl.Phys. A734 (2004) 504c.
32. V.Baran, M.Colonna, M.Di Toro, Nucl.Phys. A730 (2004) 329.
33. V.Viola et al., Phys.Rev. C31 (1985) 1550.
34. D.J.Hinde et al., Nucl.Phys. A472 (1987) 318.
35. A.Bonasera, L.P.Csernai, Phys.Rev.Lett.**59**(1987),630.
36. A.Bonasera, F.Gulminelli and J.J.Molitoris , Phys.Rep.**243** (1994),1.
37. M.Papa, T.Maruyama and A.Bonasera,Phys. Rev. **C64**,024906 (2001).
38. P.Sapienza et al., Phys.Rev.Lett.**87**(2001),072701.

PENTAQUARK STATES AND SPECTRUM

R. BIJKER
ICN-UNAM, AP 70-543, 04510 México, D.F., México.

M. M. GIANNINI AND E. SANTOPINTO
Dipartimento di Fisica dell'Università di Genova
and I. N. F. N., Sezione di Genova, via Dodecaneso 33, I-16146 Genova, Italy

We present a general classification scheme for $qqqq\bar{q}$ pentaquark states in terms of spin-flavour $SU(6)$ representations and their decomposition in terms of $SU_f(3) \otimes SU_s(2)$. Special attenction is payed to the S_4 permutational symmetry of the spin-flavour part of the four-quark wave functions. This classification is general and useful both for experimentalists and model builders. In particular, it will be useful also for the construction of higher three-quark Fock components or to compare with the nonrelativistic limits of other models. We discuss the spectroscopy of pentaquarks using a Gürsey-Radicati type mass formula, whose coefficients have been determined previously in a study of qqq baryons. Its angular momentum depends on the interplay between the spin-flavour and orbital contributions to the mass operator. The magnetic moment of the $\Theta^+(1540)$ is discussed in the constituent quark model for different values of angular momentum and parity. Finally we construct a model for the five-body problem that can be solved exactly: the properties of the pentaquark are discussed in a collective string-like model.

1. Introduction

The first report of the discovery of the pentaquark [1] has triggered an enormous amount of experimental [2] and theoretical studies of the properties of exotic baryons. The width of the Θ^+ seems to be so small that only an upper limit could be established ($<$ 20 MeV or perhaps as small as a few MeV's [3]). More recently, evidence has been found for the existence of another exotic baryon $\Xi^{--}_{3/2}(1862)$ with strangeness $S = -2$ by the NA49 Collaboration at CERN [4]. In addition, there is now the first evidence [5] for a heavy pentaquark $\Theta^+_c(3099)$ in which the antistrange quark in the Θ^+ is replaced by the anticharm quark. From these experiments there are no indications about spin and parity of these states. For a review of the experimental status we refer to [7].

However, there still exist many doubts and questions about the existence of this exotic Θ^+ states, since in addition to various confirmations there are also several experiments in which no signal has been observed [6]. Hence, it is of the utmost importance to understand the origin of these apparently contradictory results, and to have irrefutable proof for or against its existence [7]. An answer will arrive from high-statistic experiments.

The first indications of exotic baryon states have triggered an enormous amount of theoretical work. Theoretical interpretations range from chiral soliton models [8] which provided the motivation for the experimental searches, cluster models [9] and various constituent quark models [10,11]. A review of pentaquark models can be found in [12].

In this contribution we review our construction of pentaquark states based only on symmetry requirements using group theory techniques. We construct a complete classification scheme for the $q^4\bar{q}$ states.

2. Classification of pentaquark states

We consider pentaquark states to be built with five light (u, d, s) constituent quarks: the pentaquark wave functions contain contributions connected to the spatial degrees of freedom and the internal degrees of freedom (colour, flavour and spin). In the construction of the classification scheme one is guided by two conditions: the pentaquark wave function should be a colour singlet and it should be antisymmetric under any permutation of the four quarks [11]. The permutation symmetry of the four quark system is given by S_4 which is isomorphic to the tetrahedral group T_d. The labels of the latter group are used to classify the four-quark states by their permutation symmetry character: symmetric A_1, antisymmetric A_2 or mixed symmetric E, F_2 or F_1. The corresponding algebraic structure for the internal degrees of freedom is $SU_{sf}(6) \otimes SU_c(3)$. The complete spin-flavour $SU(6)$ classification of $q^4\bar{q}$ states and their full decomposition into spin and flavour states $SU_{\rm sf}(6) \supset SU_{\rm f}(3) \otimes SU_{\rm s}(2)$ can be found in Ref. [11]. In particular we use the permutational symmetry of the four-quark part of the spin-flavour states to classify the pentaquark states [11].

The spin-flavour part has to be combined with the colour part and the orbital part in such a way that the total pentaquark wave function is a colour-singlet state, and that the four quarks obey the Pauli principle, *i.e.* are antisymmetric under any permutation of the four quarks. Since the colour part of the pentaquark wave function is a $[222]_1$ singlet and that of the antiquark a $[11]_3$ anti-triplet, the colour wave function of the four-quark configuration is a $[211]_3$ triplet with F_1 symmetry under T_d. The total q^4 wave function is antisymmetric (A_2), hence the orbital-spin-flavour part is a [31] state with F_2 symmetry which is obtained from the colour part by interchanging rows and columns

$$\psi_{\rm c}(q^4) \quad [211] \quad F_1 \; , \qquad\qquad \psi_{\rm osf}(q^4) \quad [31] \quad F_2 \; . \tag{1}$$

The symmetry properties of the orbital part of the pentaquark wave function should also be discussed. If the four quarks are in a spatially symmetric S-wave ground state with A_1 symmetry, the only allowed $SU_{\rm sf}(6)$ representation is [31] with F_2 symmetry. According to Table 6 of Ref. [11], the only pentaquark configuration with F_2 symmetry that contains exotic states is $[42111]_{1134}$. On the other hand, if the four quarks are in a P-wave state with F_2 symmetry, there are several allowed $SU_{\rm sf}(6)$ representations: [4], [31], [22] and [211] with A_1, F_2, E and F_1 symmetry, respectively. The corresponding pentaquark configurations that contain exotic states are $[51111]_{700}$, $[42111]_{1134}$, $[33111]_{560}$ and $[32211]_{540}$, respectively. For each symmetry type of the orbital wave function, the corresponding symmetry of the spin-flavour wave function, as well as the associated pentaquark configurations that contain exotic states, are presented in Table 8 of Ref. [11]. In Table 7 of Ref. [11] a complete list of exotic pentaquark states is given: for each isospin multiplet the states whose combination of hypercharge Y and charge Q cannot be obtained with three-quark configurations are identified.

The constructed basis for pentaquark states makes it possible to solve the eigenvalue problem for a definite dynamical model and this is valid not only for Constituent Quark Models, but also for diquark-diquark-antiquark approaches, for which the basis is a subset of the one we have constructed. This classification scheme is complete and general, the precise ordering of the pentaquark states in the mass spectrum depends on the choice of a specific dynamical model. Moreover it will be useful also for the construction of higher Fock components of non exotic baryon states. In this respect, it could be interesting to compare our non relativistic wave functions [11] with the $SU(6)$ non relativistic limit of the Chiral Soliton Model by Diakonov [13].

2.1. *A Gürsey-Radicati mass formula*

In order to study, in a simple way the general structure of the spectrum of exotic pentaquarks, one can consider $M = M_0 + M_{\rm orb} + M_{\rm sf}$, where M_0 is a constant and $M_{\rm orb}$ describes the contribution to the pentaquark mass due to the space degrees of freedom of the constituent quarks. The last term $M_{\rm sf}$ contains the spin-flavour dependence and it is assummed to have a generalized Gürsey-Radicati form

$$M_{sf} = -AC_{2SU_{sf}(6)} + BC_{2SU_f(3)} + Cs(s+1) + DY + E[I(I+1) - \frac{1}{4}Y^2] \,. \tag{2}$$

The first two terms represent the quadratic Casimir operators of the $SU_{\rm sf}(6)$ and the $SU_{\rm f}(3)$ groups, and s, Y and I denote the spin, hypercharge and isospin, respectively. The last two terms in Eq. (2) correspond to the Gell-Mann-Okubo mass formula that describes the splitting within a flavour multiplet [15]. It was extended by Gürsey and Radicati [16] to include the terms proportional to B and C that depend on the flavour and the spin representations, which in turn was generalized further to include the spin-flavour term proportional to A [14].

The coefficients of the GR applied to the $qqqq\overline{q}$ system should be obtained from a fit of the pentaquark spectrum, but since at the moment we know at most two pentaquark states and assuming that the coefficients do not depend strongly on the structure of the quark system, we give an evaluation of the pentaquark spectrum using the coefficients taken from a prior study of qqq baryons within the hypercentral constituent quark model [17].

The average energy of any $SU_{\rm sf}(6)$ multiplet depends on the orbital part $M_{\rm orb}$ and on the term linear in the $SU_{\rm sf}(6)$ Casimir, while the terms proportional to B, C, D and E give the splittings inside the $SU_{\rm sf}(6)$ multiplet. We use the Gürsey-Radicati formula (that is without the $A\ C_{2SU(6)}$ term nor the M_{orb}) for the calculation of the energy splittings inside $SU_{\rm sf}(6)$ multiplets of the exotic pentaquark states, using the constant M_0 in order to normalize the energy scale to the observed mass of the Θ^+ and the results are given in Table 10 of Ref. [11]. The lowest pentaquark is always an $^2\overline{10}$ anti-decuplet state with isospin $I = 0$, in agreement with experimental evidence that the $\Theta^+(1540)$ is an isosinglet.

The degeneracy of the multiplets in Table 10 of Ref. [11] can be eliminated if one considers the contributions coming from the term $A\ C_{2SU(6)}$ and from the space term $M_{\rm orb}$ but for a consistent treatment of the latter, one needs a specific model that will be introduced in the next section. If we concentrate only on the effects of the term linear in A in Eq. (2) the results are reported in Table 11 of Ref [11]. Irrespective of the orbital contribution to the mass the ground state pentaquark is an anti-decuplet flavour state with spin 1/2 and isospin 0. The angular momentum and parity of the ground state exotic pentaquark depends on the relative contribution of the orbital and spin-flavour parts of the mass operator: if the splitting due to the $SU_{\rm sf}(6)$ spin-flavour term is large compared to that between the orbital states, the ground state pentaquark has positive parity, whereas for a relatively small spin-flavour splitting the parity of the lowest pentaquark state becomes negative. In case of a positive parity ground state, there is a doublet with $J^p = 1/2^+$, $3/2^+$ which, in the presence of a spin-orbit coupling term [11,18], would give rise to a pair of peaks.

3. Magnetic moments

Another unknown quantity is the magnetic moment. Although it may be difficult to determine its value experimentally, it is an essential ingredient in calculations of the photoproduction cross sections [19]. In the absence of experimental information, one has to rely on model calculations.

In Ref. [11] have analyzed the pentaquark magnetic moments of the lowest flavor antidecuplet for both positive and negative parity in the constituent quark model. The magnetic moments were obtained in closed analytic form, hence it was possible to derive generalized Coleman-Glashow sum rules for the antidecuplet magnetic moments and sum rules connecting the magnetic moments of antidecuplet pentaquarks to those of decuplet and octet baryons. The magnetic moments for negative parity $J^p = 1/2^-$ pentaquarks are an order of magnitude smaller than the proton magnetic moment, whereas for positive parity $J^p = 1/2^+$ they are even smaller due to a cancellation between orbital and spin contributions. In particular, the magnetic moment of the $\Theta(1540)$ is found to be 0.38, 0.09 and 1.05 μ_N for $J^p = 1/2^-$, $1/2^+$ and $3/2^+$, respectively. The numerical values are in qualitative agreement with those obtained in other approaches, such as correlated quark models, QCD sum rules, MIT bag model and the chiral soliton model [12].

4. String-like model

In this section, we discuss a string-like model for pentaquarks, which is a generalization to the five-body problem of the collective string-like model developed for q^3 baryons [14]. In this approach, the radial or orbital excitations of a five-body system are described in terms of a $U(13)$ spectrum-generating algebra. As a consequence of the invariance of the interations under the permutation symmetry of the four quarks, the most favorable geometric configuration is an equilateral tetrahedron in which the four quarks are located at the corners and the antiquark in its center (see Fig. 1). This configuration was also considered in [20] in which arguments based on the flux-tube model were used to suggest a nonplanar structure for the $\Theta(1540)$ pentaquark to explain its narrow width. In the flux-tube model, the strong color field between a pair of a quark and an antiquark forms a flux tube which confines them. For the pentaquark there would be four such flux tubes connecting the quarks with the antiquark.

Figure 1. Geometry of string-like pentaquarks

Hadronic spectra are characterized by the occurrence of linear Regge trajectories with almost identical slopes for baryons and mesons. Such a behavior is also expected on basis of soft QCD strings in which the strings elongate as they rotate. For this reason we use the mass squared operator $M^2 = M_0^2 + M_{\rm vib}^2 + M_{\rm rot}^2 + M_{\rm sf}^2$, where the vibrational term $M_{\rm vib}^2$ describes the vibrational spectrum corresponding to the normal modes of a tetrahedral $q^4\overline{q}$ configuration, while the rotational energies are given by a term linear in the orbital angular momentum L which is responsable for the linear Regge trajectories in baryon and meson spectra $M_{\rm rot}^2 = \alpha\, L$ and the spin-flavor part is a Gürsey-Radicati form

$$M_{\rm sf}^2 = aC_{2SU_{\rm sf}(6)} + bC_{2SU_{\rm f}(3)} + cC_{2SU_{\rm s}(2)} + dC_{1U_{\rm Y}(1)} + eC_{1U_{\rm Y}(1)}^2 + fC_{2SU_{\rm I}(2)} \,. \tag{3}$$

The coefficients α, a, b, c, d, e and f are taken from a previous study of the nonstrange and strange baryon resonances [14], and the constant M_0^2 is determined by identifying the ground state exotic pentaquark with the recently observed $\Theta(1540)$ resonance. Since the lowest orbital states with $L^p = 0^+$ and 1^- are interpreted as rotational states, for these excitations there is no contribution from the vibrational terms, so the results for the lowest Θ pentaquarks (with strangeness $S = +1$) are shown in Fig. 2.

The lowest pentaquark belongs to the flavor antidecuplet with spin $s = 1/2$ and isospin $I = 0$, in particular it is a $[42111]_{F_2}$ spin-flavor multiplet,

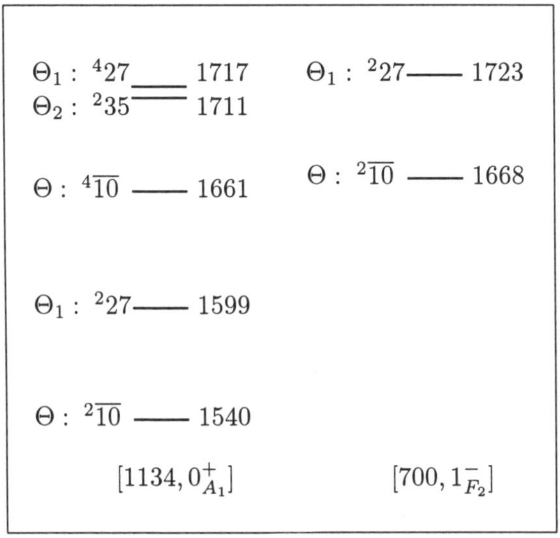

Figure 2. Spectrum of Θ pentaquarks. Masses are given in MeV

indicated in Fig. 2 by its dimension 1134, and an orbital excitation 0^+ with A_1 symmetry. Therefore, the ground state has angular momentum and parity $J^p = 1/2^-$, in agreement with recent work on QCD sum rules [12] and lattice QCD [12] but contrary to the chiral soliton model various cluster models [9] and a lattice calculation [21] that predict a ground state with positive parity.

References

1. LEPS Collaboration, T. Nakano et al., Phys. Rev. Lett. **91** (2003) 012002.
2. DIANA Collaboration, V.V. Barmin et al., Phys. Atom. Nucl. **66** (2003) 1715; SAPHIR Collaboration, J. Barth et al., Phys. Lett. B **572** (2003) 127; CLAS Collaboration,S. Stepanyan et al., Phys. Rev. Lett. **91** (2003) 252001; V. Kubarovsky et al., Phys. Rev. Lett. **92** (2004) 032001; A.E. Asratyan, A.G. Dolgolenko and M.A. Kubantsev, Phys. At. Nucl.**67** (2004) 682; HERMES Collaboration, A. Airapetian et al., Phys.Lett. B **585** (2004) 213; SVD Collaboration, A. Aleev et al., hep-ex/0401024; ZEUS Collaboration, S. Chekanov at al., Phys. Lett. B **591** (2004) 7; COSY-TOF Collaboration, M. Abdel-Bary et al., hep-ex/0403011.
3. R.A. Arndt, I.I. Strakovsky and R.L. Workman, Phys. Rev. C **68**, 042201 (2003).
4. NA49 Collaboration, C. Alt *et al.*, Phys. Rev. Lett. **92** (2004) 042003.
5. H1 Collaboration, A. Aktas et al., Phys. Lett. B **588** (2004) 17.
6. B. Aubert et al., hep-ex/0408064 and references therein.

7. Q. Zhao and F.E. Close, hep-ph/0404075; M.Karliner and H.J. Lipkin, hep-ph/0405002; hep-ph/0408001.
8. D. Diakonov, V. Petrov and M. Polyakov, Z. Phys. A **359** (1997) 305; M. Praszałowicz, in *Skyrmions and Anomalies*, Eds. M. Jezabek and M. Praszałowicz, (World Scientific; Singapore, 1987); J. Ellis, M.Karliner and M. Praszalowicz,JHEP **0405** (2004) 002.
9. R. Jaffe and F. Wilczek, Phys. Rev. Lett. **91** (2203) 232003; E. Shuryak and I. Zahed, Phys. Lett. B **589** (2004) 21; M. Karliner and H.J. Lipkin, Phys. Lett. B **575** (2003) 249.
10. Fl. Stancu, Phys. Rev. D **58** (1998) 111501; C. Helminen and D.O. Riska, Nucl. Phys. A **699** (2002) 624; A. Hosaka, Phys. Lett. B **571** (2003) 55; L.Ya. Glozman, Phys. Lett. B **575** (20003) 18; Fl. Stancu and D.O. Riska, Phys. Lett. B **575** (2003) 242; C.E. Carlson, Ch.D. Carone, H.J. Kwee and V. Nazaryan, Phys. Lett. B **579** (2004) 52.
11. R. Bijker, M. M. Giannini and E. Santopinto, hep-ph/0403029, Phys. Lett. B **595** (2004) 260; R. Bijker, M. M. Giannini and E. Santopinto, hep-ph/0310281, Eur. Phys. J. A **22** (2004) 319; R. Bijker, M. M. Giannini and E. Santopinto, hep-ph/0312380, Rev. Mex. Fís., in press .
12. B.K. Jennings and K. Maltman, Phys. Rev. D **69** (2004) 094020; S.-L. Zhu, hep-ph/0406204, to be published in Int. J. Mod. Phys. A; M. Oka, hep-ph/0406211, to be published in Progr. Theor. Phys; K. Goeke et al.,hep-ph/0411195.
13. D. Diakonov and V. Petrov, arXiv:hep-ph/0409362.
14. R. Bijker, F. Iachello and A. Leviatan, Ann. Phys. (N.Y.) **236**, 69 (1994); R. Bijker, F. Iachello and A. Leviatan, Ann. Phys. (N.Y.) **284**, 89 (2000).
15. See e.g. M. Gell-Mann and Y. Ne'eman, *The eightfold way* (W.A. Benjamin, Inc., New York, 1964).
16. F. Gürsey and L.A. Radicati , Phys. Rev. Lett. **13** (1964) 173.
17. M. M. Giannini, E. Santopinto and A. Vassallo, to be published; M. Ferraris, M. M. Giannini, M. Pizzo, E. Santopinto and L. Tiator, Phys. Lett. B **364** (1995) 231; E. Santopinto, F. Iachello, M.M. Giannini, Eur. Phys. J. A **1** (1998) 30
18. J.J. Dudek and F.E. Close, Phys. Lett. B **583** (2004) 278.
19. S.I. Nam, A. Hosaka and H.-Ch. Kim, Phys. Lett. B **579** (2004) 43; Q. Zhao, Phys. Rev. D **69** (2004) 053009; K. Nakayama and K. Tsushima, Phys. Lett. B **583** (2004) 269.
20. Xing-Chang Song and Shi-Lin Zhu, hep-ph/0403093.
21. T.-W. Chiu and T.-H. Hsieh, hep-ph/0403020

A LIGHT-FRONT QUARK MODEL FOR THE ELECTROMAGNETIC FORM FACTOR OF THE PION

J. P. B. C. DE MELO

IFT, Universidade Estadual Paulista, São Paulo, SP, Brazil

T. FREDERICO

Dep. de Física, ITA, CTA, São José dos Campos, São Paulo, Brazil

E. PACE

Dipartimento di Fisica, Università di Roma "Tor Vergata" and Istituto Nazionale di Fisica Nucleare, Sezione Tor Vergata, Via della Ricerca Scientifica 1, I-00133 Roma, Italy

G. SALMÈ

Istituto Nazionale di Fisica Nucleare, Sezione Roma I, P.le A. Moro 2, I-00185 Roma, Italy

In this contribution, an approach for a unified description of the pion electromagnetic form factor, in the space- and time-like regions, within a constituent quark model on the light front, will be reviewed. Our approach is based on i) the on-shell quark-hadron vertex functions in the valence sector, ii) the dressed photon vertex where a photon decays in a quark-antiquark pair, and iii) the emission and absorption amplitudes of a pion by a quark. Results favorably compare with the existing experimental data.

1. Introduction

Constituent quark models (CQM) developed within a light-front framework (for a general review, see, e.g., Ref. [1]) appear to be an interesting tool for investigating the electromagnetic properties of hadrons. One of the main advantage of the light-front approach is that the Fock vacuum (i.e. the vacuum of a free theory) could represent a reliable approximation of the physical, interacting vacuum, given the positivity constraint and the kinematical nature of the plus component of the total momentum, P^+. Then, expanding hadron states over a Fock basis becomes meaningful, and one can safely write for a meson

$$|meson; P^+\mathbf{P}_\perp; \zeta\rangle = \prod_{i=1,2} \int d\mathbf{k}_{i\perp}\ d\xi_i\ \delta\left(\sum_{i=1,2} \mathbf{k}_{i\perp}\right)\ \delta\left(1 - \sum_{i=1,2} \xi_i\right)\ \times$$
$$\sum_{\{\zeta_j\}} \psi_{q\bar{q}}(\mathbf{k}_{i\perp}, \xi_i, \zeta_j)\ |q\bar{q}; P^+\mathbf{P}_\perp; \mathbf{k}_{i\perp}, \xi_i, \zeta_j\rangle +$$
$$+ \prod_{i=1,4} \int d\mathbf{k}_{i\perp}\ d\xi_i\ \delta\left(\sum_{i=1,4} \mathbf{k}_{i\perp}\right)\ \delta\left(1 - \sum_{i=1,4} \xi_i\right)\ \times$$
$$\sum_{\{\zeta_j\}} \psi_{q\bar{q}q\bar{q}}(\mathbf{k}_{i\perp}, \xi_i, \zeta_j)\ |q\bar{q}q\bar{q}; P^+\mathbf{P}_\perp; \mathbf{k}_{i\perp}, \xi_i, \zeta_j\rangle +$$
$$+ \prod_{i=1,3} \int d\mathbf{k}_{i\perp}\ d\xi_i\ \delta\left(\sum_{i=1,3} \mathbf{k}_{i\perp}\right)\ \delta\left(1 - \sum_{i=1,3} \xi_i\right)\ \times$$
$$\sum_{\{\zeta_j\}} \psi_{q\bar{q}g}(\mathbf{k}_{i\perp}, \xi_i, \zeta_j)\ |q\bar{q}\ g; P^+\mathbf{P}_\perp; \mathbf{k}_{i\perp}, \xi_i, \zeta_j\rangle + ... \tag{1}$$

where a light-front spin component along the z-axis is indicated by ζ and the function ψ's are the intrinsic amplitudes for the corresponding Fock states. The possibility to include, in a given framework, states beyond the familiar CQM ones (i.e. $|q\bar{q}\rangle$ for the mesons and $|qqq\rangle$ for the baryons) allows one to address the rich phenomenology in the time-like (TL) region within a unified approach, and, in turn, to increase the constraints to be fulfilled by a chosen model.

Our model[2] has been developed for investigating the electromagnetic form factor of a pion in the whole range of momentum transfer, i.e. for positive and negative squared mass of the virtual photon. The first building block is represented by the Mandelstam formula for the matrix elements of the em current [3]. Following Ref. 3, the matrix elements for the pion in the TL region read

$$j^\mu = e\frac{2m^2}{\imath f_\pi^2} N_c \int \frac{d^4k}{(2\pi)^4} \bar{\Lambda}_\pi(k, P_\pi)\Lambda_{\bar{\pi}}(k - P_\pi, P_{\bar{\pi}})\ \times$$
$$Tr[S(k - P_\pi)\gamma^5 S(k - q)\Gamma^\mu S(k)\gamma^5] \tag{2}$$

where $S(p) = [\not{p} - m + \imath\epsilon]^{-1}$, with m the mass of the constituent quark, $\Gamma^\mu(k, q)$ is the quark-photon vertex, q^μ the virtual-photon momentum, $\Lambda_\pi(k, P_\pi)$ the pion vertex function, P^μ_π and $P^\mu_{\bar{\pi}}$ are the pion and antipion momenta, respectively. $N_c = 3$ is the number of colors and the factor 2 comes from the isospin weight, since we are dealing with a charged pion form factor. For the space-like (SL) region, P^μ_π has to be replaced by $-P^\mu_\pi$ and $\bar{\pi}$ by π'.

A key mathematical step is given by the four dimensional integration of an integrand that has, in principle, a very complicated analytical structure, due to the presence of poles from the fermion propagators and from the analytical structure of the vertex functions. A first approximation is introduced as one projects out the Mandelstam formula on the light front by a k^- integration (see Figs 1 and 2). In particular, we have assumed that: i) the meson vertex functions do not diverge in the complex plane k^- for $|k^-| \to \infty$ and ii) the contributions of their singularities are negligible. To emphasize the unified description of the em form factor in TL and SL regions, the simplifying assumption of a chiral pion ($m_\pi = 0$) has been adopted. As a matter of fact, in the SL case we have carried out our analysis in a frame where $\mathbf{P}_{\pi\perp} = \mathbf{P}_{\pi'\perp} = \mathbf{0}$, obtaining that $P^+_\pi = q^+(-1+\sqrt{1-4m^2_\pi/q^2})/2$. In the limit $m_\pi \to 0$ one has $P^+_\pi = 0$ and $P^+_{\pi'} = q^+$. Then, only the contribution of the pair-production mechanism survives, see Fig. 1(b).

In the TL case, the choice $\mathbf{P}_{\bar{\pi}\perp} = -\mathbf{P}_{\pi\perp} = \mathbf{0}$ leads to $P^+_\pi = q^+(1 \pm \sqrt{1-4m^2_\pi/q^2})/2$. In the limit $m_\pi \to 0$, in analogy with the SL case, we have adopted the choice $P^+_\pi = 0$. Then, only the contribution of the diagram (b) in Fig. 2 survives. To complete our model[2] we have to answer the following

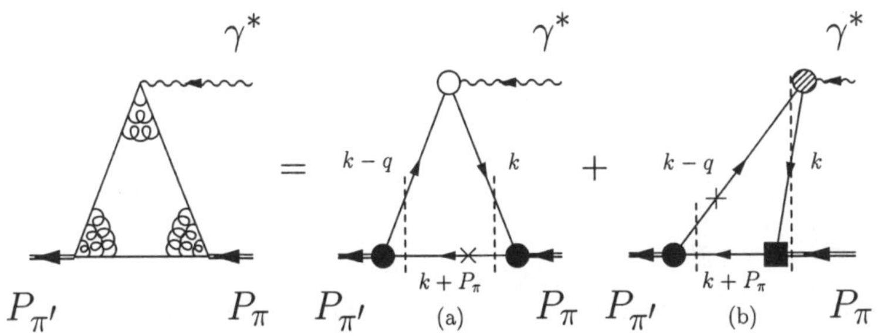

Figure 1. Light-front time-ordered diagrams contributing to the em form factor of the pion in the space-like region. In contribution (a) one has $0 < k^+ + P^+_\pi < P^+_\pi$, while in (b) $0 < k^+ < q^+$. Vertical dashed lines indicate a fixed value for the light-front time x^+ that flows from the right hand toward the left hand.
After Ref. [2]

questions : i) how to describe the $q\bar{q}$-meson vertices ? ii) how to model the dressed quark-photon vertex ? iii) how to deal with the amplitude for the emission or absorption of a pion by a quark ? In the following section, our

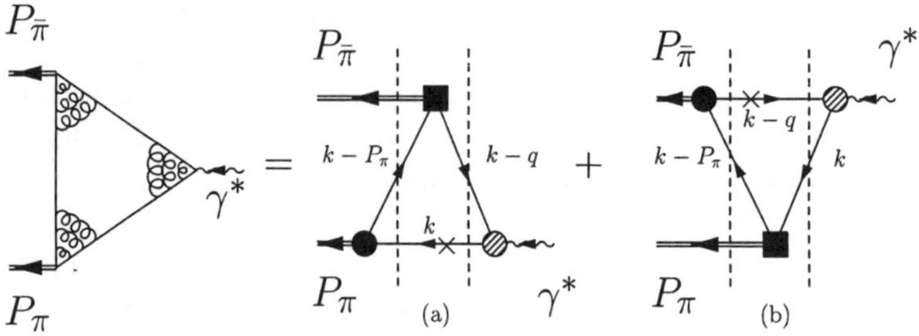

Figure 2. Light-front time-ordered diagrams contributing to the em form factor of the pion in the time-like region. In contribution (a) one has $0 < k^+ < P_\pi^+$, while in (b) $P_\pi^+ < k^+ < q^+$.
After Ref. [2]

answers will be illustrated in some detail.

2. The Model

In order to construct our model, we have first introduced an approximation for the momentum component of the pion vertex function, $\Lambda_\pi(k, P_\pi)$, when both the quarks are on their mass shell and in the interval $0 \le k^+ \le P_\pi^+$. Such a function of the quark light-front momenta, can be approximated by the "light-front pion wave function", ψ_π, obtained within the Hamiltonian light-front dynamics, through the following equation

$$(\not{k}_{on} + m)\gamma^5 \psi_\pi(k^+, \mathbf{k}_\perp; P_\pi^+, \mathbf{P}_{\pi\perp}) \left[(\not{k} - \not{P}_\pi)_{on} + m\right] =$$
$$(\not{k}_{on} + m)\gamma^5 \, \frac{m}{f_\pi} \, \frac{P_\pi^+}{[m_\pi^2 - M_0^2(k^+, \mathbf{k}_\perp; P_\pi^+, \mathbf{P}_{\pi\perp})]} \, [\Lambda_\pi(k, P_\pi)]_{[k^- = k^-_{on}]} \; \times$$
$$\left[(\not{k} - \not{P}_\pi)_{on} + m\right] \tag{3}$$

where $k^-_{on} = (\mathbf{k}_\perp^2 + m^2)/k^+$ and M_0 is the light-front free mass (see, e.g., Ref. [4]). A similar relation is also adopted for the vector mesons (VM). For the pion and the VM wave functions, we have used the eigenfunctions of the square mass operator proposed in Refs. [5, 6], within a relativistic constituent quark model which achieves a natural explanation of the "Iachello-Anisovitch law" [7,8]. The VM eigenfunctions , $\psi_n(k^+, \mathbf{k}_\perp; q^+, \mathbf{q}_\perp)$, are normalized to the probability of the lowest $(q\bar{q})$ Fock state, roughly estimated to be $\sim 1/\sqrt{2n + 3/2}$ in a simple model [2] that reproduces the "Iachello-Anisovitch law" [7,8](n is the principal quantum number).

As for the second question, following Ref. [2], a Vector Meson Dominance (VMD) approximation is applied to the quark-photon vertex $\Gamma^\mu(k, q)$, when a $q\bar{q}$ pair is produced. In particular, the plus component of the quark-photon vertex reads as follows (see Fig. 3)

$$\Gamma^+(k,q) = \sum_{n,\lambda} \left[\epsilon_\lambda \cdot \widehat{V}_n(k,k-q)\right] \Lambda_n(k,P_n) \times$$
$$\frac{\sqrt{2}[\epsilon_\lambda^+]^* f_{Vn}}{(q^2 - M_n^2 + \imath M_n \Gamma_n(q^2))} \tag{4}$$

where f_{Vn} is the decay constant of the n-th vector meson into a virtual photon, $M_n(P_n)$ the mass (four-momentum) of the VM, $\Gamma_n(q^2) = \Gamma_n q^2/M_n^2$ (for $q^2 > 0$) the corresponding total decay width, ϵ_λ the VM polarization, and $[\epsilon_\lambda \cdot \widehat{V}_n(k,k-q)\ \Lambda_n(k,q)$ the VM vertex function (for the VM Dirac structure, $\widehat{V}_n$, see below). The decay constant, f_{Vn}, is evaluated starting

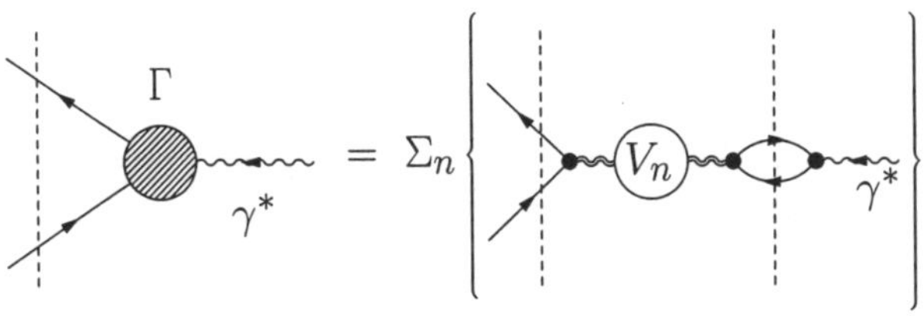

Figure 3. Diagrammatic analysis of the quark-photon vertex, $\Gamma^\mu(k,q)$. After Ref. [2]

from a four dimensional representation in terms of the VM Bethe-Salpeter vertex and integrating over k^-. In particular, one has

$$f_{Vn} = -\imath \frac{N_c}{4(2\pi)^4} \int dk^- dk^+ d\mathbf{k}_\perp \frac{Tr[\gamma^+ \ \mathcal{V}_{nz}(k,k-P_n)]\ \Lambda_n(k,P_n)}{[k^2 - m^2 + \imath\epsilon][(P_n - k)^2 - m^2 + \imath\epsilon]} =$$
$$= -\frac{N_c}{4(2\pi)^3} \int_0^{P_n^+} dk^+ \ d\mathbf{k}_\perp \frac{Tr\,[\gamma^+ \ \mathcal{V}_{nz}(k,k-P_n)]}{k^+ \ (P_n^+ - k^+)} \ \psi_n(k^+,\mathbf{k}_\perp;M_n,\vec{0}_\perp) \tag{5}$$

where $\mathcal{V}_{nz}(k,k-P_n) = (\not{k} - \not{P}_n + m)\widehat{V}_{nz}(k,k-P_n)(\not{k} + m)$.

Note that only ${}^3S_1$ isovector vector mesons have been taken into account in the calculations, and therefore the on-mass-shell spinorial part of the VM vertex must reproduce the well-known Melosh rotations for ${}^3S_1$ states, (see, e.g., Ref. [4]), viz.

$$\widehat{V}_n^\mu(k,k-q) = \gamma^\mu - \frac{k_{on}^\mu - (q-k)_{on}^\mu}{M_0(k^+,\mathbf{k}_\perp;q^+,\mathbf{q}_\perp) + 2m} \quad . \tag{6}$$

The third question raised in the Introduction is answered by describing with a constant[9] the amplitude for the emission or absorption of a pion by a quark, i.e. the pion vertex function in the non-valence sector. The value of the constant is fixed by the pion charge normalization, given our simplifying assumption of a chiral pion.

In the limit $m_\pi \to 0$ the pion form factor receives contributions only from processes where the photon decays in a $q\bar{q}$ pair. Then, by means of Eq. (4) the matrix element j^+ can be written as a sum over the vector mesons, and consequently the form factor becomes

$$F_\pi(q^2) = \sum_n \frac{f_{Vn}}{q^2 - M_n^2 + \imath M_n \Gamma_n(q^2)} \, g^+_{Vn}(q^2) \tag{7}$$

where $g^+_{Vn}(q^2)$, for $q^2 > 0$, is the form factor for the VM decay in a pair of pions.

Each VM contribution to the sum (7) is invariant under kinematical light-front boosts and can be evaluated in the rest frame of the corresponding resonance (with $q^+ = M_n$ and $\mathbf{q}_\perp = 0$).

It turns out that the same expression for $g^+_{Vn}(q^2)$ holds both in the TL and in the SL regions[2].

3. Results

In our calculations the up-down quark mass is fixed at 0.265 GeV [6] and the oscillator strength at $\omega = 1.39\ GeV^2$ [8]. For the first four vector mesons the known experimental masses and widths are used [10] (see Table I), a part the value of the mass of ρ-meson changed to $m_\rho = 0.750\ Gev$, in order to reproduce the correct position of the ρ peak in the time-like form factor of the charged pion. For the VM with $M_n >\ 2.150\ GeV$, the mass values corresponding to the model of Ref. 6 are used, while for the unknown widths we use a single value $\Gamma_n = 0.15\ GeV$. To obtain stability of the results up to $q^2 =\ 10\ (GeV/c)^2$ twenty resonances are considered.

The calculated pion form factor is shown in Fig. 4 in a wide region of square momentum transfers, from $-10\ (GeV/c)^2$ up to $10\ (GeV/c)^2$. The VM dominance ansatz for the (dressed photon)-($q\bar{q}$) vertex, within a CQ model consistent with the meson spectrum, is able to give a unified description of the pion form factor both in the SL and TL regions.

The SL form factor is notably well described, see Fig. 5, from the high-q^2 region to the low-q^2 one, as well as the charge radius. Finally, it is worth noting that the heights of the TL bumps directly depend on the calculated values of f_{Vn} and g^+_{Vn}, and for the sake of completeness a table of calculated em decay widths, $\Gamma^{th}_{e^+e^-}$, are shown in Table II.

Table I. Input values of our model, $M_n(PDG)$ and $\Gamma_n(PDG)$ (see text). For the sake of comparison, the masses evaluated in the model of Refs. [5, 6], M_n(FPZ) are also shown (the mass of the ρ is an input for this model).

VM	M_n(FPZ)	$M_n(PDG)$	$\Gamma_n(PDG)$
$\rho(770)$	*.770 GeV*	.770 GeV	0.15 GeV
$\rho(1450)$	1.408 GeV	1.465 GeV	0.37 GeV
$\rho(1770)$	1.836 GeV	1.723 GeV	0.30 GeV
$\rho(2150)$	2.182 GeV	2.149 GeV	0.18 GeV

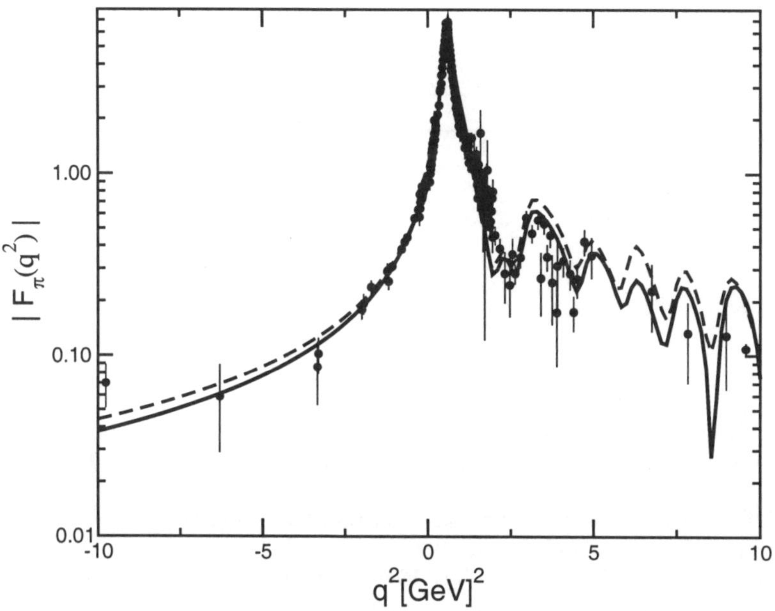

Figure 4. Pion electromagnetic form factor vs. the square momentum transfer q^2. Solid line: result obtained by using the full pion wave function. Dashed line: result obtained by using the asymptotic pion wave function (After Ref. 2).

4. Perspectives

The results obtained within our approach[2] encourage an investigation of the TL form factors of the nucleon based on a simple ansatz for the non-valence component of the nucleon state, following as a guideline the pion

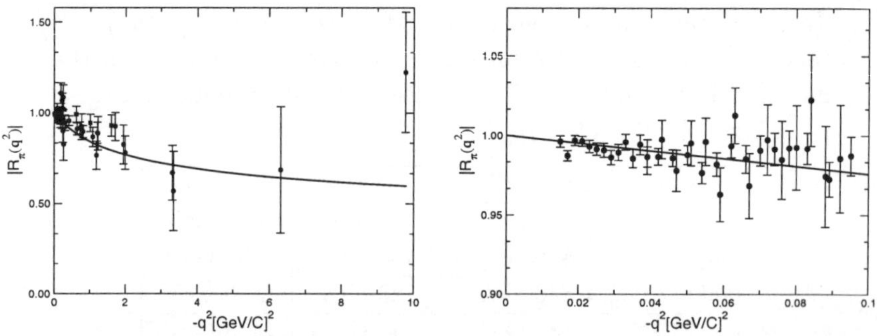

Figure 5. The ratio $R_\pi(q^2) = F_\pi(q^2)/\left[1/(1-q^2/m_\rho^2)\right]$ vs q^2, in the SL region ($m_\rho = 0.770\ GeV$). The rightmost figure shows in great detail the region $0 \leq -q^2 \leq 0.1(GeV/c)^2$, relevant for evaluating the charge radius.
Experimental data from R. Baldini et al.[11]

Table II. Em decay widths, $\Gamma^{th}_{e^+e^-}$, for the first three ρ-mesons

VM	$\Gamma^{th}_{e^+e^-}$	$\Gamma^{exp}_{e^+e^-}$
$\rho(770)$	6.37 KeV	6.77 ± 0.32 KeV
$\rho(1450)$	1.61 KeV	$> 2.30 \pm 0.50$ KeV
$\rho(1770)$	1.23 KeV	$> 0.18 \pm 0.10$ KeV

case.

References

1. S J. Brodsky, H.C. Pauli, and S.S. Pinsky, Phys. Rep. **301**, 299 (1998).
2. J.P.B.C. de Melo, T. Frederico, E. Pace and G. Salmè, Phys. Lett. **B581**,75 (2004) and to be published.
3. S. Mandelstan, Proc. Royal Soc. (London) **A233**, 248 (1956).
4. W. Jaus, Phys. Rev. **D41**, 3394 (1990).
5. T. Frederico and H.-C. Pauli, Phys. Rev. **D64**, 054007(2001).
6. T. Frederico, H.-C. Pauli and S.-G. Zhou, Phys. Rev. **D66**, 054007 (2002); ibidem **D66**, 116011 (2002).
7. F. Iachello, N.C. Mukhopadhyay and L. Zhang, Phys. Rev. **D44**, 898 (1991).
8. A.V. Anisovitch, V.V. Anisovich, and A.V. Sarantsev, Phys. Rev. **D62**, 051502(R) (2000).
9. C.-R. Ji and H.-M. Choi, Phys. Lett. **B513**, 330 (2001).
10. K. Hagiwara et al., Phys. Rev. **D66**, 010001 (2002).
11. R. Baldini, et al., Eur. Phys. J. **C11**, 709 (1999); Nucl. Phys. **A666&667**, 3 (2000) and private communication.

ELECTROMAGNETIC FORM FACTORS IN THE HYPERCENTRAL CQM

M. DE SANCTIS
Istituto Nazionale di Fisica Nucleare, Sezione di Roma, P.le A. Moro 2, I-00185 Roma, Italy

M. M. GIANNINI, E. SANTOPINTO AND A. VASSALLO
Dipartimento di Fisica dell'Università di Genova
and
Istituto Nazionale di Fisica Nucleare, Sezione di Genova, via Dodecaneso 33, I-16146 Genova, Italy
E-mail: giannini@ge.infn.it

We discuss the recent experimental results on the ratio bertween the electric and magnetic proton form factors and how they can be described by theoretical models. In particular the calculations performed using the hypercentral Constituent Quark Model are illustrated and shown to compare favourably with the data.

1. Introduction

The electromagnetic form factors are an invaluable source of information about the internal structure of the nucleon. Much attention has been devoted to this topic both by experimentalist and by model builders. The interest is however increased after the recent results of the Jefferson Laboratory, where the ratio between the electric and magnetic form factors of the proton has been measured directly by means of a double polarization experiment [1,2]. At variance with the expectations based on the widely accepted dipole fit, the ratio deviates strongly from 1, indicating that the electric and magnetic distributions of the proton are very different. Moreover, for $Q^2 \geq 1$ $(GeV/c)^2$, the ratio decreases with an almost linear behaviour, pointing towards the possible existence of a zero at $Q^2 \approx 8$ $(GeV/c)^2$, if such trend continues beyond the range of the presently available data.

These unexpected results pose some problems. The first one is the compatibility of the new data with the traditional ones obtained from a Rosenbluth plot. This point has triggered an intense discussion among theorist, proposing different explanations for such a discrepancy. Much attention has been devoted to the possible contributions coming from two-photon exchange mechanisms [3,4,5,6], which however seem to be too small for reconciling the two sets of data. A critical re-analysis of the Rosenbluth procedure is also being performed [7,8], with promising results.

The main further problem is the physical picture emerging from data. If the ratio between the electric and magnetic form factors of the proton decreases, one has to understand where this behaviour comes from. Of course the presence of a zero in the electric form factor forces the ratio to decrease, but the question is if a zero is really there and why.

The planned experiments at higher Q^2 will provide the answer about the occurrence of a zero in the electric form factor of the proton. From the theoretical point of view such zero would be a challenge for most theoretical models for the internal proton structure.

In this respect it should be mentioned that long before the double polarization data there had been phenomenological and theoretical approaches to the proton form factor implying the presence of a zero. The Vector Meson Dominance (VMD) fit by Iachello, Jackson and Landé [9] provided a good description for the form factor data available at the time, which, for the electric proton form factor were limited up to 4 $(GeV/c)^2$; however, extending the fitting formula to higher Q^2 values, one gets a zero at about 8 $(GeV/c)^2$, coming from the fact that the various VMD contributions have different signs. This value is in tremendous agreement with the one extrapolated by the presumed trend of the Jlab data.

A further example is provided by the Skyrme calculation by Holzwarth [10]. The non relativistic skyrmion has a zero in the proton form factor for low Q^2 but when a Lorentz boost is applied, in order to get a relativistic description this zero is shifted up to $Q^2 = 10$ $(GeV/c)^2$. If one compares the calculations with the Jlab data, the agreement is very good [11].

In Sect. 2 we remind briefly the features of the hypercentral Constituent Quark Model (hCQM) and its capability in describing successfully many physical properties of baryons, in Sect. 3 we report the results of recent relativistic calculations within the hCQM, with particular emphasis to the Jlab data and finally we draw some conclusions in Sect. 4.

2. The Hypercentral Constituent Quark Model

The non strange baryon spectrum can be arranged in various $SU(6)$ multiplets, whose average energies are determined by the $SU(6)$ invariant part of the three quark interaction. The internal quark motion is described by the Jacobi coordinates $\vec{\rho}$ and $\vec{\lambda}$:

$$\vec{\rho} = \frac{1}{\sqrt{2}}(\vec{r}_1 - \vec{r}_2) , \qquad \vec{\lambda} = \frac{1}{\sqrt{6}}(\vec{r}_1 + \vec{r}_2 - 2\vec{r}_3) , \tag{1}$$

or equivalently, by ρ, Ω_ρ, λ, Ω_λ. In order to describe three-quark dynamics it is convenient to introduce the hyperspherical coordinates, which are obtained substituting the absolute values $\rho = |\vec{\rho}|$ and $\lambda = |\vec{\lambda}$ by

$$x = \sqrt{\rho^2 + \lambda^2}, \qquad \xi = arctg(\frac{\rho}{\lambda}), \tag{2}$$

where x is the hyperradius and ξ the hyperangle. In this way one can use the hyperspherical harmonic formalism [12].

The three-quark potential, V, is assumed to depend on the hyperradius x only, that is to be hypercentral. It can be considered as a two-body interaction in the hypercentral approximation, that is obtained averaging the total three quark potential over the hyperangle ξ and the angles Ω_ρ, Ω_λ; this approximation has been shown to be valid, specially for the lower energy states [13]. On the other hand, the hyperradius x depends on the coordinates of all the three quarks. therefore, the interaction $V = V(x)$ is in general a three-body potential. Actually the fundamental gluon interactions, predicted by QCD, lead also to three-quark mechanisms. The situation is similar to the flux tube models, where two-body (Δ-shaped) and three-body (Y-shaped) interactions are considered.

If the potential depends on x only, one can factorize the three quark wave function as follows:

$$\Psi_{3q}(x, t, \Omega_\rho, \Omega_\lambda) = \psi_{\gamma,\nu}(x) \; Y_{[\gamma,l_\rho,l_\lambda]}(t, \Omega_\rho, \Omega_\lambda) \tag{3}$$

$Y_{[\gamma,l_\rho,l_\lambda]}(t, \Omega_\rho, \Omega_\lambda)$ are the known hyperspherical harmonics [12] and are labeled by the orbital angular momenta l_ρ and l_λ, associated to the two Jacobi coordinates, and by the grandangular quantum number $\gamma = 2n + l_\rho + l_\lambda$, n being a non negative integer. The Schrödinger equation is then reduced to a single differential equation for the hyperradial wave function $\psi_{\gamma,\nu}(x)$, which contains the dynamical information.

The hypercentral equation can be solved analytically in two cases, the h.o. potential and the 'hypercoulomb' one [13,14,15]

$$V_{hyc}(x) = -\frac{\tau}{x}, \tag{4}$$

this potential is not confining, however it has interesting properties. It leads to a power-law behaviour of the proton form factor [14] and of all the transition form factors [16] and it has a perfect degeneracy between the first 0^+ excitated state and the first 1^- states [14,15]. This degeneracy is typical of an underlying O(7) symmetry [15] and cannot be reproduced in models with only two-body forces and/or harmonic oscillator, as the excited $L = 0$ state, having one more node, lies above the $L = 1$ state. Since the first 0^+ excitated state can be identified with the Roper resonance and the first 1^- states with the negative parity resonances, the level pattern of this potential supplies a better starting point for the description of the baryon spectrum.

In the hypercentral Constituent Quark Model (hCQM) the potential is assumed to have the form [17]

$$V(x) = -\frac{\tau}{x} + \alpha x, \tag{5}$$

the $SU(6)$ violation is taking into account by adding a standard hyperfine interaction [18], treated as a perturbation. Interactions of the type linear plus Coulomb-like have been used since long time for the meson sector, e.g. the Cornell potential. This form has been obtained in recent Lattice QCD calculations [19,20] for $SU(3)$ invariant static quark sources.

The spectrum is described with $\tau = 4.59$ and $\alpha = 1.61\ fm^{-2}$ and the standard strength of the hyperfine interaction needed for the $N - \Delta$ mass difference [18].

3. The electromagnetic form factors

Having fixed the three free parameters (the quark mass is taken 1/3 of the nucleon mass), the model has been used for the prediction of various physical quantities of interest, namely the photocouplings [21], the electromagnetic transition amplitudes [22], the elastic nucleon form factors [23] and the ratio between the electric and magnetic proton form factors [24].

The calculated photocouplings are very similar to the results of other models [21], the overall behaviour is fairly reproduced but in many cases the excitation strength is underestimated. The transition form factors [22] for the negative parity resonances have in most cases the correct behaviour at medium-high Q^2, while some problems are present for low Q^2. It should be noted that the calculated r.m.s. radius of the proton, corresponding to the potential parameters quoted at the end of the last section, is about $0.5 fm$. This is actually the value which in earlier calculations has been fitted in order to reproduce the D_{13} photocouplings [25,26]. So the emerging picture is that of a quark core surrounded by meson clouds and/or quark-antiquark pairs, not explicitely included in the CQM.

The hCQM is non relativistic and one could in principle think that is the origin of the above discussed discrepancies. Actually, first order relativistic corrections have been introduced in the calculation of the helicity amplitudes [27]. The three quark states for the initial (nucleon) and the final (resonance) states have been boosted to a common reference frame and the matrix elements of the three quark current have been expanded up to first order in the quark momentum. The result is that the "relativistic" helicity amplitude can be written as

$$A_{rel}(Q^2) \;=\; F_{kin}\; A_{nonrel}(Q^2_{eff}) \tag{6}$$

where the non relativistic amplitude is calculated at the scaled momentum transfer $Q^2_{eff} \;=\; Q^2(M_N/E_N)^2$ and F_{kin} is given by calculable kinematic factors. However, these relativistic corrections do not alter significantly the behaviour of the helicity amplitudes [27].

The situation is completely different for the elastic nucleon form factors. The non relativistic calculation with the hCQM, implying, as stated above, a proton radius of $0.5 fm$, leads to nucleon form factors rather far from the data. The introduction of the first order relativistic corrections (see Eq. (6)) is very beneficial, since the theoretical curves are much closer to the experimental data. The most remarkable effect is, however, obtained in connection with the problem of the recent Jlab data. By means of a double polarisation experiment, the ratio of the electric (G^p_E) and magnetic (G^p_M) proton form factors has been directly measured [1,2]

$$R \;=\; \mu_p\, \frac{G^p_E}{G^p_M} \tag{7}$$

where μ_p is the proton magnetic moment. These new data show an unexpected decrease for increasing values of the momentum trasfer Q^2. The non relativistic calculations predict the value $R = 1$ and introducing the hyperfine interaction makes no difference ($R = 0.99$). However, the first order relativistic corrections [24] give rise to a ratio which is significantly deviating from 1 [24]. It is interesting to note that the hCQM results coincide with the dispersion relation calculation of the Mainz group [28].

Relativity is then a fundamental ingredient for the description of the elastic nucleon form factors within the hCQM and therefore we have recently reformulated the model and calculated the elastic nucleon form factors in a completely relativistic frame. First, we have introduced in the hCQM the correct relativistic kinetic energy and, using the same hypercentral potential of Eq. (5), we have obtained an equivalently good description of the baryon spectrum [29]. As for the calculation of the form factors [30], after having boosted the new three quark states to the Breit frame, we have taken into account the quark current up to any order. Finally, considering that constituent quark may acquire a finite size, we have introduced quark form factors. The free parameters in the quark form factors have been used to fit four set of experimental data, namely the ratio R, the proton magnetic form factor G_M^p, the neutron electric (G_E^n) and magnetic (G_M^n) form factors [27]. The results for the ratio R are shown in Fig. 2 and are remarkable, since the free parameters provided by the quark form factors are not sufficient by themselves to obtain a good fit, it is necessary that already the pointlike calculations provide a realistic description. In any case the size of the quarks obtained in this fit is not larger than $0.3fm$. It should be reminded that a recent analysis of the inelastic proton structure functions [31] has shown a possible evidence of the proton containing extended objects with a size of about $0.2 - 0.3fm$.

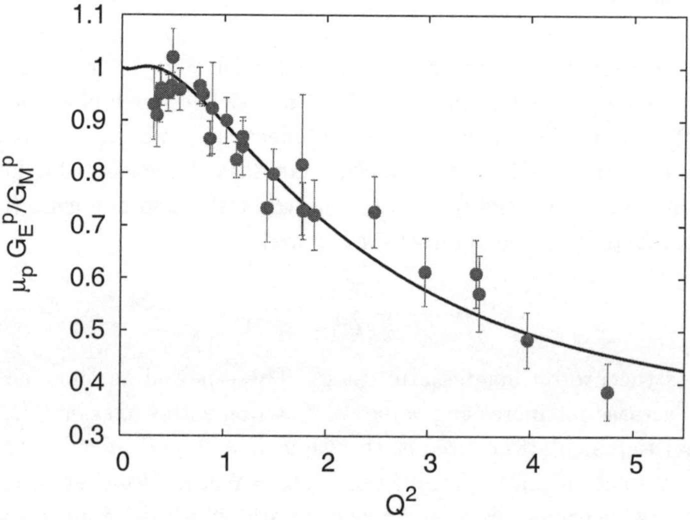

Figure 1. The ratio $\mu_p\ G_E^p/G_M^p$ calculated with the relativized Hypercentral Constituent Quark Model [30] (full curve). The data are taken from [1,2].

4. Conclusions

The recent Jlab data on the ratio $R = \mu_p\ G_E^p/G_M^p$ show a strong deviation from the value 1 predicted by the previous widely accepted dipole fit and by most models for the internal structure of the proton. Moreover, extrapolating their trend at higher Q^2, one can infer the presence of a dip. Of course, if the electric form factor of the proton has somewhere a zero, then the ratio R is forced to decrease. In any case the data show that the electric and magnetic distributions of the proton are quite different. Moreover, as mentioned in the Introduction, there is the problem of reconciling these new data with those obtained from a Rosenbluth plot.

The results obtained with the hCQM allow to state that relativity is crucial in explaining the decrease of the R. It remains to explain the eventual zero in the proton electric form factor. A dip is present in fits with Vector Meson Dominance [9], where propagators with opposite signs interfere destructively. Also models where the internal (core) region is sharply defined (Bag, Soliton [10]) lead naturally to a dip. In a CQM, like the one discussed here, the dip may be produced by the presence of different quark form factors.

The answer to the question if there is a dip in the proton form factor is very important for the understanding of the internal nucleon structure and it will be hopefully obtained by the planned experiments.

References

1. M.K. Jones et al., Phys. Rev. Lett. **84**, 1398 (2000).
2. O. Gayou et al., Phys Rev. Lett. **88**, 092301 (2002).
3. J. Arrington, hep-ph/0408261
4. P. G. Blunden, W. Melnitchouk and J. A. Tjon, Phys. Rev. Lett. **91**, 142304 (2003).
5. P. A. M. Guichon and M. Vanderhaegen, Phys. Rev. Lett. **91**, 142303 (2003).
6. M. P. Rekalo and E. Tomasi-Gustafsson, Eur.Phys.J. bf A22, 331 (2004).
7. J. Arrington, Phys. Rev. **C69**,022201 (2004).
8. E. Tomasi-Gustafsson, contribution to the International Conference Baryons 04, Palaiseau (France), 25-29 October 2004.
9. F. Iachello, A. D. Jackson and A. Landé, Phys. Lett. **43B**, 191 (1973).
10. G. Holzwarth, Z. Phys. **A356**, 339 (1996).
11. G. Holzwarth, hep-ph/0201138
12. G. Morpurgo, Nuovo Cimento **9**, 461 (1952); Yu. A. Simonov, Sov. J. Nucl. Phys. **3**, 461 (1966); J. Ballot and M. Fabre de la Ripelle, Ann. of Phys. (N.Y.) **127**, 62 (1980); M. Fabre de la Ripelle, in "Models and Methods in Few-Body Physics" (L.S. Ferreira, A. C. Fonseca and L. Streit eds.) Lecture Notes in Physics 273, Springer (Berlin), 1987, p. 283
13. M. Fabre de la Ripelle and J. Navarro, Ann. Phys. (N.Y.) **123**, 185 (1979).
14. H.J. Lipkin, Rivista Nuovo Cimento I (volume speciale), 134 (1969); J. Leal Ferreira and P. Leal Ferreira, Lett. Nuovo Cimento vol. III, 43 (1970); M.M. Giannini, Nuovo Cimento **A76**, 455 (1983); D. Drechsel, M.M. Giannini and L. Tiator, in "The Three-Body Force in the Three-Nucleon System", eds. B.L. Berman and B.F. Gibson, Lecture Notes in Physics **260**, 509 (1986); Few-Body Syst. Suppl. **2**, J.-L. Ballot and M. Fabre de la Ripelle eds. , 448 (1987).
15. E. Santopinto, M.M. Giannini and F. Iachello, in "Symmetries in Science VII", ed. B. Gruber, Plenum Press, New York, 445 (1995); F. Iachello, in

"Symmetries in Science VII", ed. B. Gruber, Plenum Press, New York, 213 (1995).

16. E. Santopinto, F. Iachello and M.M. Giannini, Nucl. Phys. **A623**, 100c (1997).
17. M. Ferraris, M.M. Giannini, M. Pizzo, E. Santopinto and L. Tiator, Phys. Lett. **B364**, 231 (1995).
18. N. Isgur and G. Karl, *Phys. Rev.* **D18**, 4187 (1978); *Phys. Rev.***D19**, 2653 (1979).
19. Gunnar S. Bali et al., Phys. Rev. **D62**, 054503 (2000);
20. Gunnar S. Bali, Phys. Rep. **343**, 1 (2001).
21. M. Aiello, M. Ferraris, M.M. Giannini, M. Pizzo and E. Santopinto, Phys. Lett. **B387**, 215 (1996).
22. M. Aiello, M. M. Giannini, E. Santopinto, J. Phys. G: Nucl. Part. Phys. **24**, 753 (1998)
23. M. De Sanctis, E. Santopinto, M.M. Giannini, Eur. Phys. J. **A1**, 187 (1998).
24. M. De Sanctis, M.M. Giannini, L. Repetto, E. Santopinto, Phys. Rev. **C62**,025208 (2000).
25. L. A. Copley, G. Karl and E. Obryk, Phys. Lett. **29**, 117 (1969).
26. R. Koniuk and N. Isgur, Phys. Rev. **D21**, 1868 (1980).
27. M. De Sanctis, E. Santopinto, M.M. Giannini, Eur. Phys. J. **A2**, 403 (1998).
28. H.-W. Hammer, U.-G. Meissner, and D. Drechsel, Phys. Lett. **B 385**, 343 (1996); P. Mergell, U.-G. Meissner, and D. Drechsel, Nucl. Phys. **A 596**, 367 (1996).
29. M.M. Giannini, E. Santopinto and A. Vassallo, to be published.
30. M. De Sanctis, M.M. Giannini, E. Santopinto and A. Vassallo, to be published.
31. R. Petronzio, S. Simula and G. Ricco, Phys. Rev. **D 67**,094004 (2003).

HADRONIC DECAYS OF BARYONS IN POINT-FORM RELATIVISTIC QUANTUM MECHANICS

T. MELDE,* W. PLESSAS, AND R. F. WAGENBRUNN

Theoretische Physik, Institut für Physik, Universität Graz
Universitätsplatz 5, A-8010 Graz, Austria

L. CANTON

INFN, Sezione di Padova, and Dipartimento di Fisica dell'Università di Padova
Via F. Marzolo 8, I-35131 Padova, Italy

We discuss strong decays of baryon resonances within the concept of relativistic constituent quark models. In particular, we follow a Poincaré-invariant approach along the point form of relativistic quantum mechanics. Here, we focus on pionic decay modes of N and Δ resonances. It is found that the covariant quark-model predictions calculated in the point-form spectator model in general underestimate the experimental data considerably. This points to a systematic defect in the used decay operator and/or the baryon wave functions. From a detailed investigation of the point-form decay operator it is seen that the requirement of translational invariance implies effective many-body contributions. Furthermore, one has to employ a normalization factor in the definition of the decay operator in the point-form spectator model. Our analysis suggests that this normalization factor is best chosen consistently with the one used for the electromagnetic and axial current operators for elastic nucleon form factors.

1. Introduction

Constituent quark models (CQMs) provide an effective description of hadrons in the low-energy regime of quantum chromodynamics (QCD). It has been of particular interest to find an appropriate type of (phenomenological) interaction between the constituent quarks, which is usually comprised of a confinement and a hyperfine part. Usually the hyperfine interaction is derived from one-gluon exchange (OGE)[1]. Beyond that also alternative types of CQMs have been suggested such as the ones based on instanton-induced (II) forces[2] or Goldstone-boson-exchange

*Work partially supported by INFN and MIUR-PRIN.

(GBE) dynamics[3]. All these variants of modern CQMs describe the overall trends of the hadronic spectra reasonably well. With regard to baryons, however, only the GBE CQM succeeds in reproducing simultaneously the level ordering of positive- and negative-parity resonances as well as the $N - \Delta$ splitting in agreement with phenomenology[4]. The typical spectra of the different types of CQMs are exemplified in fig. 1.

Another problem has been the role of relativity. Specifically, it has been found that covariant predictions of relativistic CQMs for electroweak nucleon form factors agree surprisingly well with experimental data. This is especially true for the II CQM in a Bethe-Salpeter type approach[7]. The same has been found for the GBE CQM and likewise the OGE CQM within the point-form approach of relativistic quantum mechanics[8,9,10]. In contrast, any nonrelativistic calculation fails drastically.

The covariant results obtained so far for mesonic decay widths of N and Δ resonances have shown a systematic underestimation of the experimental data for any of the three types of CQMs[11,12,13]. Clearly, this points to shortcomings in the baryon (resonance) wave functions and/or the decay operators employed. In this contribution we give a review of the current status of the investigations on hadronic decays of N and Δ resonances.

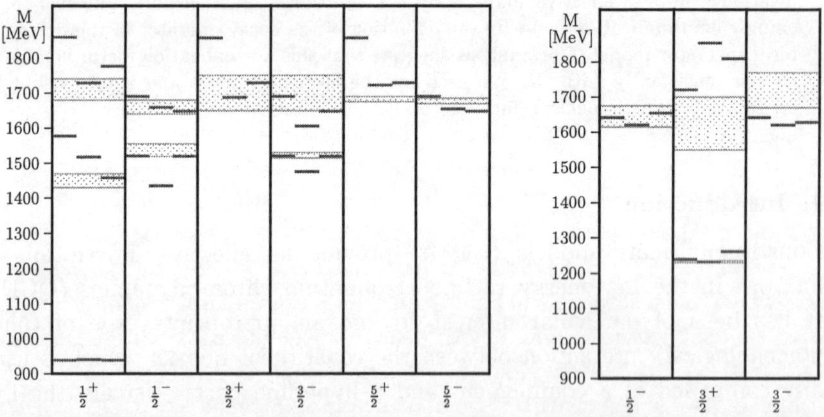

Figure 1. Nucleon excitation spectra of three different types of relativistic CQMs. The left panel shows the nucleon spectrum and the right panel the Δ spectrum. In each column the left horizontal lines represent the results of the relativized Bhaduri-Cohler-Nogami CQM[5], the middle ones of the II CQM (Version A)[2], and the right ones of the GBE CQM[3]. The shadowed boxes give the experimental data with their uncertainties after the latest compilation of the PDG[6].

2. Theory

Generally, the decay width Γ of a resonance is defined by the expression

$$\Gamma = 2\pi\rho_f \left|F\left(i \to f\right)\right|^2 , \tag{1}$$

where $F\left(i \to f\right)$ is the transition amplitude and ρ_f is the phase-space factor. In eq. (1) one has to average over the initial and to sum over the final spin-isospin projections. A common problem in nonrelativistic approximations of the transition amplitude is the ambiguity of the proper phase-space factor[14,15,16]. Here, we present a Poincaré-invariant approach, adhering to the point form of relativistic quantum mechanics[17]. In this case the generators of the Lorentz transformations remain purely kinematical and the theory is manifestly covariant[18]. This also allows to resolve the ambiguity in the phase-space factor. The interactions between the constituent quarks are introduced into the (invariant) mass operator following the Bakamjian-Thomas construction[19]. In this approach the covariant transition amplitude for mesonic decays is defined via the matrix element of the decay operator

$$\begin{aligned} F\left(i \to f\right) &= \left\langle P', J', \Sigma' \right| \hat{D}_m \left| P, J, \Sigma \right\rangle \\ &= \frac{2}{MM'} \sum_{\sigma_i \sigma'_i} \sum_{\mu_i \mu'_i} \int d^3\vec{k}_2 d^3\vec{k}_3 d^3\vec{k}'_2 d^3\vec{k}'_3 \\ &\quad \sqrt{\frac{\left(\omega_1+\omega_2+\omega_3\right)^3}{2\omega_1 2\omega_2 2\omega_3}} \sqrt{\frac{\left(\omega'_1+\omega'_2+\omega'_3\right)^3}{2\omega'_1 2\omega'_2 2\omega'_3}} \\ &\quad \Psi^\star_{M'J'\Sigma'}\left(\vec{k}'_i;\mu'_i\right) \prod_{\sigma'_i} D^{\star\frac{1}{2}}_{\sigma'_i\mu'_i}\left\{R_W\left[k'_i; B\left(V'\right)\right]\right\} \\ &\quad \left\langle p'_1, p'_2, p'_3; \sigma'_1, \sigma'_2, \sigma'_3 \right| \hat{D}_m \left| p_1, p_2, p_3; \sigma_1, \sigma_2, \sigma_3 \right\rangle \\ &\quad \prod_{\sigma_i} D^{\frac{1}{2}}_{\sigma_i\mu_i}\left\{R_W\left[k_i; B\left(V\right)\right]\right\} \Psi_{MJ\Sigma}\left(\vec{k}_i;\mu_i\right) , \end{aligned} \tag{2}$$

where overall momentum conservation, $P_\mu - P'_\mu = Q_\mu$, is explicitly satisfied; Q_μ being the meson four-momentum. Here $\left|P, J, \Sigma\right\rangle$ and $\left\langle P', J', \Sigma'\right|$ are the eigenstates of the decaying resonance and the outgoing nucleon ground state, respectively. The eigenstates are denoted by the eigenvalues P, J, Σ of the four-momentum operator $\hat{P}$, the total angular-momentum operator $\hat{J}$ and its z-component $\hat{\Sigma}$. The corresponding rest-frame baryon wave functions $\Psi^\star_{M'J'\Sigma'}$ and $\Psi_{MJ\Sigma}$ stem from the velocity-state representations of the baryon states $\left\langle P', J', \Sigma'\right|$ and $\left|P, J, \Sigma\right\rangle$. The rest-frame quark momenta $\vec{k}_i$, for which $\sum_i \vec{k}_i = \vec{0}$, are related to the individual quark four-momenta by the Lorentz boost relations $p_i = B(v)\, k_i$, with analogous relations holding for the primed variables.

In previous studies of the electroweak nucleon structure[8,9,10] a point-form spectator model (PFSM) for the electromagnetic and axial currents performed very well. Consequently, in a first investigation of mesonic decays we adopt a PFSM also for the decay operator. Assuming a pseudovector quark-meson coupling we express it by

$$\begin{aligned} &\langle p_1', p_2', p_3'; \sigma_1', \sigma_2', \sigma_3' |\, \hat{D}_m \,| p_1, p_2, p_3; \sigma_1, \sigma_2, \sigma_3 \rangle = \\ &3N \frac{i g_{qq\pi}}{2m_1 (2\pi)^{\frac{3}{2}}} \bar{u}(p_1', \sigma_1')\, \gamma_5 \gamma^\mu \lambda_m u(p_1, \sigma_1)\, Q_\mu \\ &2p_{20}\delta(\vec{p}_2 - \vec{p}_2')\, 2p_{30}\delta(\vec{p}_3 - \vec{p}_3')\, \delta_{\sigma_2 \sigma_2'} \delta_{\sigma_3 \sigma_3'}\,, \end{aligned} \tag{3}$$

with the flavor matrix λ_m characterizing the particular decay mode. Here, the incoming and outgoing momenta of the active quark are determined uniquely by the overall momentum conservation of the transition amplitude, $P_\mu - P_\mu' = Q_\mu$, together with the two spectator conditions. Generally, the momentum transferred to the active quark, $\vec{p}_1 - \vec{p}_1{}' = \vec{\tilde{q}}$, is different from the momentum transfer $\vec{Q}$ to the baryon as a whole. It has been shown that this is a consequence of translational invariance and also induces effective many-body contributions in the definition of the spectator-model decay operator[20]. Furthermore, in eq. (3) there appears an overall normalization factor N. In the PFSM electromagnetic and axial currents for the nucleon elastic form factors the factor

$$N = \left(\frac{M}{\sum_i \omega_i} \frac{M'}{\sum_i \omega_i'} \right)^{\frac{3}{2}} \tag{4}$$

was employed to recover the proper proton charge[8,9,10]. It depends on the individual quark momenta through the ω_i and the on-mass-shell conditions of the quarks. However, this is not a unique choice, if only Poincaré invariance and charge normalization are imposed. Also any other normalization factor of the asymmetric form

$$N(y) = \left(\frac{M}{\sum_i \omega_i} \right)^{3y} \left(\frac{M'}{\sum_i \omega_i'} \right)^{3(1-y)} \tag{5}$$

would be possible[21]. Below we also discuss the consequences of these further choices of N with regard to the pionic decay widths.

3. Results

In table 1 we quote the covariant predictions of the GBE and OGE CQMs for pionic decay widths calculated with the PFSM decay operator of eq. (3) with the normalization factor of eq. (4). For comparison, we also included the results obtained with the II CQM along the Bethe-Salpeter approach[22]. It is apparent that the results show a systematic underestimation of the experimental data. Only the N^*_{1535} and N^*_{1710} predictions agree with the experimental values.

We also present the numerical results for the decay widths with specific attention to the $\Delta\pi$ branching ratios, for the magnitudes of which a striking relationship is found to the underestimation of the experimental value: the larger the $\Delta\pi$ branching ratio of a resonance, the smaller the relative prediction for the π decay width. At this point it is not yet clear, if this behaviour is purely accidental or substantiates a shortcoming of the present description of baryon decays. In particular, additional degrees of freedom might be missing.

Let us now examine the influence of the normalization factor N in eq. (3). In fig. 2 we show the results for different possible choices of $N(y)$

Table 1. PFSM predictions for π decay widths of the relativistic GBE[3] and OGE[5] CQMs in comparison to the Bethe-Salpeter results[22] of the II CQM[2] and experimental data[6]. In the last three columns the theoretical results are expressed as percentage fractions of the (best-estimate) experimental values in order to be compared to the measured $\Delta\pi$ branching ratios.

Decays	Experiment	Rel. CQM			$\Delta\pi$	% of Exp. Width		
$\to N\pi$	[MeV]	GBE	OGE	II	branching ratio	GBE	OGE	II
N^*_{1440}	$(227 \pm 18)^{+70}_{-59}$	33	53	38	$20-30\%$	14	24	17
N^*_{1520}	$(66 \pm 6)^{+\ 9}_{-\ 5}$	17	16	38	$15-25\%$	26	24	58
N^*_{1535}	$(67 \pm 15)^{+28}_{-17}$	90	119	33	$< 1\%$	134	178	49
N^*_{1650}	$(109 \pm 26)^{+36}_{-\ 3}$	29	41	3	$1-7\%$	27	38	3
N^*_{1675}	$(68 \pm 8)^{+14}_{-\ 4}$	5.4	6.6	4	$50-60\%$	8	10	6
N^*_{1700}	$(10 \pm 5)^{+\ 3}_{-\ 3}$	0.8	1.2	0.1	$> 50\%$	8	12	1
N^*_{1710}	$(15 \pm 5)^{+30}_{-\ 5}$	5.5	7.7	n/a	$15-40\%$	37	51	n/a
Δ_{1232}	$(119 \pm 1)^{+\ 5}_{-\ 5}$	37	32	62	–	31	27	52
Δ_{1600}	$(61 \pm 26)^{+26}_{-10}$	0.07	1.8	n/a	$40-70\%$	≈ 0	3	n/a
Δ_{1620}	$(38 \pm 8)^{+\ 8}_{-\ 6}$	11	15	4	$30-60\%$	29	39	11
Δ_{1700}	$(45 \pm 15)^{+20}_{-10}$	2.3	2.3	2	$30-60\%$	5	5	4

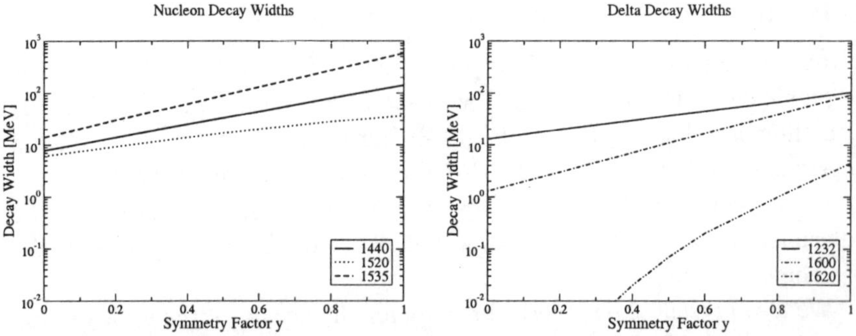

Figure 2. Dependence of the π decay widths on the asymmetry parameter y in the normalization factor of eq. (5) for selected N and Δ resonances.

according to eq. (5) in case of the GBE CQM. It is seen that in the range $0 \leq y \leq 1$ all decay widths grow rapidly with increasing y. Thus one could enhance or reduce the results as compared to the ones obtained with the symmetric factor $N(y = \frac{1}{2})$. While all other choices of $N(y)$ are also allowed, the symmetric factor was used for the electroweak nucleon form factors in refs.[8,9,10] and likewise the results given in table 1. We consider this as the best choice also for the baryon decays. It has the property that no theoretical result overshoots the corresponding experimental value. This is welcome and can be considered as reasonable, since one may expect further contributions not yet included in the present decay operator to raise the theoretical predictions.

The normalization factor N in eq. (3) effectively introduces a momentum cut-off. This can even better be seen if we investigate the form

$$N(x) = \left(\frac{M}{\sum_i \omega_i} \frac{M'}{\sum_i \omega'_i} \right)^{\frac{x}{2}} \tag{6}$$

with an arbitrary exponent x. It still represents a Poincaré-invariant construction but it does not guarantee for the proper charge normalization unless $x = 3$. It is clearly seen in fig. 3 that the choice $x = 0$ representing the bare case without normalization factor yields unreasonable results. The theoretical decay widths then evolve smoothly with increasing exponent x. For certain resonances (namely, $N^\star_{1440}$, $N^\star_{1710}$, and Δ_{1600}) the decay widths have a minimum. We notice that these are known as so-called structure-dependent resonances[23]. They are the radial excitations of the N and Δ ground states, respectively, with a corresponding nodal behaviour in their wave functions. These characteristics are quite distinct from the other reso-

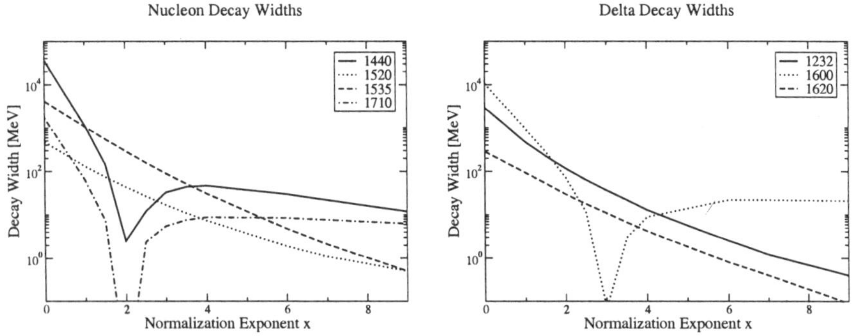

Figure 3. Dependence of the π decay widths on the exponent x of the normalization factor in eq. (6) for selected N and Δ resonances.

nances, which show a monotonous dependence on the exponent x. Through this study we get interesting insights into the causes for occurrence of narrow decay widths, at least for baryon states with nodal behaviour.

If we assume again the criterion that the theoretical predictions for decay widths with the decay operator (3) should not exceed the experimental data, we find that the case with $x = 3$ (and $y = \frac{1}{2}$) is the optimal choice. This leads to a single postulate for normalizing the PFSM operators for strong and electroweak processes.

4. Summary

We have discussed a Poincaré-invariant description of strong baryon resonance decays in point-form relativistic quantum mechanics. Covariant predictions of relativistic CQMs have been shown for π decay widths. They are considerably different from previous nonrelativistic results or results with relativistic corrections included. The covariant results calculated with a spectator-model decay operator show a uniform trend. In almost all cases the corresponding theoretical predictions underestimate the experimental data considerably. This is true in the framework of Poincaré-invariant quantum mechanics (here in point form) as well as in the Bethe-Salpeter approach[22]. Indications have been given that for a particular resonance the size of the underestimation is related to the magnitude of the $\Delta\pi$ branching ratio. This hints to systematic shortcomings in the description of the decay widths.

The investigation of different possible choices for the normalization factor occurring in the spectator-model decay operator has led to the suggestion that the symmetric choice is the most natural one. It is also consistent with the same (symmetric) choice adopted before for the spectator-model currents in the study of the electroweak nucleon form factors.

Acknowledgements

This work was supported by the Austrian Science Fund (Project P16945). T. M. would like to thank the INFN and the Physics Department of the University of Padova for their hospitality and MIUR-PRIN for financial support.

References

1. S. Capstick and N. Isgur, Phys. Rev. D **34**, 2809 (1986).
2. U. Loering, B. C. Metsch, and H. R. Petry, Eur. Phys. J. **A10**, 395 (2001); ibid. 447 (2001).
3. L. Y. Glozman, W. Plessas, K. Varga, and R. F. Wagenbrunn, Phys. Rev. D **58**, 094030 (1998).
4. L. Y. Glozman *et al.*, Phys. Rev. C **57**, 3406 (1998).
5. L. Theussl, R. F. Wagenbrunn, B. Desplanques, and W. Plessas, Eur. Phys. J. **A12**, 91 (2001).
6. S. Eidelman *et al.*, Phys. Lett. **B592**, 1 (2004).
7. D. Merten *et al.*, Eur. Phys. J. **A14**, 477 (2002).
8. R. F. Wagenbrunn *et al.*, Phys. Lett. **B511**, 33 (2001).
9. L. Y. Glozman *et al.*, Phys. Lett. **B516**, 183 (2001).
10. S. Boffi *et al.*, Eur. Phys. J. **A14**, 17 (2002).
11. T. Melde, W. Plessas, and R. F. Wagenbrunn, Few-Body Syst. Suppl. **14**, 37 (2003).
12. B. Metsch, U. Loring, D. Merten, and H. Petry, Eur. Phys. J. **A18**, 189 (2003).
13. T. Melde, W. Plessas, and R. F. Wagenbrunn, Contribution to the N*2004 Workshop, Grenoble, to appear in the Proceedings; hep-ph/0406023 (2004).
14. P. Geiger and E. S. Swanson, Phys. Rev. D **50**, 6855 (1994).
15. S. Kumano and V. R. Pandharipande, Phys. Rev. D **38**, 146 (1988).
16. R. Kokoski and N. Isgur, Phys. Rev. D **35**, 907 (1987).
17. B. D. Keister and W. N. Polyzou, Adv. Nucl. Phys. **20**, 225 (1991).
18. W. H. Klink, Phys. Rev. C **58**, 3587 (1998).
19. B. Bakamjian and L. H. Thomas, Phys. Rev. **92**, 1300 (1953).
20. T. Melde, L. Canton, W. Plessas, and R. F. Wagenbrunn, hep-ph/0411322 (2004).
21. T. Melde, L. Canton, W. Plessas, and R. F. Wagenbrunn, Contribution to the Mini-Workshop on Quark Dynamics, Bled, to appear in the Proceedings; hep-ph/0410274 (2004).
22. B. Metsch, Contribution to the 10th International Conference on Hadron Spectroscopy, Aschaffenburg, 2003; hep-ph/0403118 (2004).
23. R. Koniuk and N. Isgur, Phys. Rev. D **21**, 1868 (1980).

PARTONIC STRUCTURE OF THE NUCLEON IN QCD AND NUCLEAR PHYSICS: NEW DEVELOPMENTS FROM OLD IDEAS

M. RADICI

Istituto Nazionale di Fisica Nucleare, Sezione di Pavia, and Dipartimento di Fisica Nucleare e Teorica, Università di Pavia, via Bassi 6, 27100 Pavia, Italy
E-mail: radici@pv.infn.it

The nucleon is an ideal laboratory to solve QCD in the nonperturbative regime. There are several experimental observations that still lack a rigorous interpretation; they involve the nucleon as a (polarized) target as well as a beam (in collisions and Drell-Yan processes). These data look like big azimuthal and spin asymmetries, related to the transverse polarization and momentum of the nucleon and/or the final detected particles. They suggest internal reaction mechanisms that are suppressed in collinear perturbative QCD but that are "natural" in Nuclear Physics: quark helicity flips, residual final state interactions, etc.. In my talk, I will give a brief survey of the main results and I will flash the most recent developments and measurements.

1. Introduction

The spin structure of the nucleon can be best studied by using the socalled spin asymmetry, defined as the ratio between the difference and the sum of differential cross sections obtained by flipping the spin of one of the particles involved in the reaction. Spin asymmetries are known since a long time; historically, the first one was obtained at FERMILAB, where an anomalous large transverse polarization of the Λ produced in proton-nucleon annihilations was measured [1], surviving even at large values of the Λ transverse momentum. More recently, similar anomalously large asymmetries have been observed, for example by the STAR [2] collaboration in inclusive pion production from collisions of transversely polarized protons, as well as in Deep-Inelastic Scattering (DIS) of lepton probes on polarized protons by the HERMES [3] and CLAS [4] collaborations.

Assuming that partons are collinear with their parent hadrons and that a factorization theorem exists for the process at hand, QCD relates the spin asymmetry for transversely polarized objects to the off-diagonal components of the elementary scattering amplitude in the parton helicity basis; actually, to the imaginary part of products of such components [5]. But with the above assumptions any mechanism flipping the helicity of the parton is suppressed in QCD up to terms proportional to its mass. Since we are considering here mainly protons, which contain only *up* and *down* valence quarks, we can assume that deviations to this rule are negligible. Therefore, QCD in collinear approximation is not capable to explain the above mentioned amount of experimental observations.

Indeed, since in polarized DIS the structure function $g_T = g_1 + g_2$ (related to the partonic transverse spin distribution) appears at subleading twist, a common prejudice has always driven people to consider transverse spin effects as suppressed because inextricably associated to off-shellness, higher-order quark-gluon interactions, etc.. But in perturbative QCD (pQCD), longitudinal- and transverse-spin effects can be described on the same footing provided that the appropriate helicity and transverse basis are selected, respectively [6]. As a consequence, the spin structure of the proton at leading twist is not fully exploited by just the well known momentum and helicity distributions $f_1(x)$ and $g_1(x)$ (or, $q(x)$ and $\Delta q(x)$, in another common notation). A third one, the transversity $h_1(x)$ (or, $\delta q(x)$), is necessary which is basically unknown because it is related to helicity-flip mechanisms, as it should be clear from the above discussion. Since helicity and chirality coincide at leading twist, it is usually referred to as a chiral-odd distribution. Observables, like the cross section, must be chiral-even; hence, the transversity needs another chiral-odd partner to be extracted from a spin asymmetry. This is why, for example, it is suppressed in simple inclusive DIS (for a review on the transversity distribution, see Ref. [7]).

From this short introduction, it should hopefully be clear that the standard framework in which pQCD is usually calculated, namely collinear approximation neglecting transverse spin components, is not adequate to interpret the wealth of data collected over the years under the form of spin (azimuthal) asymmetries. In the following, I will review the main flowing ideas about improving such framework.

2. Intrinsic transverse momentum distribution of partons

Let consider the annihilation process $pp^\uparrow \to \pi X$, where a pion is semi-inclusively detected. If a factorization proof holds for the elementary process, in collinear approximation the differential cross section can be schematically written as

$$\frac{d\sigma}{d\eta\, dP_\pi\, d\phi_S} \propto \sum_{abcd} \int dx_a dx_b dz_c dz_d\, \phi(x_a, Q^2)\, \phi(x_b, Q^2)\, d\hat{\sigma}(ab^\uparrow \to c^\uparrow d) \times \chi(z_c, Q^2)\, \delta(z_c - \bar{z}_c)\, \delta(z_d - 1)\,, \tag{1}$$

where the functions ϕ are the distributions for the two annihilating partons carrying the fractions x_a and x_b of the two corresponding proton momenta, respectively, while χ is the fragmentation function for the pion with momentum P_π and carrying a fraction z_c of the fragmenting parton. Momentum conservation at the partonic level implies that z_c is constrained to $\bar{z}_c$, which is a function of x_a, x_b, of the center-of-mass (cm) energy s and of the rapidity η. The elementary cross section $d\hat{\sigma}$ with transversely polarized partons can be deduced from Ref. [8]. The angle ϕ_S is formed between the directions of the polarization and of the momentum of the polarized proton.

If this framework is not appropriate to describe experimental observations of spin asymmetries involving transversely polarized objects, a possible generalization consists in assuming (and trying to prove) that the factorization holds also when an explicit dependence upon the parton transverse momenta is introduced in Eq. (1), namely

$$\frac{d\sigma}{d\eta\, dP_\pi\, d\phi_S} \propto \sum_{abcd} \int dx_a dx_b dz_c dz_d d\mathbf{p}_{Ta} d\mathbf{p}_{Tb} d\mathbf{p}_{T\pi}\, \delta(z_c - \bar{z}_c)\, \delta(\mathbf{p}_{T\pi} \cdot \hat{\mathbf{p}}_c) \times \phi(x_a, \mathbf{p}_{Ta}, Q^2)\, \phi(x_b, \mathbf{p}_{Tb}, Q^2)\, d\hat{\sigma}(ab^\uparrow \to c^\uparrow d) \times \chi(z_c, \mathbf{p}_{T\pi}, Q^2)\, \delta(z_d - 1)\,. \tag{2}$$

At present, this factorization scheme has been proven only for Drell-Yan and e^+e^- processes [9], as well as for semi-inclusive DIS in some kinematical regions [10]. Therefore, for the considered process Eq. (2) is no more than an assumption, but it leads to several interesting consequences.

When partons have an intrinsic transverse momentum with respect to the direction of the parent hadron momentum, the list of the leading-twist distribution and fragmentation functions is far longer than in the collinear case [11]. Several (chiral-odd) functions appear with a specific number density interpretation, that can originate new interesting spin effects. In the initial state, when the proton is transversely polarized, we have

$$f(q/p^\uparrow) = f_1^q(x, \mathbf{p}_T) - f_{1T}^{\perp q}(x, \mathbf{p}_T)\, \frac{\hat{\mathbf{P}} \times \mathbf{p}_T \cdot \mathbf{S}_T}{M}\,, \tag{3}$$

where the function $f_{1T}^{\perp}(x, \mathbf{p}_T) \propto f(q/p^{\uparrow}) - f(q/p^{\downarrow})$ describes how the distribution gets distorted by the proton transverse polarization. This is possible because $f_{1T}^{\perp}$ appears weighted by the correlation $\mathbf{P} \times \mathbf{p}_T \cdot \mathbf{S}_T$ between the momentum $\mathbf{P}$ and transverse spin $\mathbf{S}_T$ of the proton with mass M, and the transverse momentum $\mathbf{p}_T$ of the parton (see Fig. 1-a). As a consequence, the final pion emerging from the fragmentation can be deflected into different directions according to the transverse polarization of the initial proton, the socalled Sivers effect [12]. From the field-theoretical point of view, $f_{1T}^{\perp}$ can be represented as in Fig. 1-a, i.e. it is diagonal in the parton helicity basis but not in the hadron one. As such, it is chiral-even but does not fullfil the constraints of time-reversal invariance (in jargon, it is T-odd); a possible interpretation for the latter feature is that in the considered process $pp^{\uparrow} \to \pi X$, a sort of initial state interaction occurs before the collision [13], which prevents the S matrix from being invariant under time-reversal transformations. Interestingly, $f_{1T}^{\perp}$ can be linked to the helicity-flip Generalized Parton Distribution (GPD) E, where the correlation between $\mathbf{S}_T$ and $\mathbf{p}_T$ can be directly interpreted as due to the orbital angular momentum of the partons themselves [14].

When the proton is not polarized, we can have

$$f(q^{\uparrow}/p) = \frac{1}{2} \left(f_1^q(x, \mathbf{p}_T) - h_1^{\perp q}(x, \mathbf{p}_T) \, \frac{\hat{\mathbf{P}} \times \mathbf{p}_T \cdot \mathbf{S}_{qT}}{M} \right) , \qquad (4)$$

where the function $h_1^{\perp}(x, \mathbf{p}_T) \propto f(q^{\uparrow}/p) - f(q^{\downarrow}/p)$ describes the influence of the parton transverse polarization $\mathbf{S}_{qT}$ on its momentum distribution inside an unpolarized proton. Alternatively to the Sivers effect, the final pion can be deflected according to the direction of $\mathbf{S}_{qT}$ entering the correlation factor (see Fig. 1-b). In the corresponding handbag diagram, the function is diagonal in the hadron helicity but not in the parton one, hence it is chiral-odd and it represents a natural partner of transversity for the considere process [15]. Extraction of $h_1^{\perp}$ is of great importance, because it is believed to be responsible for the well known violation of the Lam-Tung sum rule [16], an anomalous big azimuthal asymmetry of the unpolarized Drell-Yan $pp \to \mu^+\mu^- X$ process, that neither Next-to-Leading Order (NLO) QCD calculations [17], nor higher twists or factorization-breaking terms in NLO QCD [18,19,20] are able to justify.

In the final state, a transversely polarized/unpolarized parton can fragment into a hadron with mass M_h and carrying a fraction z of the momentum. As for the former, since we are considering a semi-inclusive process with a pion in the final state, we can have

$$D(h/q^{\uparrow}) = D_1^q(z, \mathbf{P}_{hT}) + H_1^{\perp q}(z, \mathbf{P}_{hT}) \frac{\hat{\mathbf{p}} \times \mathbf{P}_{hT} \cdot \mathbf{S}_{qT}}{z M_h} , \qquad (5)$$

where $D_1^q(z, \mathbf{P}_{hT})$ is the probability for a quark q to fragment into a hadron with transverse momentum $\mathbf{P}_{hT}$, while $H_1^{\perp q}(z, \mathbf{P}_{hT}) \propto D(h/q^{\uparrow}) - D(h/q^{\downarrow})$ is the analogue for a quark with transverse polarization $\mathbf{S}_{qT}$, i.e. the Collins function [21]. Again, also the polarization of the fragmenting quark can be responsible for an asymmetric distribution of the detected pion via the correlation $\hat{\mathbf{p}} \times \mathbf{P}_{hT} \cdot \mathbf{S}_{qT}$ (see Fig. 1-c). In the same figure, the diagram corresponding to $H_1^{\perp}$ displays a helicity flip for the fragmenting quark; the Collins function is both chiral-odd and T-odd, and it represents another possible partner for extracting the transversity through the Collins effect. Several experimental collaborations are pursuing this goal [2,3,4], with the help of theoretical calculations as well [22,23].

Finally, the last combination is given by

$$D(h^{\uparrow}/q) = \frac{1}{2} \left(D_1^q(z, \mathbf{P}_{hT}) + D_{1T}^{\perp q}(z, \mathbf{P}_{hT}) \frac{\hat{\mathbf{p}} \times \mathbf{P}_{hT} \cdot \mathbf{S}_h}{z M_h} \right) . \qquad (6)$$

The function $D_{1T}^{\perp q}(z, \mathbf{P}_{hT}) \propto D(h^{\uparrow}/q) - D(h^{\downarrow}/q)$ describes the fragmentation of an unpolarized parton into a hadron with polarization $\mathbf{S}_h$ like, e.g., the Λ: it is not pertinent to the process we have here selected, but it is anyway important because the correlation $\hat{\mathbf{p}} \times \mathbf{P}_{hT} \cdot \mathbf{S}_h$ is believed to be the mechanism responsible for the observed asymmetric production of Λ in unpolarized proton collisions [24,13] (see Fig. 1-d). From the related diagram, the function $D_{1T}^{\perp}$ is related to a helicity flip of just the polarized hadron; it is then chiral even, but T-odd.

3. Interference fragmentation functions

From previous section, it emerges that in the non-collinear factorized framework of Eq. (2) at least three different interpretations arise at leading twist for the spin asymmetry observed in the $pp^{\uparrow} \to \pi X$ process: the Sivers effect, related to the convolution $f_1^a \otimes f_{1T}^{\perp b} \otimes D_1^c$ for the elementary process $a + b \to c + d$; the Collins effect, related to the convolution $f_1^a \otimes h_1^b \otimes H_1^{\perp c}$ for $a + b^{\uparrow} \to c^{\uparrow} + d$; the effect related to the convolution $h_1^{\perp a} \otimes h_1^b \otimes D_1^c$ for $a^{\uparrow} + b^{\uparrow} \to c + d$. An intense experimental and theoretical activity is ongoing in this field in order to unravel the physics contained in the asymmetry data, and also in related topics, like for example the new project of extracting transversity from Drell-Yan using (polarized) antiprotons [25].

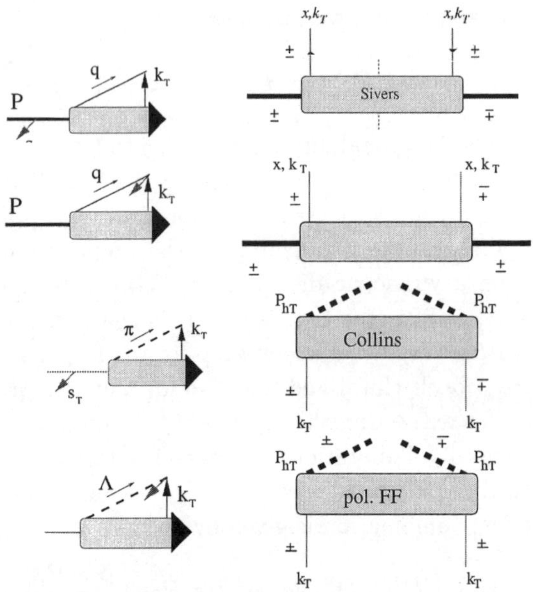

Figure 1. On the left, from top to bottom: a- the nonperturbative correlation giving rise to the Sivers effect; b- the same for transversely polarized quarks in unpolarized protons; c- the same for the Collins effect; d- the same for the polarized fragmentation. On the right, from top to bottom, the corresponding field-theoretical interpretations.

But all the above mechanisms require an explicit dependence of the distribution and fragmentation functions upon an intrinsic tranverse momentum of partons. We already stressed that in such a non-collinear framework the factorization leading to Eq. (2) has not yet been proven for hadron-hadron collisions. A lot of work is being done also in this specific, though quite general, subject [26,27,28]. Nevertheless, it would be desirable to find out a mechanism that leads to a spin asymmetry without invoking an explicit dependence on $\mathbf{p}_T$. Since we are considering here final states made of unpolarized hadrons only, we need an additional 4-vector that can be provided by semi-inclusively detecting a second hadron inside the same jet of the first one. In fact, in this case the system of two unpolarized hadrons with momenta P_1 and P_2 is described by two 4-vectors, the cm one, $P_h = P_1 + P_2$ and the relative one, $R = (P_1 - P_2)/2$, by which we can construct the correlation $\mathbf{S}_{qT} \cdot \mathbf{R} \times \mathbf{P}_h$ [29,30,31] (see Fig. 2).

Indeed, for the DIS lepton-induced production from a transversely po-

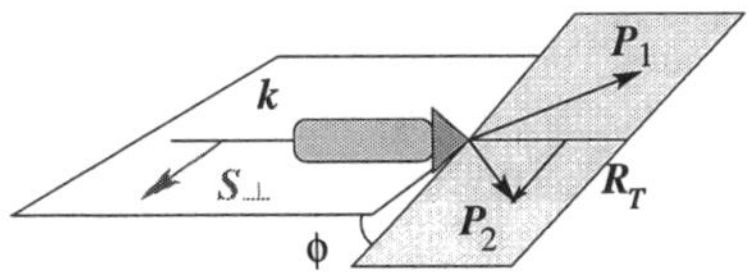

Figure 2. The mechanism leading to a spin asymmetry for a transversely polarized quark fragmenting into two unpolarized hadrons.

larized proton of two pions with the cm momentum aligned to the jet axis (i.e., with $\mathbf{P}_{hT} = 0$), it has been shown [32] that at leading twist the cross section contains a term like $(\mathbf{S}_{qT} \times \mathbf{R}_T)\, h_1(x)\, H_1^{\sphericalangle}(z, M_h^2, \mathbf{R}_T^2)$, where the chiral-odd (T-odd) $H_1^{\sphericalangle}$ is representative of a new class of fragmentation functions, the Interference Fragmentation Functions (IFF). A spin asymmetry can be built by flipping the spin of the transversely polarized proton target, which isolates the transversity h_1 at leading twist via $H_1^{\sphericalangle}$. A thorough analysis of IFF has been carried out up to twist-3 [33], and $H_1^{\sphericalangle}$ is the only one surviving at leading twist in the collinear situation, since it is related to the azimuthal position of the plane containing the two pions: the latter ones are differently distributed in the azimuthal angle ϕ of Fig. 2 according to the direction of the transverse polarization of the fragmenting quark. A partial-wave analysis of the two-pion system allows to isolate the main channels, where the two pions are produced in a relative s or p wave. Interference between the two or between different components of p waves is responsible of the T-odd nature of IFF [34]. These functions can be extracted in principle from the process $e^+e^- \to (\pi\pi)_{jet1}(\pi\pi)_{jet2}X$ [35]. Measurements aiming at the extraction of IFF are under way at HERMES (DIS) and BELLE (e^+e^-) and could be possible at COMPASS and BABAR too. However, it has been recently proposed [36] that a consistent extraction of h_1 and $H_1^{\sphericalangle}$ could be achieved by considering again the collision of (un)polarized protons leading to one or two pion pairs semi-inclusively detected in the final state. For the case $pp^\uparrow \to (\pi\pi)X$, the cross section contains the convolution $f_1^a \otimes h_1^b \otimes H_1^{\sphericalangle c}$ for the elementary process $a+b^\uparrow \to c^\uparrow d$. On the other hand, for the corresponding unpolarized process $pp \to (\pi\pi)_{jet1}(\pi\pi)_{jet2}X$ the cross section contains $f_1^a \otimes f_1^b \otimes H_1^{\sphericalangle c} \otimes H_1^{\sphericalangle d}$ for $a + b \to c^\uparrow + d^\uparrow$. By combining the two measurements, information on the two unknowns h_1 and $H_1^{\sphericalangle}$ can be extracted consistently in the same experiment. Finally, the very same formalism offers the possibility of observing for the first time the transverse linear polarization of gluons, even using spin-$\frac{1}{2}$ targets [36].

References

1. G. Bunce *et al.*, *Phys. Rev. Lett.* **36**, 1113 (1976).
2. J. Adams *et al.* (STAR), *Phys. Rev. Lett.* **92**, 171801 (2004).
3. A. Airapetian *et al.* (HERMES), `hep-ex/0408013`.
4. H. Avakian (CLAS), `hep-ex/0301005`
5. G. L. Kane, J. Pumplin and W. Repko, *Phys. Rev. Lett.* **41**, 1689 (1978).
6. R. L. Jaffe, proc. *Ettore Majorana International School on the Spin Structure of the Nucleon*, Erice (Italy), 3-10 Aug. 1995, `hep-ph/9602236`.
7. V. Barone, A. Drago and P. G. Ratcliffe, *Phys. Rep.* **359**, 1 (2002).
8. R. Gastmans and T. T. Wu, *The Ubiquitous photon: Helicity methods for QED and QCD* (Clarendon, Oxford, UK, 1990).
9. J. Collins, D. Soper and G. Sterman, *Nucl. Phys.* **B250**, 199 (1985).
10. X. Ji, J.-P. Ma and F. Yuan, *Phys. Lett.* **B597**, 299 (2004).
11. P. J. Mulders and R. D. Tangermann, *Nucl. Phys.* **B461**, 197 (1996).
12. D. W. Sivers, *Phys. Rev.* **D43**, 261 (1991).
13. M. Anselmino *et al.*, `hep-ph/0408356`.
14. Burkardt, *Nucl. Phys.* **A735**, 185 (2004).
15. D. Boer, *Phys. Rev.* **D60**, 014012 (1999).
16. J. S. Conway *et al.*, *Phys. Rev.* **D39**, 92 (1989).
17. A. Brandenburg, O. Nachtmann and E. Mirkes, *Z. Phys.* **C60**, 697 (1993).
18. A. Brandenburg, S. J. Brodsky, V. V. Khoze and D. Mueller, *Phys. Rev. Lett.* **73**, 939 (1994).
19. K. J. Eskola, P. Hoyer, M. Vanttinen and R. Vogt, *Phys. Lett.* **B333**, 526 (1994).
20. E. L. Berger and S. J. Brodsky, *Phys. Rev. Lett.* **42**, 940 (1979).
21. J. C. Collins, *Nucl.Phys.* **B396**, 161 (1993).
22. A. V. Schweitzer, K. Goeke and P. Schweitzer, *Eur. Phys. J.* **C32**, 337 (2003).
23. A. Bacchetta, R. Kundu, A. Metz and P. J. Mulders, *Phys. Rev.* **D65**, 094021 (2002).
24. M. Anselmino *et al.*, *Phys, Rev.* **D63**, 054029 (2001).
25. PAX Coll., Letter of Intent for *Antiproton-proton Scattering Experiments with Polarization* (2003), `http://www.fz-juelich.de/ikp/pax`.
26. A. Metz, *Phys. Lett.* **B549**, 139 (2002).
27. D. Boer, P. J. Mulders and F. Pijlman, *Nucl. Phys.* **B667**, 201 (2003).
28. J. C. Collins and A. Metz, `hep-ph/0408249`.
29. J. C. Collins and G. A. Ladinski, `hep-ph/9411444`.
30. R. L. Jaffe, X. Jin and J. Tang, *Phys. Rev. Lett.* **80**, 1166 (1998).
31. A. Bianconi, S. Boffi, R. Jakob and M. Radici, *Phys. Rev.* **D62**, 034008 (2000).
32. M. Radici, R. Jakob and A. Bianconi, *Phys. Rev.* **D65**, 074031 (2002).
33. A. Bacchetta and M. Radici, *Phys. Rev.* **D69**, 074026 (2004).
34. A. Bacchetta and M. Radici, *Phys. Rev.* **D67**, 094002 (2003).
35. D. Boer, R. Jakob and M. Radici, *Phys. Rev.* **D67**, 094003 (2003).
36. A. Bacchetta and M. Radici, *Phys. Rev.* **D** in press, `hep-ph/0409174`.

A GENERAL FORMALISM FOR SINGLE AND DOUBLE SPIN ASYMMETRIES IN INCLUSIVE HADRON PRODUCTION

U. D'ALESIO*, S. MELIS†, F. MURGIA‡

Istituto Nazionale di Fisica Nucleare, Sezione di Cagliari and Dipartimento di Fisica, Università di Cagliari C.P. 170,I-09042 Monserrato (CA),Italy

We present a calculation of single and double spin asymmetries for inclusive hadron production in hadronic collisions. Our approach is based on Leading Order (LO) perturbative QCD and generalized factorization theorems, with full account of intrinsic parton momentum, $\boldsymbol{k}_\perp$, effects. This leads to a new class of spin and $\boldsymbol{k}_\perp$-dependent distribution and fragmentation functions. Limiting ourselves to consider leading twist functions, we show how they could play a relevant role in producing non-vanishing spin asymmetries.

1. Introduction

It has been experimentally known since a long time that transverse single spin asymmetries (SSA) in hadronic collisions can be, in particular kinematical regions, very large. Two relevant examples are: the transverse Λ polarization, P_T^Λ, measured in unpolarized hadron collisions; the SSA, A_N, observed in $p^\uparrow p \to \pi + X$. P_T^Λ and A_N are defined as:

$$P_T^\Lambda = \frac{d\sigma^{AB\to\Lambda^\uparrow X} - d\sigma^{AB\to\Lambda^\downarrow X}}{d\sigma^{AB\to\Lambda^\uparrow X} + d\sigma^{AB\to\Lambda^\downarrow X}} \quad A_N = \frac{d\sigma^{A^\uparrow B\to CX} - d\sigma^{A^\downarrow B\to CX}}{d\sigma^{A^\uparrow B\to CX} + d\sigma^{A^\downarrow B\to CX}} \quad (1)$$

where $d\sigma$ stands for the corresponding invariant differential cross section and $\uparrow,\downarrow$ denote transverse polarization with respect to the hadron production plane. Both these observables can reach in size values up to 30%-40%. These results are at first puzzling in the context of perturbative QCD (pQCD) if, as usual, one assumes a collinear partonic kinematics. In fact,

*E-mail: umberto.dalesio@ca.infn.it
†E-mail: stefano.melis@ca.infn.it
‡E-mail: francesco.murgia@ca.infn.it

it is easy to see that in this case pQCD predicts almost vanishing SSA at large energy scales. Writing a transverse spin state as a combination of helicity states, $|\uparrow/\downarrow\rangle = \frac{1}{\sqrt{2}}[|+\rangle \pm i|-\rangle]$, we have schematically:

$$A_N \sim \frac{\langle\uparrow|\uparrow\rangle - \langle\downarrow|\downarrow\rangle}{\langle\uparrow|\uparrow\rangle + \langle\downarrow|\downarrow\rangle} \sim \frac{2\mathrm{Im}\langle+|-\rangle}{\langle+|+\rangle + \langle-|-\rangle}, \tag{2}$$

where $\langle\uparrow|\uparrow\rangle$ stands for the partonic cross section $\mathrm{d}\hat{\sigma}^{\uparrow}$. The presence of a term like $\mathrm{Im}\langle+|-\rangle$ in the numerator of Eq. (2) requires both helicity flip at the partonic level and a relative phase between helicity amplitudes. Because massless QCD conserves helicity and Born level amplitudes are real, A_N should be proportional to the quark mass times a factor α_s coming from the required higher order contributions. This implies $A_N \propto \alpha_s m_q/E_q$, which would be negligible at large energies. A possible way out is to take into account partonic transverse momenta, $\boldsymbol{k}_\perp$, in parton distribution and fragmentation functions and in the elementary scattering process. This in turn leads to the introduction of a new class of spin and $\boldsymbol{k}_\perp$ dependent partonic distribution (PDF) and fragmentation (FF) functions that, at least in principle, are able to generate spin asymmetries.

2. General formalism

In the usual collinear pQCD approach at leading order (LO) and leading twist, the unpolarized cross section for the process $AB \to CX$ reads:

$$\begin{aligned} \mathrm{d}\sigma^{AB\to CX} &= \sum_{a,b,c,d} f_{a/A}(x_a, Q^2) \otimes f_{b/B}(x_b, Q^2) \\ &\otimes \mathrm{d}\hat{\sigma}^{ab\to cd}(\hat{s}, \hat{t}, \hat{u}, x_a, x_b) \otimes D_{C/c}(z, Q^2), \end{aligned} \tag{3}$$

where $f_{a/A}(x_a, Q^2)$ denotes the PDF for parton a inside hadron A carrying a fraction x_a of the parent hadron light-cone momentum; $D_{C/c}(z, Q^2)$ is the FF for parton c fragmenting into hadron C with a fraction z of the parton light-cone momentum; $\mathrm{d}\hat{\sigma}^{ab\to cd}$ denotes the partonic cross section for the elementary scattering $ab \to cd$; $\hat{s}$, $\hat{t}$, $\hat{u}$ are the Mandelstam variables for the partonic process.

In order to include $\boldsymbol{k}_\perp$ effects we have to generalize standard collinear PDF and FF to $\boldsymbol{k}_\perp$-dependent functions; for instance $f_{a/A}(x_a)$ is generalized to $\hat{f}_{a/A}(x_a, \boldsymbol{k}_{\perp a})$, where $\boldsymbol{k}_{\perp a}$ is the parton momentum perpendicular to the hadron momentum, with the condition:

$$f_{a/A}(x_a) = \int \mathrm{d}^2\boldsymbol{k}_{\perp a} \hat{f}_{a/A}(x_a, \boldsymbol{k}_{\perp a}). \tag{4}$$

Analogously, the fragmentation function $D_{C/c}(z)$ is generalized to $\hat{D}_{C/c}(z, \boldsymbol{k}_{\perp C})$, where $\boldsymbol{k}_{\perp C}$ is the transverse momentum of the observed hadron C with respect to the fragmenting parton c momentum.

The next step is to extend the pQCD expression for the cross section, Eq. (3), including $\boldsymbol{k}_\perp$ effects:

$$\begin{aligned} \mathrm{d}\sigma^{AB\to CX} &= \sum_{a,b,c,d} \hat{f}_{a/A}(x_a, \boldsymbol{k}_{\perp a}; Q^2) \otimes \hat{f}_{b/B}(x_b, \boldsymbol{k}_{\perp b}; Q^2) \\ &\otimes \mathrm{d}\hat{\sigma}^{ab\to cd}(\hat{s}, \hat{t}, \hat{u}, x_a, x_b) \otimes \hat{D}_{C/c}(z, \boldsymbol{k}_{\perp C}; Q^2)\,, \qquad (5) \end{aligned}$$

where now $\hat{s}$, $\hat{t}$ and $\hat{u}$ depend on the full $\boldsymbol{k}_\perp$ kinematics.[1]

A factorization theorem with the inclusion of transverse momenta has not been formally proved in general,[2] but only for the Drell-Yan process, for two-particle inclusive production in e^+e^- annihilation[3] and, recently, for SIDIS processes in particular kinematical regions.[4]

In order to study spin asymmetries we have to include explicit spin dependences into Eq (5). This can be done by introducing in the factorization scheme the parton helicity density matrices describing the parton spin states. In this way, starting from Eq. (5) we can write the polarized cross section as

$$\mathrm{d}\sigma^{(A,S_A)+(B,S_B)\to C+X} = \sum_{a,b,c,d,\{\lambda\}} \rho^{a/A,S_A}_{\lambda_a,\lambda'_a}\, \hat{f}_{a/A,S_A}(x_a, \boldsymbol{k}_{\perp a}) \otimes$$
$$\rho^{b/B,S_B}_{\lambda_b,\lambda'_b}\, \hat{f}_{b/B,S_B}(x_b, \boldsymbol{k}_{\perp b}) \otimes \hat{M}_{\lambda_c,\lambda_d;\lambda_a,\lambda_b}\, \hat{M}^*_{\lambda'_c,\lambda_d;\lambda'_a,\lambda'_b} \otimes \hat{D}^{\lambda_C,\lambda_C}_{\lambda_c,\lambda'_c}(z, \boldsymbol{k}_{\perp C}) \qquad (6)$$

where $\{\lambda\}$ is a shorthand for the sum over all helicity indices involved. In Eq. (6), $\rho^{a/A,S_A}_{\lambda_a,\lambda'_a}$ is the helicity density matrix of parton a inside the hadron A with generic polarization S_A. Similarly for parton b inside hadron B. The $\hat{M}_{\lambda_c,\lambda_d;\lambda_a,\lambda_b}$'s are the helicity amplitudes for the elementary process $ab \to cd$, normalized so that the unpolarized cross section, for a collinear collision, is given by

$$\frac{d\hat{\sigma}^{ab\to cd}}{d\hat{t}} = \frac{1}{16\pi\hat{s}^2}\frac{1}{4} \sum_{\lambda_a,\lambda_b,\lambda_c,\lambda_d} |\hat{M}_{\lambda_c,\lambda_d;\lambda_a,\lambda_b}|^2\,. \qquad (7)$$

The $\hat{M}_{\lambda_c,\lambda_d;\lambda_a,\lambda_b}$'s in Eq. (6) are defined in the hadronic c.m. frame; they are related to the usual helicity amplitudes, defined in the "canonical' partonic c.m. frame, $\hat{M}^0_{\lambda_c,\lambda_d;\lambda_a,\lambda_b}$, in a non trivial way by proper phases coming from the rotations and the boost connecting the two reference frames. These azimuthal phases are crucial in determining, when integrating over partonic phase space, the size of each allowed contribution to Eq. (6).[5,6,7]

Finally, $\hat{D}^{\lambda_C,\lambda'_C}_{\lambda_c,\lambda'_c}(z,\boldsymbol{k}_{\perp C})$ is the product of *fragmentation amplitudes* for the $c \to C + X$ process

$$\hat{D}^{\lambda_C,\lambda'_C}_{\lambda_c,\lambda'_c} = \sum\!\!\!\!\!\!\!\int_{X,\lambda_X} \hat{\mathcal{D}}_{\lambda_X,\lambda_C;\lambda_c}\,\hat{\mathcal{D}}^*_{\lambda_X,\lambda'_C;\lambda'_c}\,, \tag{8}$$

where $\sum\!\!\!\!\!\!\int_{X,\lambda_X}$ stands for a spin sum and phase space integration over all undetected particles, considered as a system X. The usual unpolarized fragmentation function $D_{C/c}(z)$, *i.e.* the number density of hadrons C resulting from the fragmentation of an unpolarized parton c and carrying a light-cone momentum fraction z, is given by

$$D_{C/c}(z) = \frac{1}{2}\sum_{\lambda_c,\lambda_C}\int d^2\boldsymbol{k}_{\perp C}\,\hat{D}^{\lambda_C,\lambda_C}_{\lambda_c,\lambda_c}(z,\boldsymbol{k}_{\perp C})\,. \tag{9}$$

One can also give a more explicit connection between hadron and parton polarizations. Interpreting the partonic distribution, at LO, as the inclusive cross section for the process $A \to a + X$ we can write:

$$\begin{aligned}\rho^{a/A,S_A}_{\lambda_a,\lambda'_a}\,\hat{f}_{a/A,S_A}(x_a,\boldsymbol{k}_{\perp a}) &= \sum_{\lambda_A,\lambda'_A}\rho^{A,S_A}_{\lambda_A,\lambda'_A}\sum\!\!\!\!\!\!\!\int_{X_A,\lambda_{X_A}}\hat{\mathcal{F}}_{\lambda_a,\lambda_{X_A};\lambda_A}\,\hat{\mathcal{F}}^*_{\lambda'_a,\lambda_{X_A};\lambda'_A}\\ &= \sum_{\lambda_A,\lambda'_A}\rho^{A,S_A}_{\lambda_A,\lambda'_A}\,\hat{F}^{\lambda_a,\lambda'_a}_{\lambda_A,\lambda'_A}\,,\end{aligned} \tag{10}$$

where $\rho^{A,S_A}_{\lambda_A,\lambda'_A}$ is the hadron helicity density matrix,

$$\hat{F}^{\lambda_a,\lambda'_a}_{\lambda_A,\lambda'_A} \equiv \sum\!\!\!\!\!\!\!\int_{X_A,\lambda_{X_A}}\hat{\mathcal{F}}_{\lambda_a,\lambda_{X_A};\lambda_A}\,\hat{\mathcal{F}}^*_{\lambda'_a,\lambda_{X_A};\lambda'_A}\,, \tag{11}$$

and the $\hat{\mathcal{F}}$'s are the *helicity distribution amplitudes* for the $A \to a + X$ process. Inserting Eq. (10) into Eq. (6) we obtain our master formula for a collision between polarized hadrons:

$$\begin{aligned}d\sigma^{(A,S_A)+(B,S_B)\to C+X} = &\sum_{a,b,c,d,\{\lambda\}}\rho^{A,S_A}_{\lambda_A,\lambda'_A}\,\hat{F}^{\lambda_a,\lambda'_a}_{\lambda_A,\lambda'_A}\otimes\rho^{B,S_B}_{\lambda_B,\lambda'_B}\,\hat{F}^{\lambda_b,\lambda'_b}_{\lambda_B,\lambda'_B}\\ &\otimes\hat{M}_{\lambda_c,\lambda_d;\lambda_a,\lambda_b}\,\hat{M}^*_{\lambda'_c,\lambda_d;\lambda'_a,\lambda'_b}\otimes\hat{D}^{\lambda_C,\lambda_C}_{\lambda_c,\lambda'_c}\,.\end{aligned} \tag{12}$$

This expression contains all allowed combinations of spin and $\boldsymbol{k}_\perp$-dependent distribution and fragmentation functions: at LO and leading twist these functions have a simple partonic interpretation and are related to the spin and $\boldsymbol{k}_\perp$-dependent functions discussed in other papers.[8,9] As an example, the relation between the definitions of the Sivers, the Boer-Mulders, and the Collins functions in two widely adopted notations is the following:[10]

$$f_{q/h^\uparrow} - f_{q/h^\downarrow} \equiv \Delta^N f_{q/h^\uparrow} = 4\mathrm{Im}(\hat{F}^{+,+}_{+,-}) \propto f_{1T}^\perp \tag{13}$$

$$f_{q^\uparrow/h} - f_{q^\downarrow/h} \equiv \Delta^N f_{q^\uparrow/h} = 2\mathrm{Im}(\hat{F}^{+,-}_{+,+}) \propto h_1^\perp \tag{14}$$

$$D_{h/q^\uparrow} - D_{h/q^\downarrow} \equiv \Delta^N D_{h/q^\uparrow} = 2\mathrm{Im}(D^h_{+-}) \propto H_1^\perp \,. \tag{15}$$

$\Delta^N f_{q/h^\uparrow}$ in Eq. (13) gives the probability to find an unpolarized parton inside a transversely polarized hadron. $\Delta^N f_{q^\uparrow/h}$ in Eq. (14), is the probability to find a transversely polarized quark inside an unpolarized hadron. Finally, $\Delta^N D_{h/q^\uparrow}$ is the probability for a transversely polarized quark to fragment into an unpolarized hadron.

2.1. $A_N(A^\uparrow B \to \pi + X)$

As an application of this formalism let us consider the process $A^\uparrow B \to \pi + X$, where A and B are spin one-half hadrons. In this case the helicity density matrices for A, B, take the form:

$$\rho^{A,\uparrow/\downarrow}_{\lambda_A,\lambda'_A} = \frac{1}{2}\begin{pmatrix} 1 & \mp i \\ \pm i & 1 \end{pmatrix} \qquad \rho^{B,0}_{\lambda_B,\lambda'_B} = \frac{1}{2}\begin{pmatrix} 1 & 0 \\ 0 & 1 \end{pmatrix} . \tag{16}$$

By performing explicitly the sum over hadron helicity indices in Eq. (12) and using Eq. (16) we obtain:

$$\begin{aligned} d\sigma^{A^{\uparrow/\downarrow}+B\to\pi+X} = & \sum_{a,b,c,d,\{\lambda\}} [\hat{F}^{\lambda_a,\lambda'_a}_{+,+} + \hat{F}^{\lambda_a,\lambda'_a}_{-,-} \mp i(\hat{F}^{\lambda_a,\lambda'_a}_{+,-} - \hat{F}^{\lambda_a,\lambda'_a}_{-,+})] \\ & \otimes [\hat{F}^{\lambda_b,\lambda'_b}_{+,+} + \hat{F}^{\lambda_b,\lambda'_b}_{-,-}] \otimes \hat{M}_{\lambda_c\lambda_d;\lambda_a\lambda_b} \hat{M}^*_{\lambda'_c\lambda_d;\lambda'_a\lambda'_b} \otimes \hat{D}^\pi_{\lambda_c,\lambda'_c} \,, \end{aligned} \tag{17}$$

where $\{\lambda\}$ stands for a sum over partonic helicity indices.

In Eq. (17) there are terms that change their sign when changing the sign of the corresponding hadron polarization. These terms survive in the numerator of the asymmetry and involve, depending also on the partonic subprocess considered, different combinations of the Sivers, Boer-Mulders and Collins functions.

As an explicit example let us now consider a particular partonic subprocess: $qq \to qq$. Then, using known symmetry properties of the helicity distribution functions[5] we can write:

$$\hat{F}^{\lambda_a,\lambda'_a}_{\lambda_A,\lambda'_A}(x_a, \boldsymbol{k}_{\perp a}) = F^{\lambda_a,\lambda'_a}_{\lambda_A,\lambda'_A}(x_a, k_{\perp a}) \exp[i(\lambda_A - \lambda'_A)\phi_a] \,, \tag{18}$$

$$F^{-\lambda_a,-\lambda'_a}_{-\lambda_A,-\lambda'_A} = (-1)^{(\lambda_A-\lambda_a)+(\lambda'_A-\lambda'_a)} F^{\lambda_a,\lambda'_a}_{\lambda_A,\lambda'_A} \,, \tag{19}$$

where ϕ_i $(i = a, b, c, d)$ are the azimuthal angles of parton i three-momentum in the AB c.m. frame. Analogously for fragmentation functions we have:

$$\hat{D}^{\pi}_{\lambda_c,\lambda'_c} = D^{\pi}_{\lambda_c,\lambda'_c} \exp[i(\lambda_c - \lambda'_c)\phi^H_\pi] \,, \tag{20}$$

where ϕ^H_π is the azimuthal angle of the pion three-momentum as seen in the parton c helicity frame.[5]

As mentioned above, the helicity amplitudes $\hat{M}_{\lambda_c\lambda_d;\lambda_a\lambda_b}$ in Eq. (17) are defined in the hadronic c.m. frame; we can relate these amplitudes to those given in the canonical partonic c.m. frame, $\hat{M}^0_{\lambda_c\lambda_d;\lambda_a\lambda_b}$, where Z is the direction of the colliding partons and the XZ-plane coincides with the scattering plane. In this frame the $\hat{M}^0_{\lambda_c\lambda_d;\lambda_a\lambda_b}$'s exhibit explicit symmetry properties.
To reach the canonical partonic c.m. frame, from the hadronic c.m. frame, we have first to perform a boost along the $\boldsymbol{p}_a + \boldsymbol{p}_b$ direction, so that the boosted three-vector $\boldsymbol{p}'_a + \boldsymbol{p}'_b$ is vanishing. In this frame (S') partons a and b are in the "head-on" configuration, but not aligned along the Z-axis direction. We then perform a rotation to align the colliding initial partons with the Z-axis. We call this new frame S''. In this frame parton c three-momentum does not lie in the XZ-plane but has a transverse component, with an azimuthal angle ϕ''_c. A final rotation around Z by ϕ''_c leads to the canonical configuration. The relationship between $\hat{M}_{\lambda_c\lambda_d;\lambda_a\lambda_b}$ and $\hat{M}^0_{\lambda_c\lambda_d;\lambda_a\lambda_b}$ is then given by:[5]

$$\begin{aligned}\hat{M}_{\lambda_c,\lambda_d;\lambda_a,\lambda_b} &= \hat{M}^0_{\lambda_c,\lambda_d;\lambda_a,\lambda_b} \\ &\quad \times e^{-i[\lambda_a(\xi_a+\tilde{\xi}_a)+\lambda_b(\xi_b+\tilde{\xi}_b)-\lambda_c(\xi_c+\tilde{\xi}_c)-\lambda_d(\xi_d+\tilde{\xi}_d)]}\, e^{i(\lambda_a-\lambda_b)\phi''_c} \\ &= \hat{M}^0_{\lambda_c,\lambda_d;\lambda_a,\lambda_b} \\ &\quad \times e^{-i(\lambda_a\xi_a+\lambda_b\xi_b-\lambda_c\xi_c-\lambda_d\xi_d)}\, e^{-i[(\lambda_a-\lambda_b)\tilde{\xi}_a-(\lambda_c-\lambda_d)\tilde{\xi}_c]}\, e^{i(\lambda_a-\lambda_b)\phi''_c} \,,\end{aligned} \tag{21}$$

where ξ_i, $\tilde{\xi}_i$ $(i = a, b, c, d)$ are phases due to the behaviour of helicity states under the Lorentz transformations connecting the hadronic and the canonical partonic c.m. frame.[5,11]

Using Eqs. (18), (19), (20), (22) into Eq. (17) we obtain the contribution to the numerator of A_N coming from the $qq \to qq$ partonic subprocess:

$$
\begin{aligned}
&\mathrm{d}\sigma^{A^\uparrow B\to\pi X}_{qq\to qq} - \mathrm{d}\sigma^{A^\downarrow B\to\pi X}_{qq\to qq} \propto \\
&\quad 2\,\mathrm{Im}[\hat{F}^{++}_{+-}(x_a,k_{\perp a})]\cos\phi_a \otimes \hat{f}_{b/B}(x_b,k_{\perp b}) \\
&\qquad\quad \otimes \left(|\hat{M}^0_{++;++}|^2 + |\hat{M}^0_{-+;-+}|^2 + |\hat{M}^0_{-+;+-}|^2\right) \otimes \hat{D}^\pi_{++} \\
&- \Big\{\hat{F}^{+-}_{+-}(x_a,k_{\perp a})\cos(2\phi''_c - \xi_a - \tilde{\xi}_a + \xi_b + \tilde{\xi}_b + \phi_a) \\
&\qquad - \hat{F}^{-+}_{+-}(x_a,k_{\perp a})\cos(2\phi''_c - \xi_a - \tilde{\xi}_a + \xi_b + \tilde{\xi}_b - \phi_a)\Big\} \\
&\qquad\quad \otimes 4\,\mathrm{Im}[\hat{F}^{+-}_{++}(x_b,k_{\perp b})] \otimes \hat{M}^0_{-+;+-}\hat{M}^0_{-+;-+} \otimes \hat{D}^\pi_{++} \qquad (22) \\
&+ \Big\{\hat{F}^{+-}_{+-}(x_a,k_{\perp a})\cos(\phi''_c + \phi^H_\pi - \xi_a - \tilde{\xi}_a + \xi_c + \tilde{\xi}_c + \phi_a) \\
&\qquad - \hat{F}^{-+}_{+-}(x_a,k_{\perp a})\cos(\phi''_c + \phi^H_\pi - \xi_a - \tilde{\xi}_a + \xi_c + \tilde{\xi}_c - \phi_a)\Big\} \\
&\qquad\quad \otimes \hat{f}_{b/B}(x_b,k_{\perp b}) \otimes \hat{M}^0_{++;++}\hat{M}^0_{-+;-+} \otimes 2\,\mathrm{Im}(\hat{D}^\pi_{+-}) \\
&- 8\,\mathrm{Im}[\hat{F}^{++}_{+-}(x_a,k_{\perp a})]\cos\phi_a \otimes \mathrm{Im}[\hat{F}^{+-}_{++}(x_b,k_{\perp b})] \\
&\qquad \times\cos(\phi''_c - \phi^H_\pi + \xi_b + \tilde{\xi}_b - \xi_c - \tilde{\xi}_c) \otimes \hat{M}^0_{++;++}\hat{M}^0_{-+;+-} \otimes \mathrm{Im}(\hat{D}^\pi_{+-})\,.
\end{aligned}
$$

With the help of Eqs. (13), (14) and (15) one can identify in this expression the Sivers, Boer-Mulders and Collins mechanisms. The terms $\hat{F}^{+-}_{+-}$ and $\hat{F}^{-+}_{+-}$ are in turn related to the distribution of transversely polarized quarks inside a transversely polarized hadron, the well-known transversity function.

3. Final hadron polarization in hadronic collisions

Within the same formalism we are able to calculate polarized cross sections for processes in which the final, spin one-half hadron (e.g. a Λ hyperon), is polarized. Our master formula, Eq. (12), is modified by the introduction of the helicity density matrix $\rho^\Lambda_{\lambda_\Lambda,\lambda'_\Lambda}$ of the observed hadron, describing its polarization state. This way, Eq. (12) becomes:

$$
\begin{aligned}
&\rho^\Lambda_{\lambda_\Lambda,\lambda'_\Lambda}\mathrm{d}\sigma^{(A,S_A)(B,S_B)\to\Lambda X} \propto \not\sum_{\lambda_X,X}\sum_{\lambda_c,\lambda'_c}\hat{\mathcal{D}}_{\lambda_\Lambda,\lambda_X,\lambda_c}\hat{\mathcal{D}}^*_{\lambda'_\Lambda,\lambda_X,\lambda'_c}\rho^{\Lambda/c}_{\lambda_c,\lambda'_c} \\
&= \sum_{a,b,c,d,\{\lambda\}}\rho^{A,S_A}_{\lambda_A,\lambda'_A}\hat{F}^{\lambda_a,\lambda'_a}_{\lambda_A,\lambda'_A}\otimes\rho^{B,S_B}_{\lambda_B,\lambda'_B}\hat{F}^{\lambda_b,\lambda'_b}_{\lambda_B,\lambda'_B}\otimes\hat{M}_{\lambda_c\lambda_d;\lambda_a\lambda_b}\hat{M}^*_{\lambda'_c\lambda_d;\lambda'_a\lambda'_b}\otimes\hat{D}^{\lambda_\Lambda,\lambda'_\Lambda}_{\lambda_c,\lambda'_c}\,.
\end{aligned}
\qquad (23)
$$

By choosing S_A and S_B, and performing the sum over partonic helicity indices (as discussed above), Eq. (23) allows one to compute all polarization states (longitudinal and transverse) for a final spin-1/2 hadron produced in (un)polarized hadron-hadron collisions. Notice also the appearance in the fragmentation sector of new terms with respect to the pion case, depending on the final hadron helicities.

4. Conclusions

Spin effects in inclusive high-energy hadronic reactions play an important role in our understanding of strong interactions.
We have presented a general approach to describe, within pQCD factorization schemes and using the helicity formalism, polarized inclusive particle production in high-energy hadronic collisions. By taking into account intrinsic motion of partons in the distribution and fragmentation functions and in the partonic process, this approach allows one to give explicit expressions for single and double spin asymmetries. This requires the introduction of a new class of spin and $\boldsymbol{k}_\perp$-dependent functions. The combined study of single and double spin asymmetries for different particles and in different kinematical situations may help in gathering information on these basically unknown functions. As an example, we have briefly discussed two interesting applications of this approach, namely $A_N(A^\uparrow B \to \pi + X)$ and final hadron polarization in (un)polarized hadronic collisions.

References

1. U. D'Alesio and F. Murgia, *Phys. Rev.* **D70** (2004) 074009
2. J.C. Collins, *Nucl. Phys.* **B396** (1993) 161
3. J.C. Collins, D.E. Soper and G. Sterman, *Nucl. Phys.* **B250** (1985) 199; J.C. Collins and D.E. Soper, *Nucl. Phys.* **B193** (1981) 381
4. X. Ji, J.-P. Ma and F. Yuan, e-Print Archive: hep-ph/0404183; *Phys. Lett.* **B597** (2004) 299
5. M. Anselmino, M. Boglione, U. D'Alesio, E. Leader and F. Murgia e-Print Archive: hep-ph/0408356 (*Phys. Rev. D*, in press)
6. M. Anselmino, M. Boglione, U. D'Alesio, E. Leader, S. Melis and F. Murgia in preparation
7. M. Anselmino, M. Boglione, U. D'Alesio, E. Leader and F. Murgia, *Phys. Rev.* **D70** (2004) 074025
8. D. Boer, P. Mulders and F. Pijlman, *Nucl. Phys.* **B667** (2003) 201
9. V. Barone, A. Drago and P. Ratcliffe, *Phys. Rep.* **359** (2002) 1
10. A. Bacchetta, U. D'Alesio, M. Diehl, C. Andy Miller, e-Print Archive: hep-ph/0410050 (*Phys. Rev. D*, in press)
11. For a pedagogical introduction to all the basics of helicity formalism, see, *e.g.*, E. Leader, *Spin in Particle Physics*, Cambridge University Press, 2001

^{3}HE STRUCTURE FROM COHERENT HARD EXCLUSIVE PROCESSES

S. SCOPETTA

Dipartimento di Fisica, Università degli Studi di Perugia, via A. Pascoli 06100 Perugia, Italy
and INFN, sezione di Perugia

Hard exclusive processes, such as deep electroproduction of photons and mesons off nuclear targets, could give access, in the coherent channel, to nuclear generalized parton distributions (GPDs). Here, a realistic microscopic calculation of the unpolarized quark GPD H_q^3 of the ^{3}He nucleus is reviewed. In Impulse Approximation, H_q^3 is obtained as a convolution between the GPD of the internal nucleon and the non-diagonal spectral function, describing properly Fermi motion and binding effects. The obtained formula has the correct limits. Nuclear effects, evaluated by a modern realistic potential, are found to be larger than in the forward case. In particular, they increase with increasing the momentum transfer and the asymmetry of the process. Another feature of the obtained results is that the nuclear GPD cannot be factorized into a Δ^2-dependent and a Δ^2-independent term, as suggested in prescriptions proposed for finite nuclei. The dependence of the obtained GPDs on different realistic potentials used in the calculation shows that these quantities are sensitive to the details of nuclear structure at short distances.

1. Introduction

Generalized Parton Distributions (GPDs)[1] parametrize the non-perturbative hadron structure in hard exclusive processes. Their measurement would represent a unique way to access several crucial features of the nucleon (for a comprehensive review, see, e.g., Ref. 2). According to a factorization theorem derived in QCD[3], GPDs enter the long-distance dominated part of exclusive lepton Deep Inelastic Scattering (DIS) off hadrons. In particular, Deeply Virtual Compton Scattering (DVCS), i.e. the process $eH \longrightarrow e'H'\gamma$ when $Q^2 \gg m_H^2$, is one of the the most promising to access GPDs (here and in the following, Q^2 is the momentum transfer between the leptons e and e', and Δ^2 the one between the hadrons H and H') [4]. Therefore, relevant experimental efforts to measure GPDs by means of DVCS off hadrons are likely to take place in the next few years. Recently, the issue of measuring GPDs for nuclei has been addressed. The first paper on this subject [5], concerning the deuteron, contained already the crucial observation that the knowledge of GPDs would permit the investigation of the short light-like distance structure of nuclei, and thus the interplay of nucleon and parton degrees of freedom in the nuclear wave function. In standard DIS off a nucleus with four-momentum P_A and A nucleons of mass M, this information can be accessed in the region where $Ax_{Bj} \simeq \frac{Q^2}{2M\nu} > 1$, being $x_{Bj} = Q^2/(2P_A \cdot q)$ and ν the energy transfer in the laboratory system. In this region measurements are difficult, because of vanishing cross-sections. As explained in Ref. 5, the same physics can be accessed in DVCS at lower values of x_{Bj}. Since then, DVCS has been extensively discussed for nuclear targets. Calculations, have been performed for the deuteron[6] and for finite nuclei [7]. The study of GPDs for ^{3}He is interesting for many aspects. In fact, ^{3}He is a well known nucleus, for which realistic studies are possible, so that conventional nuclear effects can be safely calculated. Strong deviations from the predicted behaviour could be ascribed to exotic effects, such as the ones of non-nucleonic degrees of freedom, not included in a realistic wave function. Besides, ^{3}He is extensively used as an effective neutron target, in DIS, in particular in the polarized case [8,9]. Polarized ^{3}He will be the first candidate for experiments aimed at the study of GPDs of the free neutron, to unveil details of its angular momentum content. In this talk, the results of an impulse approximation (IA) calculation[10] of the quark unpolarized GPD H_q^3 of ^{3}He are reviewed. A convolution formula is discussed and numerically evaluated using a realistic non-diagonal spectral function, so that Fermi motion and binding effects are rigorously estimated. The proposed scheme is valid for $\Delta^2 \ll Q^2, M^2$ and despite of this it permits to calculate GPDs in the kinematical range relevant to the coherent, no break-up channel of deep exclusive processes off ^{3}He. In fact, the latter channel is the most interesting one for its theoretical implications, but it can be hardly observed at large Δ^2, due to the vanishing cross section. The main result of this investigation is not the size and shape of the obtained H_q^3 for ^{3}He, but the size and nature of nuclear effects on it. This will permit to test directly, for the ^{3}He target at least, the accuracy of prescriptions which have been proposed to estimate nuclear GPDs[7], providing a tool for the planning of future experiments and for their correct interpretation.

2. Formalism

The formalism introduced in Ref. 11 is adopted. If one thinks to a spin 1/2 hadron target, with initial (final) momentum and helicity $P(P')$ and $s(s')$, respectively, two GPDs $H_q(x, \xi, \Delta^2)$ and $E_q(x, \xi, \Delta^2)$, occur. If one works in a system of coordinates where the photon 4-momentum, $q^\mu = (q_0, \vec{q})$, and $\bar{P} = (P + P')/2$ are collinear along z, ξ is the so called "skewedness", parametrizing the asymmetry of the process, is defined by the relation

$$\xi = -\frac{n \cdot \Delta}{2} = -\frac{\Delta^+}{2\bar{P}^+} = \frac{x_{Bj}}{2 - x_{Bj}} + O\left(\frac{\Delta^2}{Q^2}\right) , \tag{1}$$

where n is a light-like 4-vector satisfying the condition $n \cdot \bar{P} = 1$. One should notice that the variable ξ is completely fixed by the external lepton kinematics. The values of ξ which are possible for a given value of Δ^2 are $0 \leq \xi \leq \sqrt{-\Delta^2}/\sqrt{4M^2 - \Delta^2}$. The well known natural constraints of $H_q(x, \xi, \Delta^2)$ are: i) the so called "forward" limit, $P' = P$, i.e., $\Delta^2 = \xi = 0$, where one recovers the usual PDFs $H_q(x, 0, 0) = q(x)$; ii) the integration over x, yielding the contribution of the quark of flavour q to the Dirac form factor (f.f.) of the target: $\int dx H_q(x, \xi, \Delta^2) = F_1^q(\Delta^2)$; iii) the polynomiality property[11].

In Ref. 10, specifying to the ^{3}He target the procedure developed in Ref. 12, an IA expression for $H_q(x, \xi, \Delta^2)$ of a given hadron target, for small values of ξ^2, has been obtained.

$$H_q^3(x, \xi, \Delta^2) = \sum_N \int dE \int d\vec{p} [P_N^3(\vec{p}, \vec{p} + \vec{\Delta}, E) + O(\vec{p}^2/M^2, \vec{\Delta}^2/M^2)] \\ \times \frac{\xi'}{\xi} H_q^N(x', \xi', \Delta^2) + O\left(\xi^2\right) . \tag{2}$$

In the above equation, the kinetic energies of the residual nuclear system and of the recoiling nucleus have been neglected, and $P_N^3(\vec{p}, \vec{p} + \vec{\Delta}, E)$ is the one-body off-diagonal spectral function for the nucleon N in ^{3}He:

$$P_N^3(\vec{p}, \vec{p} + \vec{\Delta}, E) = \frac{1}{(2\pi)^3} \frac{1}{2} \sum_M \sum_{R,s} \langle \vec{P}' M | (\vec{P} - \vec{p}) S_R, (\vec{p} + \vec{\Delta}) s \rangle \\ \times \langle (\vec{P} - \vec{p}) S_R, \vec{p} s | \vec{P} M \rangle \, \delta(E - E_{min} - E_R^*) . \tag{3}$$

Besides, the quantity $H_q^N(x', \xi', \Delta^2)$ is the GPD of the bound nucleon N up to terms of order $O(\xi^2)$, and in the above equation use has been made of the relations $\xi' = -\Delta^+/2\bar{p}^+$, and $x' = (\xi'/\xi)x$.

The delta function in Eq. (2) defines E, the so called removal energy, in terms of $E_{min} = |E_{^3He}| - |E_{^2H}| = 5.5$ MeV and E_R^*, the excitation energy of the two-body recoiling system. The main quantity appearing in the definition Eq. (3) is the overlap integral

$$\langle \vec{P}M|\vec{P}_R S_R, \vec{p}s\rangle = \int d\vec{y}\, e^{i\vec{p}\cdot\vec{y}} \langle \chi^s, \Psi_R^{S_R}(\vec{x})|\Psi_3^M(\vec{x},\vec{y})\rangle \ , \tag{4}$$

between the eigenfunction Ψ_3^M of the ground state of ^{3}He, with eigenvalue $E_{^3He}$ and third component of the total angular momentum M, and the eigenfunction $\Psi_R^{S_R}$, with eigenvalue $E_R = E_2 + E_R^*$ of the state R of the intrinsic Hamiltonian pertaining to the system of two interacting nucleons[13]. Since the set of the states R also includes continuum states of the recoiling system, the summation over R involves the deuteron channel and the integral over the continuum states. Eq. (2) can be written in the form

$$H_q^3(x,\xi,\Delta^2) = \sum_N \int_x^1 \frac{dz}{z} h_N^3(z,\xi,\Delta^2) H_q^N\left(\frac{x}{z},\frac{\xi}{z},\Delta^2\right) \ , \tag{5}$$

where

$$h_N^3(z,\xi,\Delta^2) = \int dE \int d\vec{p}\, P_N^3(\vec{p},\vec{p}+\vec{\Delta})\delta\left(z+\xi-\frac{p^+}{\bar{P}^+}\right) \ . \tag{6}$$

In Ref. 10, it is discussed that Eqs. (5) and (6) or, which is the same, Eq. (2), fulfill the constraint $i) - iii)$ previously listed.

3. Numerical Results

$H_q^3(x,\xi,\Delta^2)$, Eq. (2), has been evaluated in the nuclear Breit Frame.

The non-diagonal spectral function Eq. (3), appearing in Eq. (2), has been calculated along the lines of Ref. 14, by means of the overlap Eq. (4), which exactly includes the final state interactions in the two nucleon recoiling system, the only plane wave being that describing the relative motion between the knocked-out nucleon and the two-body system [13]. The realistic wave functions Ψ_3^M and $\Psi_R^{S_R}$ in Eq. (4) have been evaluated using the AV18 interaction[15] and taking into account the Coulomb repulsion of protons in ^{3}He. In particular Ψ_3^M has been developed along the lines of Ref. 16. The other ingredient in Eq. (2), i.e. the nucleon GPD H_q^N, has been modelled in agreement with the Double Distribution representation[17]. In this model, whose details are summarized in Ref. 10, the Δ^2-dependence of H_q^N is given by $F_q(\Delta^2)$, i.e. the contribution of the quark of flavour q to the nucleon form factor. It has been obtained from the experimental values of the proton, F_1^p, and of the neutron, F_1^n, Dirac form factors. For the u and d flavours, neglecting the effect of the strange quarks, one has $F_u(\Delta^2) = \frac{1}{2}(2F_1^p(\Delta^2)+F_1^n(\Delta^2))$, $F_d(\Delta^2) = 2F_1^n(\Delta^2)+F_1^p(\Delta^2)$. The contributions of

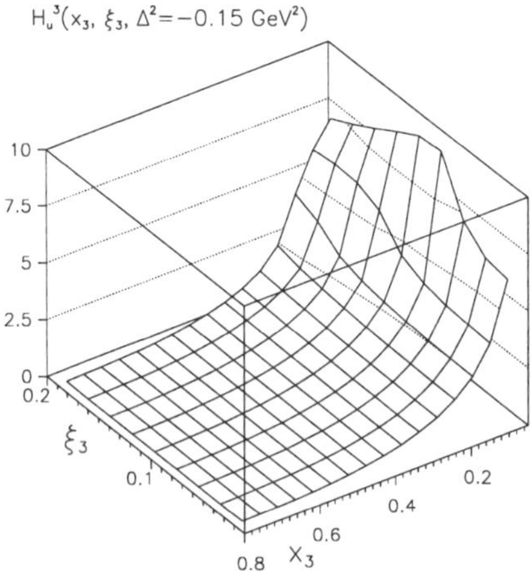

Figure 1. For the ξ_3 values which are allowed at $\Delta^2 = -0.15$ GeV^2, $H_u^3(x_3, \xi_3, \Delta^2)$, evaluated using Eq. (5), is shown for $0.05 \le x_3 \le 0.8$.

the flavours u and d to the proton and neutron f.f. are therefore $F_u^p(\Delta^2) = \frac{4}{3}F_u(\Delta^2)$, and $F_d^p = -\frac{1}{3}F_d(\Delta^2)$ and $F_u^n(\Delta^2) = \frac{2}{3}F_d(\Delta^2)$, $F_d^n(\Delta^2) = -\frac{2}{3}F_u(\Delta^2)$, respectively. For the numerical calculations, use has been made of the parametrization of the nucleon Dirac f.f. given in Ref. 18. Now the ingredients of the calculation have been completely described, so that numerical results can be presented. If one considers the forward limit of the ratio

$$R_q(x, \xi, \Delta^2) = \frac{H_q^3(x, \xi, \Delta^2)}{2H_q^p(x, \xi, \Delta^2) + H_q^n(x, \xi, \Delta^2)} , \tag{7}$$

where the denominator clearly represents the distribution of the quarks of flavour q in ^{3}He if nuclear effects are completely disregarded, i.e., the interacting quarks are assumed to belong to free nucleons at rest, the behaviour which is found, shown in Ref. 10, is typically EMC−like, so that, in the forward limit, well-known results are recovered. In Ref. 10 it is also shown that the x integral of the nuclear GPD gives a good description of ff data of ^{3}He, in the relevant kinematical region, $-\Delta^2 \le 0.25$ GeV^2. As an illustration, the result of the evaluation of $H_u^3(x, \xi, \Delta^2)$ by means of Eq. (2) is shown in Fig. 1, for $\Delta^2 = -0.15$ GeV^2 as a function of $x_3 = 3x$ and $\xi_3 = 3\xi$. The GPDs are shown for the ξ_3 range allowed and in the $x_3 \ge 0$

region. Let us now discuss the quality and size of the nuclear effects. The

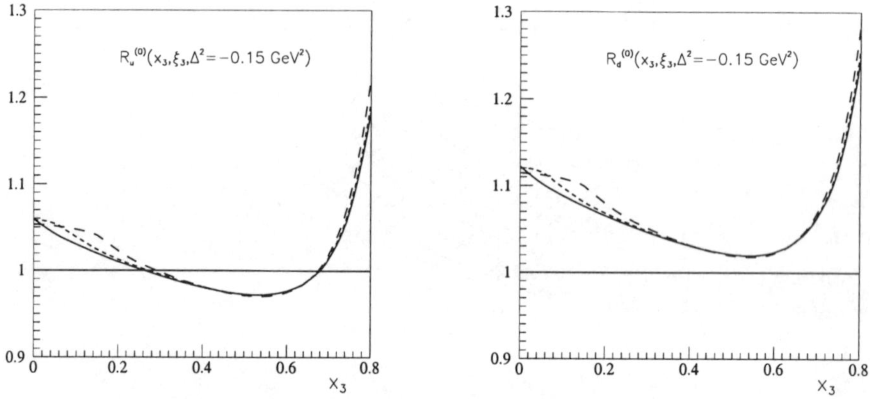

Figure 2. In the left panel, the ratio Eq. (9) is shown, for the u flavour and $\Delta^2 = -0.15$ GeV^2, as a function of x_3. The full line has been calculated for $\xi_3 = 0$, the dashed line for $\xi_3 = 0.1$ and the long-dashed one for $\xi_3 = 0.2$. The symmetric part at $x_3 \leq 0$ is not presented. In the right panel, the same is shown, for the flavour d.

full result for the GPD H_q^3, Eq. (2), will be now compared with a prescription based on the assumptions that nuclear effects are completely neglected and the global Δ^2 dependence can be described by the f.f. of ^{3}He:

$$H_q^{3,(0)}(x,\xi,\Delta^2) = 2H_q^{3,p}(x,\xi,\Delta^2) + H_q^{3,n}(x,\xi,\Delta^2)\ , \tag{8}$$

where the quantity $H_q^{3,N}(x,\xi,\Delta^2) = \tilde{H}_q^N(x,\xi)F_q^3(\Delta^2)$ represents the flavor q effective GPD of the bound nucleon $N = n,p$ in ^{3}He. Its x and ξ dependences, given by the function $\tilde{H}_q^N(x,\xi)$, is the same of the GPD of the free nucleon N, while its Δ^2 dependence is governed by the contribution of the quark of flavor q to the ^{3}He f.f., $F_q^3(\Delta^2)$.

The effect of Fermi motion and binding can be shown through the ratio

$$R_q^{(0)}(x,\xi,\Delta^2) = \frac{H_q^3(x,\xi,\Delta^2)}{H_q^{3,(0)}(x,\xi,\Delta^2)} \tag{9}$$

i.e. the ratio of the full result, Eq. (2), to the approximation Eq. (8). The latter is evaluated by means of the nucleon GPDs used as input in the calculation, and taking $F_u^3(\Delta^2) = \frac{10}{3}F_{ch}^3(\Delta^2)$, $F_d^3(\Delta^2) = -\frac{4}{3}F_{ch}^3(\Delta^2)$, where

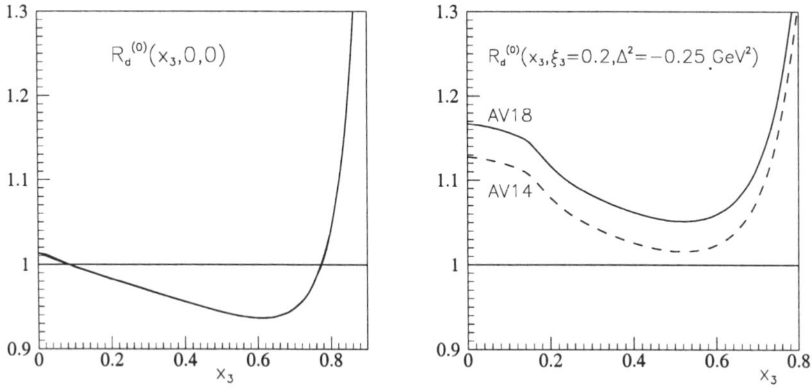

Figure 3. Left panel: the ratio $R^{(0)}$, for the d flavor, in the forward limit $\Delta^2 = 0, \xi = 0$, calculated by means of the AV18 (full line) and AV14 (dashed line) interactions, as a function of $x_3 = 3x$. The results obtained with the different potentials are not distinguishable. Right panel: the same as in the left panel, but at $\Delta^2 = -0.25$ GeV^2 and $\xi_3 = 3\xi = 0.2$. The results are now clearly distinguishable.

$F^3_{ch}(\Delta^2)$ is the f.f. which is calculated within the present approach. The coefficients 10/3 and −4/3 are simply chosen assuming that the contribution of the valence quarks of a given flavour to the f.f. of ^{3}He is proportional to their charge. The choice of calculating the ratio Eq. (9) to show nuclear effects is a very natural one. As a matter of fact, the forward limit of the ratio Eq. (9) is the same of the ratio Eq. (7), yielding the EMC-like ratio for the parton distribution q and, if ^{3}He were made of free nucleon at rest, the ratio Eq. (9) would be one. This latter fact can be immediately realized by observing that the prescription Eq. (8) is exactly obtained by placing $z = 1$, i.e. no Fermi motion effects and no convolution, into Eq. (2).

Results are presented in Fig. 2, where the ratio Eq. (9) is shown for $\Delta^2 = -0.15$ GeV^2 as a function of x_3, for three different values of ξ_3, for the flavours u and d.

Some general trends of the results are apparent:

i) nuclear effects, for $x_3 \leq 0.7$, are as large as 15 % at most.

ii) Fermi motion and binding have their main effect for $x_3 \leq 0.3$, at variance with what happens in the forward limit.

iii) nuclear effects increase with increasing ξ and Δ^2, for $x_3 \leq 0.3$.

iv) nuclear effects for the d flavour are larger than for the u flavour.

The behaviour described above is discussed and explained in Ref. 10. It is known that the point $x = \xi$ gives the bulk of the contribution to hard exclusive processes, since at leading order in QCD the amplitude for DVCS and for meson electroproduction just involve GPDs at this point. In Ref. 10 it is shown that also in this crucial region nuclear effects are systematically underestimated by the approximation Eq. (8). In Fig. 3, it is shown that nuclear effects are found to depend on the choice of the NN potential[19], at variance with what happens in the forward case. Nuclear GPDs turn out therefore to be strongly dependent on the details of nuclear structure.

The issue of applying the obtained GPDs to calculate DVCS off ^{3}He, to estimate cross-sections and to establish the feasibility of experiments, is in progress. Besides, the study of polarized GPDs will be very interesting, due to the peculiar spin structure of ^{3}He and its implications for the study of the angular momentum of the free neutron.

References

1. D. Müller, D. Robaschik, B. Geyer, F.M. Dittes, and J. Hořejši, Fortsch. Phys. 42, 101 (1994); hep-ph/9812448; A. Radyushkin, Phys. Lett. B 385, 333 (1996); X. Ji, Phys. Rev. Lett. 78, 610 (1997).
2. M. Diehl, Phys. Rept. 388, 41 (2003).
3. J.C. Collins, L. Frankfurt and M. Strikman Phys. Rev. D 56, 2892 (1997).
4. P.A. Guichon and M. Vanderhaeghen, Prog. Part. Nucl. Phys. 41, 125 (1998).
5. E.R. Berger et al., Phys. Rev. Lett. 87, 142302 (2001).
6. F. Cano and B. Pire, Nucl. Phys. A711, 133c (2002); Nucl. Phys. A721, 789 (2003); Eur. Phys. J. A19, 423 (2004).
7. V. Guzey and M.I. Strikman, Phys. Rev. C 68, 015204 (2003); A. Kirchner and D. Müller, Eur. Phys. J. C 32, 347 (2003).
8. J.L. Friar et al. Phys. Rev.C 42, 2310 (1990).
9. C. Ciofi degli Atti *et al.* Phys. Rev. C 48, 968 (1993).
10. S. Scopetta, Phys. Rev. C 70, 015205 (2004).
11. X. Ji, J. Phys. G 24, 1181 (1998).
12. S. Scopetta and V. Vento, Phys. Rev. D 69, 094004 (2004).
13. C. Ciofi degli Atti, E. Pace, and G. Salmè, Phys. Lett. B 141, 14 (1984).
14. A. Kievsky, E. Pace, G. Salmè, and M. Viviani, Phys. Rev. C 56, 64 (1997).
15. R.B. Wiringa, V.G.J. Stocks, and R. Schiavilla, Phys. Rev. C 51, 38 (1995).
16. A. Kievsky, M. Viviani, and S. Rosati, Nucl. Phys. A 577, 511 (1994).
17. A.V. Radyushkin, Phys. Lett. B 449, 81 (1999);
18. M. Gari and and W. Krümpelmann, Phys. Lett. B 173, 10 (1986).
19. S. Scopetta, nucl-th/0410057

INSTANTON-INDUCED CORRELATIONS IN HADRONS

P. FACCIOLI[12], M. CRISTOFORETTI[123] AND M. TRAINI[123]

[1] *E.C.T.*, Villazzano (Trento),* [2] *INFN, Gruppo Collegato di Trento,* [3] *Dipartimento di Fisica Università degli Studi di Trento.*

We discuss the role of non-perturbative spin- and flavor- dependent instanton-induced correlations in light hadrons. We show that the Instanton Liquid Model can reproduce the available data on proton and pion form factors at large momentum transfer and explain the delay of the onset of the perturbative regime in several exclusive reactions. The strong attraction generated by instantons in the $\bar{\mathbf{3}}_{\mathbf{c}}$ 0^+ diquark channel leads to a quantitative description of non-leptonic decays of hyperons and provides a microscopic dynamical explanation of the $\Delta I = 1/2$ rule.

1. Non-perturbative correlations in hadrons

Important information about the structure of the non-perturbative quark-quark interaction can be inferred from a comparative analysis of different electro-magnetic reactions involving light hadrons. Several recent experiments have shown that non-perturbative interactions survive also at relatively large momentum transfer, $Q^2 \gtrsim 2-6$ GeV^2. This means that there are *short-range* non-perturbative correlations in hadrons. Evidence for this fact has come, for example, from the measurement of the pion space-like form factor up to $Q^2 \simeq 2$ GeV^2 [1], and from the direct determination of the proton $G_E(Q^2)/G_M(Q^2)$ ratio up to $Q^2 \simeq 6$ GeV^2 [2]. In these reactions, the data are very far from the asymptotic perturbative QCD prediction.

A second piece of information that can be inferred from experiment is that such non-perturbative forces are very much channel dependent: they are effective in some process, but they are almost absent in some other. For example, the $\gamma\gamma^* \rightarrow \pi_0$ transition form factor and the Deeply Inelastic Scattering structure functions, follow the perturbative prediction, already for $Q^2 \gtrsim 1-2$ GeV^2.

The evidence for non-perturbative correlations at short distances (of the order of the 0.1 fm), rules-out models in which the quarks move essentially as free particle inside a confining bag and feel the non-perturbative color field only when they approach the edge of the hadron, i.e. for distances of the order $\sim$ 1 fm.

The channel dependence of the non-perturbative dynamics has been known for a long time: another example is the "Zweig rule" forbidding flavor-mixing, which works very well in the vector and axial vector meson channels, but it is violated in the scalar and pseudo-scalar channels. A strong channel-dependence of the non-perturbative correlations implies that the quark-quark interaction at low-energy has much more structure than a simple radial potential. It must at least be spin-dependent. Moreover, the rather large flavor asymmetry observed in DIS parton distributions suggests that it is also flavor-dependent. As a consequence, quarks are more correlated in some spin and flavor configuration than in other. Indeed, a number of phenomenological studies seem to indicate that quarks preferably correlate to form color- and flavor- anti-triplet scalar diquarks (see [3] and references therein).

An important testing ground for the spin-flavor structure of the non-perturbative quark-quark interaction is represented by weak-decays of hadrons. The natural scale of weak processes –set by W boson mass– is much larger than all other scales involved in the hadron internal dynamics. Hence, weak interactions can be regarded as effectively local and can resolve small structures inside hadrons. Moreover, their explicit dependence on flavor and chirality can be exploited to probe the spin and flavor structure of the non-perturbative QCD interaction. Among the large variety of weak hadronic processes, a prominent role is played by the non-leptonic decays of kaons and hyperons, which are characterized by the famous $\Delta\, I = 1/2$ rule. Neither electro-weak nor *perturbative* QCD interactions can account for the dramatic relative enhancement of the $\Delta\, I = 1/2$ decay channels. Its origin must therefore reside in the non-perturbative sector of QCD. Stech, Neubert and collaborators observed that the experimental data on both hyperon and kaon decays could be understood by assuming strong diquark correlations in the scalar, color anti-triplet channel [4].

The evidence for short-range, spin-dependent non-perturbative correlations in hadrons naturally leads to the problem of identifying their dynamical origin. To this end, it is instructive to analyze the non-perturbative scales in QCD. We know of at least two non-perturbative phenomena which occur at a momentum scale higher than Λ_{QCD}: the dynamical breaking of chiral symmetry and the anomalous breaking of the axial symmetry. The natural scale for the interactions related to chiral symmetry breaking is set by the mass of the lightest vacuum excitation which is not protected by chiral symmetry, the ρ-meson. Similarly, the typical scale of topological interactions is given by the mass of the η' meson. Hence, we should not be surprised to find non-perturbative effects at the GeV scale. The physical properties of the pion are certainly strongly influenced by the interactions responsible for the breaking of chiral symmetry. Presumably, these forces play an important role in the lightest baryons.

Instantons have been proposed long ago as the dynamical mechanism driving both the saturation of the axial anomaly [5] and the spontaneous breaking of chiral symmetry [6 7] (for a review see [8]). Physically, instantons are gluon fields which are generated during tunneling events between degenerate QCD vacua. Mathematically, they are non-perturbative solutions of the Euclidean Yang-Mills equation of motion. Being minima of the Yang-Mills action, they have been used in the context of a saddle-point (semi-classical) analysis of the Euclidean QCD path integral [5].

In the Instanton Liquid Model (ILM) the functional integral over all possible gluon field configurations is replaced by a sum over the configurations of an ensemble of instantons and anti-instantons. The path integral can then be solved by exploiting the formal analogy between the Euclidean QCD generating functional and the partition function of a grand-canonical statistical ensemble. The two phenomenological parameters in the model are the average instanton density $\bar{n} \simeq 1 \text{ fm}^{-4}$– which relates to the rate of tunneling in the vacuum – and the average instanton size $\bar{\rho} \simeq 1/3$ fm – which determines how long each tunneling event lasts for – . These values were extracted more than two decades ago from the global vacuum properties [6] and indicate that the diluteness (or "packing fraction") of the instanton ensemble is a small parameter: $\kappa = \bar{n}\,\bar{\rho}^4 \simeq 0.01$. Tunnelings are fairly rare events.

The instanton field induces an effective vertex between quarks ('t Hooft interaction) which, for $Q^2 \ll 1/\bar{\rho}^2$, reduces to a 2 N_f-leg contact interaction. For example, for $N_f = 2$ it reads:

$$\mathcal{L}_{'t\ H} = G_{\bar{\rho},\bar{n}} \left([(\psi^\dagger \tau_a^- \psi)^2 - (\psi^\dagger i\gamma_5 \tau^- \psi)^2] + \frac{1}{2(2N_c - 1)} (\psi^\dagger \sigma_{\mu\nu} \tau_a^- \psi)^2 \right), \tag{1}$$

where $\tau^- = (\vec{\tau}, i)$ ($\vec{\tau}$ are isospin Pauli matrices), and $G_{\bar{n}\,\bar{\rho}}$ is a coupling constant depending on the typical density and size of instantons in the vacuum. The finite size of the instanton field provides a natural cut-off scale for the 't Hooft interaction.

The non-perturbative instanton-induced interaction (1) has two important characteristic features: (i) it involves quarks of different flavor, (ii) it is chirality-mixing, i.e. quarks must flip their chirality any time they cross the field of an instanton. These two properties distinguish the instanton-induced interaction from the perturbative quark-gluon vertex, which is flavor-blind and chirality-conserving.

The chirality-flipping nature of the 't Hooft vertex allows to understand why *instanton effects are strongly channel dependent.* In fact, processes in which a quark undergoes a chirality-flip can occur any time there is an instanton nearby. In this case, the corresponding matrix elements will get a contribution from the instanton-induced interaction at the leading order in the instanton packing fraction, i.e. $o(\kappa)$. Conversely, transitions in which quarks do not change their chirality get contribution from instantons only at the next-to-leading order in the packing fraction, i.e. $o(\kappa^2)$. The physical reason is that the probability for two tunneling events to occur during the same scattering process is suppressed, if tunnelings are sufficiently rare events.

This mechanism explains the suppression of non-perturbative effects in several channels. A good example is the suppression of flavor-mixing in the vector and axial-vector meson channels (Zweig rule). The Lagrangian (1) induces flavor-mixing. Due to its chirality-flipping structure, its contribution to the scalar and pseudo-scalar channels is of $o(\kappa)$, so maximally enhanced. On the other hand, its contribution to the vector and axial-vector channels is of $o(\kappa^2)$, so much smaller.

In the following sections we shall show that the ILM quantitatively reproduces the JLAB data on electro-magnetic form factors and explains why the perturbative regime sets-in very early in DIS and in the $\gamma^* \gamma \to \pi_0$ transition form factor. Moreover, we will also show that instantons lead to a quantitative understanding of non-leptonic weak decays of hyperons and to an explanation of the $\Delta I = 1/2$ rule.

2. Instantons and the electro-magnetic structure of hadrons

Let us first discuss the instanton contribution to the electro-magnetic form factors. The framework to compute momentum-dependent hadronic matrix elements, from vanishing to large momentum transfer, has been recently developed in a number of papers [10 11 12 13 14]. At large momentum transfer the calculations of hadronic matrix elements in the ILM can even be carried-out analytically, by means of the Single Instanton Approximation (SIA) [15]. This approach exploits the fact that small-sized correlation functions are dominated by the interaction of the quarks with a single instanton. At small or vanishing momentum transfer, many-instanton effects are important and one has to rely on numerical Monte Carlo methods to compute the path integral and extract the matrix elements from appropriate ratio of correlation functions [14].

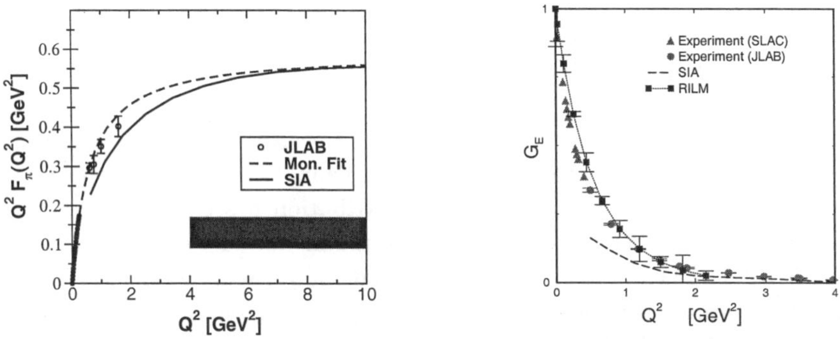

Figure 1.
LEFT PANEL: The JLAB data for $Q^2 F_\pi(Q^2)$ in comparison with the asymptotic perturbative QCD prediction (thick bar, for a typical $\alpha_s \sim 0.2 - 0.4$), the monopole fit (dashed line), and the SIA calculation (solid line). The SIA calculation is not reliable below $Q^2 \sim 1\,\text{GeV}^2$. The solid circles denote the SLAC data.
RIGHT PANEL: The electric form factor of the proton in the ILM and from experiment. Triangles are low-energy SLAC data, which follow a dipole fit. Circles are experimental data obtained from the recent JLAB result for G_E/G_M, assuming a dipole fit for the magnetic form factor. Squares are result of many-instanton simulations in the ILM, and the dashed line is the SIA curve.

The result of our analytic calculation of the pion form factor at moderate and large momentum transfer [13] is reported in Fig.1. The instanton contribution is $o(\kappa)$, i.e. maximally enhanced. The ILM can quantitatively reproduce the available data and explain the deviation from the perturbative regime at large momentum transfer.

Conversely, we have observed that in the $\gamma\gamma^* \to \pi_0$ transition form factor the instanton effects are of order $o(\kappa^2)$, hence parametrically suppressed by an additional power of the packing fraction. This explains the early onset of the perturbative regime in such a form factor. Moreover, a calculation of the pion light-cone distribution amplitude in the ILM was performed in [16]. It was found that instantons can explain the behavior of the low-energy experimental data ($Q^2 < 2$ GeV^2) for the $\gamma\gamma^* \to \pi^0$ transition form factor.

Our ILM calculation of the proton electric form factor $G_E(Q^2)$ is reported in Fig.1 (right panel), where it is compared with a SLAC data at low Q^2 and of the JLAB data at large-Q^2. Again, the short-range correlations induced by instantons quantitatively explain the experimental data. Similar results have been obtained also for the magnetic form factor [14].

Physically, the fact that instantons give very hard hadronic form factors can be interpreted as follows. Due to the strong zero-mode attraction, the hadron wave-function in coordinate space is very narrow and peaked around the origin. As a consequence, the charge distribution changes very rapidly at short distances, leading to large hard-momentum components in its Fourier transform (i.e. the form factor). However, due to the finite size of the instanton, the zero-mode attraction cannot transfer infinitely large momenta. Hence, the electric charge distribution has to become eventually flat, for distances much smaller than the instanton size. Indeed, when $Q^2 \gtrsim 10-15$ GeV^2, we have observed that the zero-mode contribution to the electro-magnetic correlation function rapidly dies-out and the perturbative regime is finally free to set-in. In conclusion, the ILM predicts that the perturbative limit will not be reached, in the kinematic region accessible to the forthcoming JLAB experiments.

Instanton also provide an explanation of why DIS perturbative evolution equations works so well, already for $Q^2 \gtrsim 1$ GeV^2. Lee, Weiss and Goeke [17] have shown that the instanton-contribution to the twist 2 and 3 operators in the Operator Product Expansion is parametrically suppressed by powers of the packing fraction κ, in complete analogy with what we found in the $\gamma\gamma^* \to \pi_0$ transition form factor. Moreover, Ostrovsky and Shuryak have recently shown that the ILM can explain the available data on azimuthal spin asymmetries in DIS [18].

3. Instantons and the $\Delta I = 1/2$ rule for hyperon decays

As we have already mentioned, non-leptonic weak decays of hyperons are good testing grounds for spin-dependent, non-perturbative instanton correlations.

The decay amplitudes can be parametrized in terms of two constants, corresponding to parity-violating and parity-conserving transitions:

$$\langle B' \, \pi | H_{eff} | B \rangle = i \, \bar{u}_{B'} \, [A - B\gamma_5] \, u_B. \tag{2}$$

H_{eff} is the effective Hamiltonian, which incorporates the electro-weak and the hard-gluon contributions, B (B') denotes the initial (final) baryon, and A and B are respectively called *S-wave* and *P-wave* amplitudes, each of

Table 2. Random Instanton Liquid Model prediction and experimental results for P-wave and S-wave amplitudes for non-leptonic weak decays of hyperons. Following the standard notation, B_q^Q corresponds to Amp($B^Q \to B' + \pi^q$).

Amplitude ($\times\ 10^7$)	P-wave (theory)	P-wave (experiment)	S-wave (theory)	S-wave (experiment)
Λ^0_0	-10.9 ± 1.17	-15.61 ± 1.4	-1.75 ± 0.34	-2.36 ± 0.03
Λ^0_-	17.71 ± 1.66	22.40 ± 0.54	2.25 ± 0.57	3.25 ± 0.02
Σ^+_0	22.4 ± 3.55	26.74 ± 1.32	-3.55 ± 0.64	-3.25 ± 0.02
Σ^+_+	31.84 ± 4.81	41.83 ± 0.17	0	0.14 ± 0.03
Σ^-_-	-1.52 ± 0.30	-1.44 ± 0.17	4.34 ± 0.90	$4..27 \pm 0.01$
Ξ^-_-	14.15 ± 2.75	17.45 ± 0.58	-4.22 ± 0.82	-4.49 ± 0.02
Ξ^0_0	-10.42 ± 1.95	-12.13 ± 0.71	3.20 ± 0.58	3.43 ± 0.06

which can be decomposed in $\Delta\ I = 1/2$ and $\Delta\ I = 3/2$ components. The $\Delta\ I = 1/2$ transition amplitudes are found to be typically $\sim$ 20 times larger that the $\Delta\ I = 3/2$ amplitudes ("$\Delta\ I = 1/2$" rule).

We have calculated these amplitudes in the ILM [19]. Our results are reported and compared with the experiment in table 1. We have found that not only instantons can explain the $\Delta\ I = 1/2$ rule, but also that the theoretical predictions for the amplitudes were in quantitative agreement with experiment. Note that a $\sim$ 20% discrepancy between theory and experiemt is of the same order of the uncertainty which was introduced by assuming the flavor $SU(3)$ limit in the calculation.

4. Conclusions

We have reported on the results of a series of investigations of the role played by instanton-induced correlations in the physics of light hadrons. We have shown that the ILM provides a qualitative and quantitative understanding on the global electro-weak structure of the pion, the proton and hyperons. In particular, it can explain the fact that the short-range non-perturbative correlations are very strong in some reactions and almost absent in some others.

References

1. J. Volmer *et al.* [The Jefferson Laboratory F_π Collaboration], Phys. Rev. Lett. **86** (2001) 1713.
2. M.K. Jones, *et al.*, Phys. Rev. Lett. **84** (2000) 1398. O. Gayou, *et al.*, Phys. Rev. Lett. **88** (2002) 092301.

3. R.L. Jaffe, hep-ph/0409065
4. B. Stech, Phys. Rev. D **36** (1987) 975. B. Stech and Q.P. Xu, Z. Phys. C **49** (1991) 491. H.D. Dosch, M. Jamin and B. Stech, Z. Phys. C **42** (1989) 167. M. Neubert, Z. Phys. C **50** (1991) 243. M. Neubert and B. Stech, Phys. Rev. D **44** (1991) 775.
5. G.'t Hooft, Phys. Rev. Lett., **37** (1976) 8. G.'t Hooft, Phys. Rev., **D14** (1976) 3432. G. 't Hooft, Phys. Rept. **142**, 357 (1986).
6. E.V. Shuryak, Nucl. Phys. **B214** (1982) 237.
7. D.J. Dyakonov and V.Yu. Petrov, Nucl. Phys., **B272** (1986) 457.
8. T. Schäfer and E.V. Shuryak, Rev. Mod. Phys. **70** (1998) 323.
9. R. L. Jaffe and F. Wilczek, Phys. Rev. Lett. **91**(2003) 232003.
10. P. Faccioli and E. V. Shuryak, Phys. Rev. D **65** (2002) 076002.
11. P. Faccioli, A. Schwenk and E. V. Shuryak, Phys. Lett. B **549** (2002) 93.
12. P. Faccioli, Phys. Rev. D **65**(2002) 094014.
13. P. Faccioli, A. Schwenk and E.V. Shuryak, Phys. Rev. **D67** (2003) 113009.
14. P. Faccioli, Phys. Rev. C **69** (2004) 065211.
15. P. Faccioli and E.V. Shuryak, Phys. Rev. **D64** (2001) 114020.
16. A. E. Dorokhov, JETP Lett. **77** (2003) 63 [Pisma Zh. Eksp. Teor. Fiz. **77**, 68 (2003)].
17. N. Y. Lee, K. Goeke and C. Weiss, Phys. Rev. D **65** (2002) 054008.
18. D. Ostrovsky and E. Shuryak, hep-ph/0409253.
19. M. Cristoforetti, P. Faccioli, E. V. Shuryak and M. Traini, Phys. Rev. **D70**(2004) 054016.

NUCLEAR DYNAMICS

G. POLLAROLO
Dipartimento di Fisica Teorica, Università di Torino
and INFN Sezione di Torino, Via Pietro Giuria 1, 10125 Torino, Italy
E-mail: nanni@to.infn.it

Some aspects of nuclear dynamics relevant for nuclear structure studies, like the excitation of the double giant resonances and the multi-nucleon transfer reactions, are summarized in some details by using a semi-classical approach that allows a clear separation between the relative motion and the intrinsic degrees of freedom.

1. Introduction

Nuclear dynamics deals with the different phenomena that appear when two heavy ions are brought in contact so that the strong nuclear forces that hold together a nucleus are felt by the nucleons of the other nucleus. What make this field of research so reach is that the nucleus presents both the degrees of freedom associated with the single-particle motion and the collective degrees of freedom associated to strong surface vibrations and rotations.

At moderate bombarding energies (up to 10-15 MeV/A) it is the interplay of these two classes of degrees of freedom that governs the evolution of the heavy-ion reaction from the Coulomb excitations and quasi-elastic transfer to the more complex fusion and deep inelastic reactions[1]. At higher bombarding energies (close to the Fermi energy or higher) the intrinsic degrees of freedom play a minor role and new phenomena, like multi-fragmentation, enter into play and allow to study the equation of state of nuclear matter[2]. Even at this high bombarding energies, by selecting very forward events, one can exploit the strong Coulomb force to study collective states at high excitation energies like the giant isoscalar and isovector modes[3]. By appropriately selecting targets and projectiles one may be able to excite different degrees of freedom and search for new one. For instance from the study of multi-nucleon transfer reactions one should be able to understand the role of pair correlations in the nuclear media and thus to learn more on the paring interaction.

Many different theoretical models have been developed for the analysis of heavy ion reactions. The most successful in giving an overall description employ the Fokker-Plank diffusion equation where the exchange of nucleons is described in term of transport and diffusion coefficients (for a review see ref.[4]). In these models no special treatment of the grazing reactions is done. This kind of reactions are incorporated by, simply, using a radial dependence of the transport coefficients. But there is a great body of evidence showing that the quasi-elastic reactions form a well defined class where the structure of the individual partners of the collision plays an important role and where a coherent description should be possible.

This contribution will summarize some of the main researches in the field of nuclear dynamics that have been carried out in Italy in the past few years and it will be mostly concerned on Coulomb excitation and quasi-elastic transfer reactions. After a short introduction of the model used in the analysis of the data, a more detail discussion will be given to topics about the double phonons excitations of the giant isoscalar and isovector modes and to some results on transfer reactions discussing in particular the multi-nucleon transfer reactions for their relevance in the understanding of pair correlation in the nuclear medium.

2. The model

The grazing collisions of two heavy ions are well described, in a coupled channel treatment, by using as base vectors the product of the asymptotic states wave functions of the two nuclei. By utilizing the fact that the wave length associated to the relative motion is much shorter than the interaction region one can use the semi-classical approximation and write the following system of coupled equations:

$$i\hbar\dot{c}_\beta(t) = \sum_\alpha < \beta|V_{int}|\alpha > c_\alpha(t) e^{\frac{i}{\hbar}(E_\beta - E_\alpha)t + i(\delta_\beta - \delta_\alpha)} \tag{1}$$

where the coefficients $c_\beta(t)$ represent the amplitude that the system is, at time t, in the channels $\beta \sim (b, B)$. This is described by the product of the intrinsic states wave function ψ_b and ψ_B that are solution of the Schödinger equations:

$$\hat{H}_b\psi_b = \left[\sum_i \epsilon_i a_i^\dagger a_i + \sum_{\lambda\mu} \hbar\omega_\lambda \Gamma^\dagger_{\lambda\mu}(b)\Gamma_{\lambda\mu}(b)\right]\psi_b = E_b\psi_b \tag{2}$$

and similarly for the other nucleus B. The operators $a_i^\dagger$ (a_i) create(annihilate) a particle in the single particle level i of energy ϵ_i and the operators $\Gamma^\dagger_{\lambda\mu}$ $(\Gamma_{\lambda\mu})$ create(annihilate) a phonon of energy $\hbar\omega_\lambda$. The first term in the Hamiltonian thus describes the single particle motion while the second the surface vibrations.

The semi-classical phases δ_α and δ_β have to be introduced to take into account that the relative motion of the two ions is not a straight line but follows a classical trajectory in a nuclear plus Coulomb field. It is trough these phases that the semi-classical description incorporates the recoil effects.

The coupling Hamiltonian V_{int}, responsible for the excitation on the surface modes and for the exchange of nucleons among the reactants, is written as:

$$\hat{V}_{int}(t) = \hat{V}_{tr}(t) + \hat{V}_{in}(t) + \Delta U_{bB}(t) \tag{3}$$

where the transfer part $(\hat{V}_{tr})$ has the form:

$$\hat{V}_{tr} = \sum_{(\nu\pi)j,k} f^{kj'}(\mathbf{r}) a_k^\dagger a_{j'} \; + \; h.c. \tag{4}$$

with the summation extended to all active single particle states for neutrons (ν) and protons (π) in target and projectile, with $f^{kj'}$ the form-factor for the transfer of a nucleon from the single particle state j' in the projectile to the single particle state k in the target. The inelastic excitation term $(\hat{V}_{in})$, for the case of the target (B), is written as:

$$\hat{V}_{in} = \sum_{\lambda\mu} f^B_{\lambda\mu}(\mathbf{r}) \Big(\Gamma^\dagger_{\lambda\mu}(B) + \Gamma_{\lambda\mu}(B) \Big) \; + \; (b) \tag{5}$$

where $f^B_{\lambda\mu}(\mathbf{r})$ is the inelastic form-factor for the excitation of the surface modes. The last term in the coupling Hamiltonian, ΔU_{bB}, takes into account the re-normalization of the ion-ion potential due to the transfer processes. In the above expressions with $\mathbf{r}$ we have indicated the ion-ion vector distance. The time dependence of the different operator is obtained by solving the newtonian equations for the relative motion of the two ions in a Coulomb plus nuclear potentials.

For the inelastic excitation one usually uses the form factor derived from the macroscopic model thus for the isoscalar modes (of the target) one has the expression:

$$f^B_{\lambda\mu}(\mathbf{r}) = \left(\beta_B \frac{\partial U_{bB}(r)}{\partial r} + \frac{4\pi Z_b e}{2\lambda+1} \frac{\sqrt{B(E2)}}{r^{\lambda+1}} \right) Y^*_{\lambda\mu}(\hat{r}) \tag{6}$$

where β_B is the deformation length:

$$\beta_B = R_0 < 1_{\lambda\mu}|\alpha^B_{\lambda\mu}|0 > \tag{7}$$

and $U_{bB}(r)$ is the nuclear part of the potential that defines the trajectory of relative motion. The last term in (6) is the well known matrix element of the Coulomb excitation.

The form factor for stripping of neutrons or protons are calculated from the single particle wave function in projectile and target, namely:

$$f_{i'j}(\mathbf{r}) \sim \int d^3r' \psi^\dagger_{i'}(\bar{r}') V_a(|\bar{r}'|) \psi_j(\bar{\mathbf{r}} - \bar{r}') \tag{8}$$

where $V_b(|\bar{r}'|)$ is the shell model potential that bound the transferred nucleon to the projectile. For the pick-up reaction, very similar expressions hold.

The cross section for a given transition to the final channels β is written from the amplitude, solution of the system (1), and from the elastic scattering angular distribution as:

$$\frac{d\sigma_{\alpha\beta}}{d\Omega} = \frac{d\sigma_{el}}{d\Omega} \left|c_{\alpha\beta}(\theta)\right|^2 \tag{9}$$

being α the entrance channel. The amplitude has to be calculated along the trajectory leading to the scattering angle θ.

In what follows the above system of coupled equation is used to analyze the Coulomb excitation of the double giant states at relativistic energy and to analyze some multi-nucleon transfer reactions.

3. Multi-phonon excitation

The giant resonances (GR) are collective states at high excitation energy that involve a coherent motion of many nucleons (in first approximation they are a superposition of particle-hole states) that are characterized by a spreading width much larger than the decay width (γ, particle emission and fission). This spreading width accounts for coupling to more complicated configurations. In the collective model these modes are seen as surface vibrations that, due to their small amplitude, should have an harmonic character. In this approximation the two phonons state has an energy that is twice the one of the one-phonon and a width larger by a factor $\sqrt{2}$. The coupling Hamiltonian (5) may excite the double phonon state via a successive mechanism with a probability related to the one of the one-phonon.

While the double phonon states build on the low lying collective surface vibrations have been recognized in several nuclei it is only recently that the excitation of the double giant resonances (DGR) have been identified. These states have been seen in pion scattering[5] via double charge-exchange reactions (for the dipole isovector mode) and in Coulomb excitation[6]. Some evidence for the excitation of the double isoscalare modes is also been established in heavy-ion collision at intermediate energy[7].

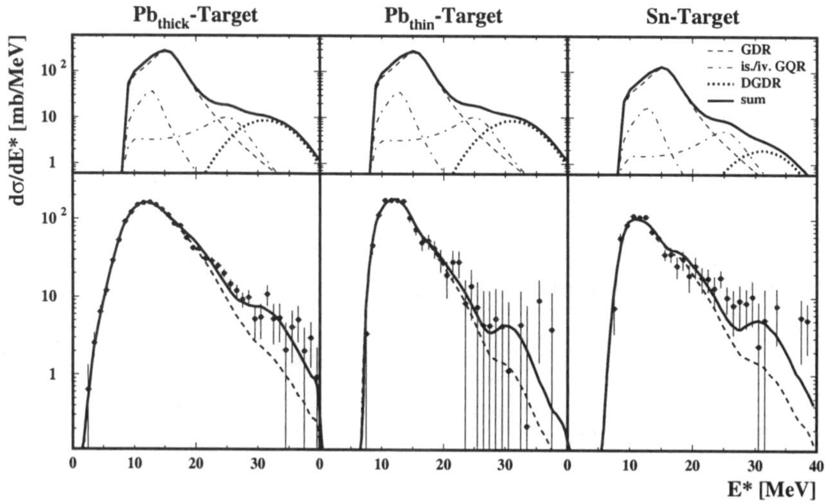

Figure 1. In the lower panel the experimental energy spectra for the very forward events in the collision of ^{136}Xe at 700 MeV/A on the indicated target. The dash line indicates an analysis of the data where only the one phonon states of the GDR, isoscalar and isovector GQR are included. The full line indicates when the two phonon states contribution are added. In the upper panel the contribution of the different giant states is shown. For more details cfr. ref.[6]

To summarize the properties of these double phonon states, in Fig. 1, are shown the energy spectra of very forward scattering events for the reaction of ^{136}Xe on several targets at a bombarding energy of 700 MeV/A[6]. The forward-angle selection has been done to ensure that only Coulomb excitations took place. The analysis of the data has been carried out with the relativistic Coulomb excitation model of ref.[3] by including the contributions from the dipole isovectore mode and the quadrupole isovector and isoscalar modes. It is clear that the high energy part of the spectra could be fitted only by adding the contribution of the double phonon of the isovectore dipole mode. A systematic analysis of all the experiments leads to the conclusion that the double phonon has an energy in reasonable agreement with the harmonic approssimation (the energy is in some cases 2-3 MeV lower) and its width is a factor 1.5~2 larger than the one phonon states but the harmonic approximation underpredicts the cross section by a factor ranging from 1.3 to 2.

To overcome these discrepancies the Hamiltonian of eqs. (2) and (5) have been modified to include anharmonic terms in the intrinsic Hamiltonian and non linear terms in the couplings[8,9,10] thus allowing a direct path for the excitation of double phonon states. This is done by mapping the RPA phonons into boson operators[12]. As it is well known, in the RPA, the excited states of a nucleus are described as a superposition of particle-hole states build on the nuclear ground state $|0>$, thus the one phonon state is written:

$$|1>= \Gamma_\nu^\dagger |0> \tag{10}$$

where the quasi-phonon operator $\Gamma_\nu^\dagger$ is:

$$\Gamma_\nu^\dagger = \sum_{(ph)} \left(X_{ph}^\nu a_p^\dagger a_h + Y_{ph}^\nu a_h^\dagger a_p \right) \tag{11}$$

with the forward and backward amplitudes solutions of the RPA equations.

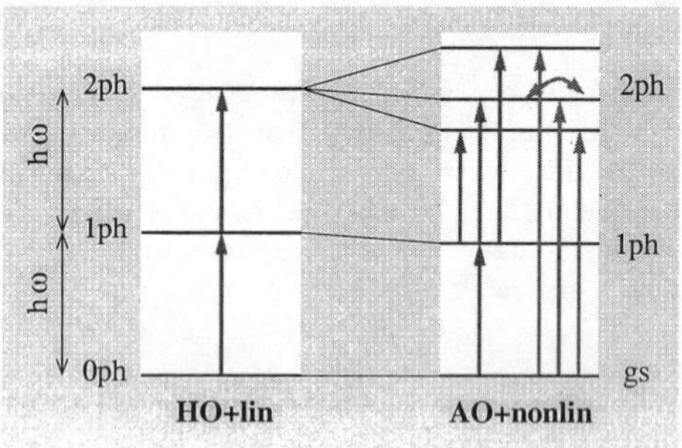

Figure 2. Schematic rapresentation of the mapping procedure that generalizes the harmonic plus linear couplingn to the anharmonic oscillator with non linear couplings.

By mapping the RPA phonons in the boson operators space:

$$a_p^\dagger a_h \to B_{ph}^\dagger + (1-\sqrt{2}) \sum_{p'h'} B_{p'h'}^\dagger B_{p'h}^\dagger B_{ph'} + \dots \tag{12}$$

one obtains for the intrinsic Hamiltonian:

$$H = \sum_{\mu} E_{\mu} Q^{\dagger}_{\mu} Q_{\mu} + \left[\sum_{\mu_1,\mu_2\mu_3} V^{21}_{\mu_1\mu_2\mu_3} Q^{\dagger}_{\mu_1} Q^{\dagger}_{\mu_2} Q_{\mu_3} + \sum_{\mu_1,\mu_2\mu_3\mu_4} V^{22}_{\mu_1\mu_2\mu_3\mu_4} \cdots \right. \tag{13}$$

Here beside the harmonic component (the first term) appear non linear terms containing matrix elements that couple one- with two-phonons states (V^{21}) and two- with two-phonons (V^{22}). The eigenstates of this Hamiltonian are mixed states of one- and two- phonons and their corresponding eigenvalues are not harmonic.

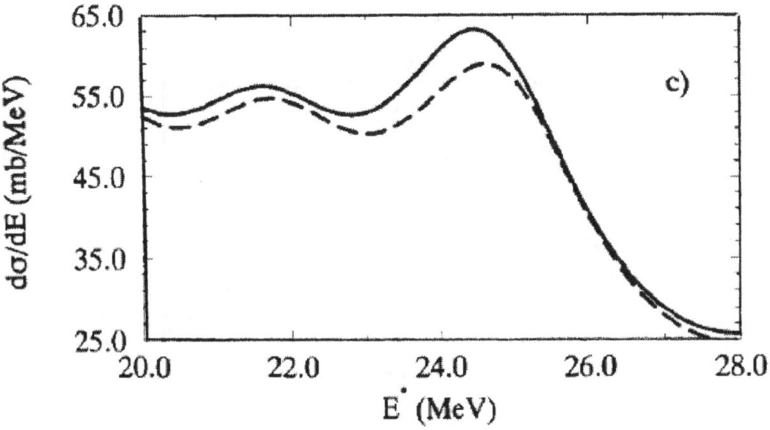

Figure 3. Relativistic Coulomb target excitation for the ^{208}Pb+^{208}PB system at 641 MeV/A as a function of the excitation energy in the region of the DGDR. The full line corresponds to the calculation where anharmonicity and non linearity are taken into account. The dash line corresponds to the harmonic approximation with linear coupling. The cross section for each state $|\psi_\alpha>$ has been smoothed with a Lorentzian with a 3 MeV width.

In the above expression we have introduced the new operators $Q^{\dagger}_{\nu}$ e Q_{ν} defined as:

$$Q^{\dagger}_{\nu} = \sum_{(ph)} \left(X^{\nu}_{ph} B^{\dagger}_{ph} + Y^{\nu}_{ph} B^{\dagger}_{hp} \right) \tag{14}$$

and similarly for Q_{ν}

By applying the same mapping procedure to the interaction Hamiltonian (5) one obtains the following results

$$V_{int} = W^{00} + \sum_{\mu} W^{10}_{\mu} Q^{\dagger}_{\mu} + \sum_{\mu\mu'} W^{11}_{\mu\mu'} Q^{\dagger}_{\mu} Q_{\mu'} + \sum_{\mu\mu'} W^{20}_{\mu\mu'} Q^{\dagger}_{\mu} Q^{\dagger}_{\mu'} + \dots \quad (15)$$

The first term represents the ground-to ground state interaction between the two ions. The W^{10} terms connect states differing by one phonon, the W^{11} connect states with the same number of phonons while the W^{20} terms allow transitions from the ground state to the two phonons states (cfr. Fig. 2). Clearly this direct path (not present in the linear coupling) may increase the cross section for the two-phonons state.

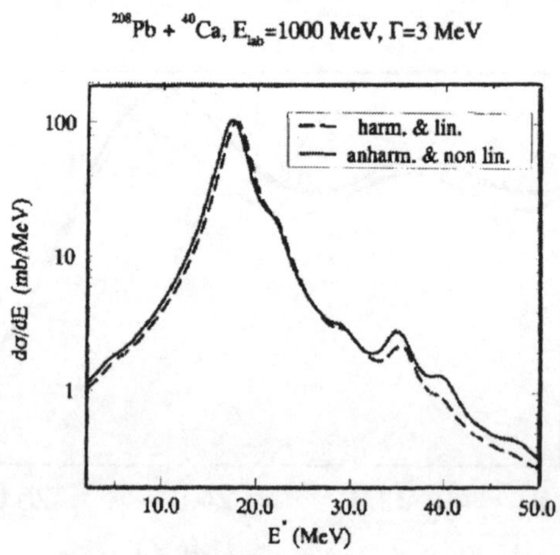

Figure 4. Relativistic Coulomb target excitation for the ^{208}Pb+^{40}Ca system at 1GeV/A as a function of the excitation energy. The cross section for each state $|\psi_\alpha>$ has been smoothed with a Lorentzian with a 3 MeV width.

In this model the calculation proceeds by first running an HF+RPA calculation with an SGI interaction to construct the base vectors for the diagonalization of the intrinsic Hamiltonian (13). All the roots of the RPA equation with angular momentum less or equal to 3 and with an energy weighted sum rule (EWSR) larger than 5% are used to construct all possible combination of two-phonon states and in this combined space the Hamiltonian (13) is diagonalized to obtain the intrinsic states $|\psi_\alpha>$ of the two ions.

The cross section are calculated by solving the semiclassical system of coupled equations (1) constructed by expanding the solution of the

Schrödinger equation in the base vector $|\psi_\alpha >$ for each impact parameter b:

$$\sigma_\alpha = 2\pi \int_0^{+\infty} |c_\alpha(b, t = +\infty)|^2 T(b) b db \tag{16}$$

where the transmission coefficient $T(b)$ has been taken equal to a sharp cutoff function $\theta(b - b_{min})$ being b_{min} chosen to ensure that the nuclear interaction does not contribute. Since these calculations are performed at very high energy the relative motion is approximated by a straight line.

To discuss the application of the above model to actual case let us start with the excitation of ^{208}Pb in a collision with ^{208}Pb at 641 MeV/A. Since there are two different aspects, the anharmonicity and non linearity let start with few words about anharmonicity. In all cases the anharmonicities predicted by the microscopic Hamiltonian (13) are quite small, the energy shift being of the order of few hundreds keV. In Fig. 3 is shown the result for the excitation of ^{208}Pb in the region of the double giant dipole resonance (DGDR) in comparison with the traditional calculation. As it is seen this model predicts a 10% increase of the cross section. In the case of ^{208}Pb+^{40}Ca system at 1GeV/A (cfr. Fig. (4)) the increase is of the order of a 20% bringing the calculation much closer to the experimental data.

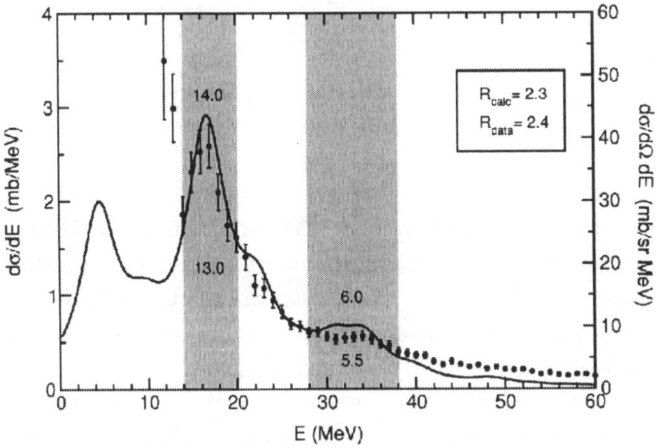

Figure 5. Comparison between experimental coincidence inelastic spectrum (right scale) with theoretical calculations (left scale). For more details cfr. ref. [10].

The above formalism can be generalized for reaction at lower energy. In this case the coupling term have to be generalized to include the nuclear component of the interaction as in (6). Also the trajectory calculation has to be done explicitly by solving the Newtonian equation of motion. Along the trajectory an imaginary part of the interaction has to be included to avoid the uncertainty of the integration over small impact parameters. Fig. (5) shows the comparison for the experimental coincidence inelastic spectra for the collision of ^{40}Ca + ^{40}Ca at 50 MeV/N. The theoretical curve, calculated with the above model, has been smoothed-out with a Lorentzian of 5 MeV width. Notice that the experimental data have no background subtraction.

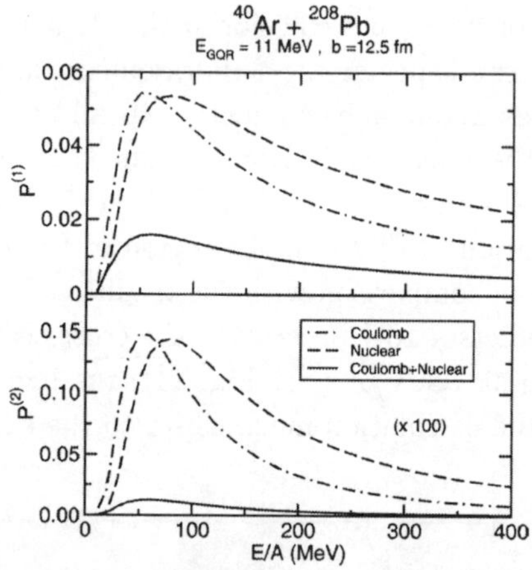

Figure 6. Excitation probability as a function of bombarding energy of one- and two-phonon states of a GQR at 11 MeV for the indicated reaction and at the indicated impact parameter.

The role of the nuclear coupling in the excitation of the GRs and DGRs and its interplay with the long range Coulomb component have been investigated in grait detail in ref.[13] where also the relation of line-shape of the excitation function to the spreading width of these states is discussed. This analysis has been done by studying the inelastic cross section as a function of the different parameters that specify the intrinsic states by utilizing the semiclassical model discussed above. Instead of solving the system of coupled equations (1) a perturbative approach has been used. The amplitude to excite the μ component of the phonon of multipolarity λ is written as:

$$a^{(1)}_{\lambda\mu} = -\frac{i}{\hbar}\int_{-\infty}^{+\infty} dt f_{\lambda\mu}(\mathbf{r}(t)) e^{\frac{i}{\hbar}E_1 t} \quad (17)$$

where the integral has to be evaluated along the classical trajectory $\mathbf{r}(t)$,

E_1 is the energy of the GR and $f_{\lambda\mu}$ is the inelastic form factor (6). In a similar way the probability to excite the two-phonons state (DGR) with angular momentum L and projection M :

$$a_{LM}^{(2)} = \frac{1}{\hbar^2} \sum_{\mu} G_{M\mu} \int_{-\infty}^{+\infty} dt f_{\lambda M-\mu}(\mathbf{r}) e^{\frac{i}{\hbar}(E_2-E_1)t} \int_{-\infty}^{t} dt' f_{\lambda\mu}(\mathbf{r}) e^{\frac{i}{\hbar}(E_1)t'} \tag{18}$$

being E_2 the energy of the two-phonons state and $G_{M\mu} = \sqrt{(1+\delta_{\mu,M-\mu})}$ a geometrical factor. Once the amplitude are known the cross sections are calculated in the usual manner by using (9) with the introduction of an absorption to take into account the de-population on the initial mass partition due to other reaction channels.

The bombarding energy dependence of the probabilities for the excitation of one- and two-phonons states is shown in Fig. 6. After a rapid increase of the probabilities of excitation that reach their maximum at around 50 MeV/A a gradual decline sets in until an almost constant behavior is reached at the higher energies. At still higher energies a relativistic approach should be used but the trend will not be altered. This beaviour is quite clear since the interaction time reduces with the increase of the bombarding energy up to a point where all the states are less and less favored (cfr. Fig. 7 for the Coulomb interaction).

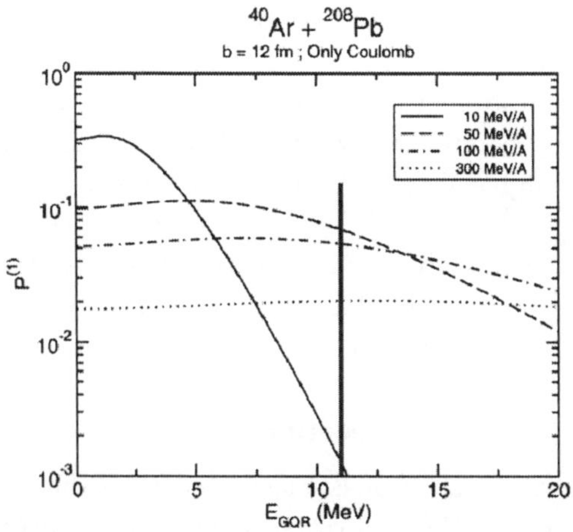

Figure 7. Probability for the excitation of one phonon as a function of the phonon energy for several incident energies. The vertical line indicates the actual position of the GQR.

An interesting aspect of the inelastic process emerges by looking at the bombarding energy dependence of the excitation of the GR and DGR shown in Fig. 8. In fact one should notice that at all bombarding energies the one-phonon state is dominated by the Coulomb formfactor while the two-phonons state is dominated by the nuclear coupling.

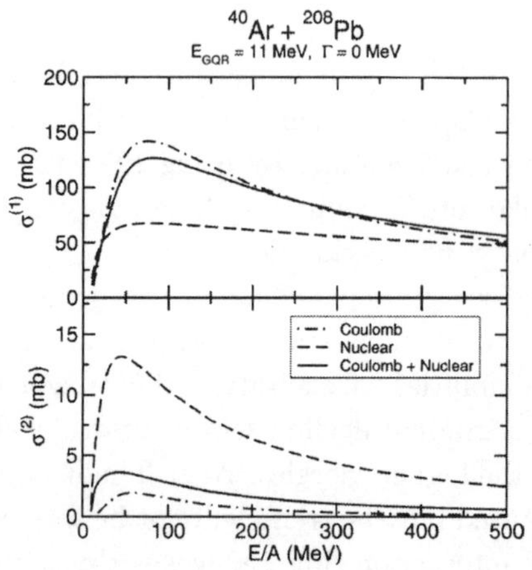

Figure 8. Excitation cross section for the GR and the DGR as a function of incident energy.

Since Q-value considerations, as seen in Fig. 7, have a pronounce effect on the excitation probabilities it is expected that they will play an important role in the excitation mechanism since the GR have a sizable width and the line-shape may be greitly altered. This is seen clearly in Fig. 9 where the Q-value effects give also rise to a shift of the maxima of the excitation function.

From this detailed analysis it is clear that the excitation process of the GR and DGR needs to be further studied before any relaible conclusions on the importance of anharmonic terms or non linear coupling are drown.

4. Multi-nucleon transfer reactions

Among the different kinds of quasi-elastic reactions the exchange of nucleons are peculiar in that they convey specific informations on nuclear properties like single particle levels and correlations, of the two colliding nuclei. For example, the pairing model received considerable inputs from the extensive experimental work with (p, t) and (t, p) reactions. With heavy-ions

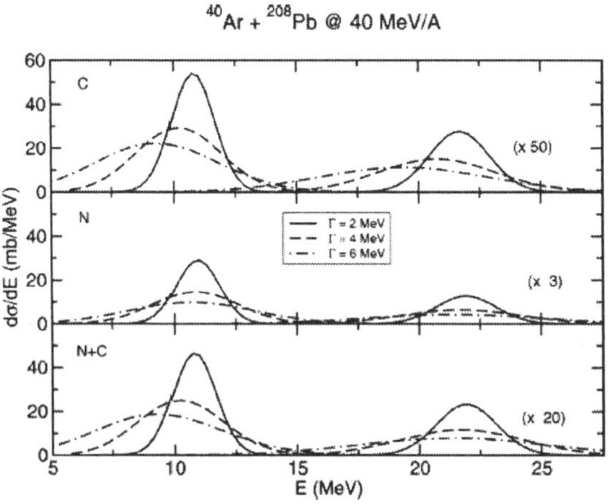

Figure 9. Excitation function for the indicated reaction for three values of the width of the states. The contributions of the Coulomb (C), nuclear (N) and total (C+N) are shown.

these studies may be further extended since the two partners may exchange many nucleons (neutrons and protons) and thus one should be able, at least in principle, to measure the pair density in the nuclear medium.

From the reaction mechanism point of view the study of multi-nucleon transfer reactions is also very important since one learns about which degrees of freedom have to be included in any model in order to describe the evolution of the heavy ion reaction from the quasi elastic to the deep-inelastic regimes and to fusion. Discussions are still going on about the role played by transfer degrees of freedom in the enhancement of the fusion cross section at very low energies.

To illustrate the problematic encountered in the analysis of multi-nucleon transfer reactions we use the results of a recent experiment performed at the INFN National Laboratory of Legnaro (Italy)[14,15] where the isotopic distribution of multi-nucleon transfer reaction has been measured for the ^{40}Ca + ^{208}Pb reaction at several incident energies. In Fig. 10 are shown the total cross sections for pure proton stripping and pure neutron pick-up channels obtained by integrating the angular and Q-value distributions for each isotope. The cross sections for the neutron pick-up drop by almost a constant factor for each transferred neutron, as an independent particle mechanism would suggest. The pure proton cross sections behave differently, with the population of the $-2p$ channel as strong as the $-1p$.

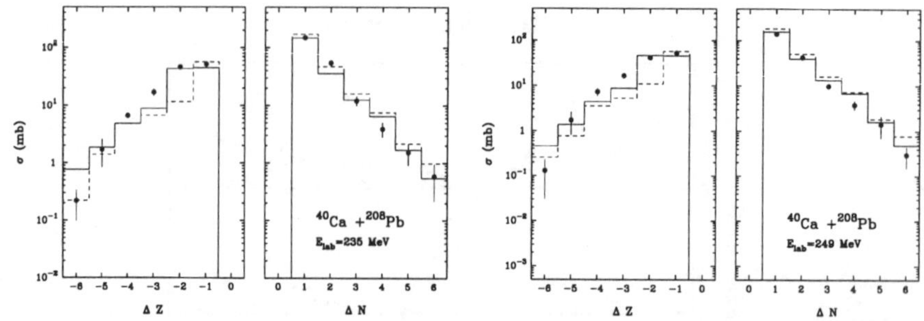

Figure 10. Inclusive cross sections for pure proton stripping and pure neutron pick-up channels for the ^{40}Ca+^{208}Pb reaction at the indicated energies. The histograms indicate theoretical calculation (see text).

This may, at first, indicates the contribution of processes involving the direct transfer of proton pairs in addition to the successive transfer of single protons.

This proton and neutron asymmetric behavior, present in all the reactions insofar studied, should not lead us to think that pairing correlations are more important for protons then for neutrons. Nuclear structure calculations have, in fact, shown that the paring interaction has the same strength. Since the one-neutron transfer cross section is almost one order of magnitude larger than the one-proton transfer the contribution of pair transfer mode is masked, in the neutron sector, by the successive mechanism. It is clear that any model that want to analyze these reactions must incorporate both proton-pair and neutron pair modes together with the one nucleon transfer. The model should also be able to incorporate evaporation since there is evidence that this process plays an important role in altering the isotopic distribution of the fragments toward the lighter one. Here is worth to remember that the transfer process is dominated by Q-values consideration and these, for stable nuclei, favor only neutron pick-up and proton stripping reactions.

The data are analyzed by using a semiclassical model that generalizes the one described above in that it uses a WKB approximation for the description of relative motion[16]. The formalism uses the same approximation utilized to calculate the absorbitive and polarization potentials. It incorporates all the one-particle transfer channels connecting all the single particle levels around the Fermi surface of projectile and target. It calculates the multi-nucleon transfer by a successive approximation and incorporates the pair transfer mode by utilizing the macroscopic approximation of the form-

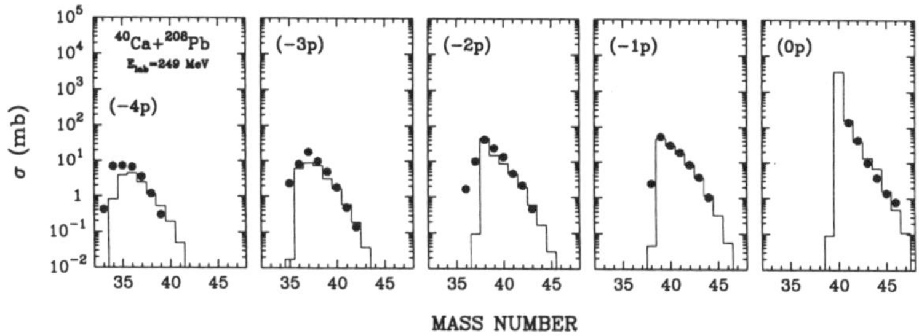

Figure 11. Inclusive cross sections for all the isotopes for the indicated charge transfer channels for the ^{40}Ca+^{208}Pb reaction at 249 MeV. The histograms are calculations.

factor. The results of such calculations are shown with histograms on the same figure. The full line indicates the results when a pair transfer mode is included both for protons and for neutrons (they have the same strength), the dash line is the results of a calculation when the contribution of the pair-modes has been neglected. We have also to remind that the shown cross sections are corrected by evaporation. This can be done since the model is able to calculate the excitation energy of the fragments and their intrinsic angular momentum.

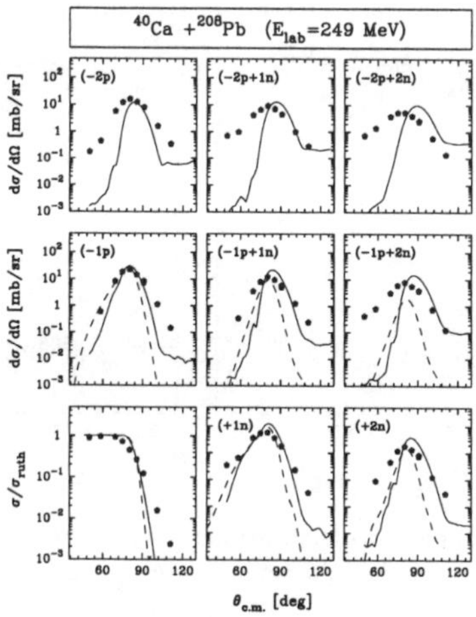

Figure 12. Angular distribution of the dominant transfer channels for the ^{40}Ca+^{208}Pb reaction at 249 MeV. Lines are calculations (see text).

To appreciate an overall description of the reaction we show in Fig. 11 the total cross section measured for all the isotopes for all the channels with a charge transfer less or equal to 4. The angular distribution for the dominant transfer channels is shown in Fig. 12, here in the lower left panel is also shown the quasi elastic (ratio to Rutherford) angular distribution. The angular distributions display the well known bell-shape, that is characteristic of the direct reactions, with a maximum at the grazing angle and a width that increase for the more massive transfer. The increase of the width indicates that during the transfer process the two nuclei undergo large deformations so to remain in contact for a longer time (neck formation). While the theory describes white well the shapes of the $(+1n)$ and $(-1p)$ channels it misses the widening of the angular distribution since it does not incorporate the surface modes. To show that this is the case we repeated the calculation with the semicalssical model of ref.[17,18,19] that incorporates both transfer and collective degrees of freedom. This model calculates the different observables by solving, in an approximate way, the system of coupled equations (1). The model is not, at present, able to include the pair-transfer modes, here the multi-nucleon is calculated simply via a successive mechanism. The results are shown with a dot line in the same figures, notice that the angular distributions are wider in the forward direction. To see if the correct treatment of the inelastic processes is

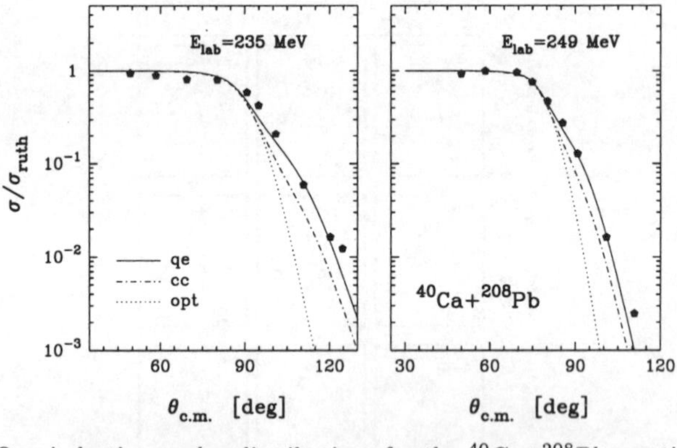

Figure 13. Quasi elastic angular distributions for the $^{40}Ca+^{208}Pb$ reaction at the indicated energies. Lines are coupled channels calculations..

important a coupled channel calculation, for the quasi-elastic angular distribution, has been done with the PTOLEMY code. The real part of the optical potential we have used is the one of ref.[1] while the imaginary part has been calculated microscopically by using the same set of single particle states as the calculations of the cross section above. The inelastic coupling to the low lying 2+ and 3- states of ^{40}Ca and ^{208}Pb have been included in the calculation. The results are shown in Fig. 13. The full lines correspond to the sum of the true elastic plus all the inelastic channels (i.e. the quasi-elastic cross section) in comparison with the true elastic. The other two lines correspond to the true elastic and the one obtained with an optical model calculation. From the nice fit to the data we get confirmation that a correct treatment of the deformations is essential in the treatment of heavy ion reactions.

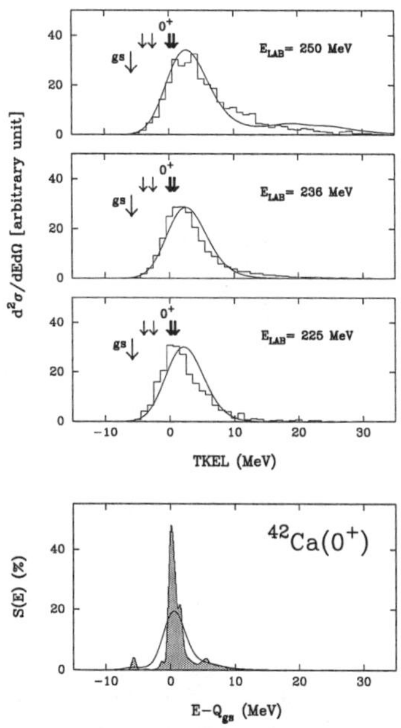

Figure 14. Experimental (histograms) and theoretical (curves) total kinetic energy loss distributions of the two neutron pick-up channels at the indicated energies. The arrows correspond to the energies of 0^+ states in ^{42}Ca with an excitation energy lower than 7 MeV. Bottom panel shows the strength function $S(E)$ from SM calculations (see text) after convoluting with Gaussian of two different widths: 300 keV and 1.5 MeV (close to the experimental energy resolution - curve). The represented strength function has been obtained after 200 Lanczos iterations to allow a correct convergence of all eigenstates.

In Fig. 14 are shown the total kinetic energy loss (TKEL) distributions measured at the grazing angle for the (+2n) channel in comparison with the theoretical predictions calculated for a partial wave close to the grazing one. As can be appreciated all the distribution pick at optimum Q-value (for this channel it is close to 0) leaving the ground state unpopulated. The calculation gives an overall nice description of the data. By looking at the final population of the single particle levels we can infer that the maxima for the +2n channel is essentially due to two neutrons in the $p_{3/2}$ orbital. This fact together with the known low energy spectra of ^{42}Ca suggests to interpret this maximum as the excited 0^+ states that were interpreted as pair mode[20,21]. To strengthen this interpretation a calculation

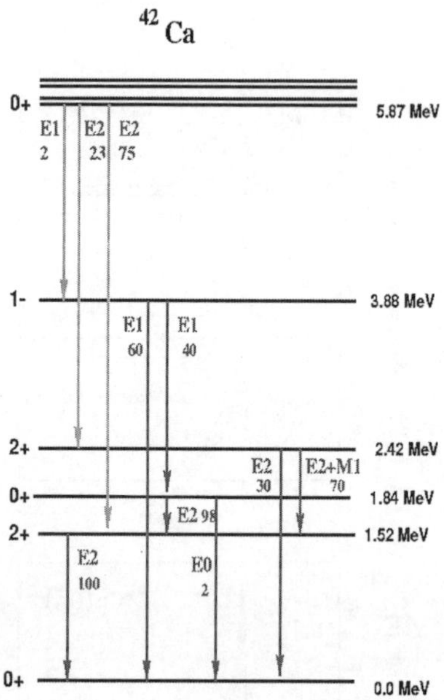

Figure 15. Energy levels for ^{42}Ca below 6 MeV. The arrows indicate the decay patterns (notice that the one of the $^{+}0$ are not known). The number close to each vertical line indicates the expected number of γ transition in a 7 days run.

has been performed to obtain the strength distribution of these peculiar 0^+ states (with two neutrons in the $p_{3.2}$ orbitals). This has been done in the framework of large scale shell model (SM) calculations by using the same model space and interaction as in a recent publication concerning various spectroscopic features of calcium isotopes [22]. To extract the channel strength function the Lanczos method has been used with as pivot state $\psi_1 = (a^+_{p_{3/2}} a^+_{p_{3/2}})^0 |0^+\rangle$ that corresponds to the creation of two neutrons, coupled to 0, on the $|0^+\rangle$ ground state of ^{40}Ca. The valence space, used in these SM calculations consists of a ^{28}Si inert core and of the $2s_{1/2}$, $1d_{3/2}$, $1f_{7/2}$ and $2p_{3/2}$ sub-shells for both protons and neutrons. The strength distribution $S(E)$, shown in the lower panel of Fig. 14, displays, clearly, a strong concentration near ~ 6 MeV of excitation energy, an energy very close to that of a configuration where a $p_{3/2}$ neutron pair is coupled to a closed shell of ^{40}Ca ground state ($E \sim 5.9$ MeV). This calculation demonstrates the dominant character of the $p_{3/2}$ orbitals and the predicted very narrow energy distribution suggests its interpretation as a pair mode.

With the present energy resolution of the experiment would be impossible to arrive at a definite assignment of these states but the obtained results and the large measured cross section will enable the study of the decay pattern (cfr. Fig. 15) of these states with the new large angle spectrometer (PRISMA) that has been installed at the Legnaros's Laboratory and coupled with a large clover array (CLARA). This new apparatus will allow to measure particle-γ coincidence and if the statistics is good all the $\gamma - \gamma$ coincidence for the determination of the decay pattern.

5. Conclusions

In this contribution I attempted an overview of some of the researches concerning the reaction mechanism of heavy ion collisions that have been done in Italy in the past few years. To arrive at a reasonable consistent description I decided to focus on researches relevant for nuclear structure.

References

1. R. Broglia and A. Winther, *Heavy Ion Reactions*, Addison-Wesley Pub. Co., Redwood City CA, 1991
2. A. Bonasera, in this proceedings.mtablex.datmtablex.dat
3. A. Winther and K. Alder, Nucl. Phys. **A319** (1979) 518.
4. H. Feldmeier, Rep. Prog. Phys. **50** (1987) 915.
5. S. Mondechai, H. Fortune, J.O'Donell et. al., Phys. Rev **C41** (1990) 202.
6. K. Boretzky, A. Grünschloff, S. Llievski, et al,, Phys. Rev. **C68** (2003) 024317.
7. P. Chomaz and N. Frascaria, Phys. Rep. **252** (1995) 275.
8. E.G. Lanza, M.V. Andrés, F. Catara et al., Nucl. Plys. **A613** (1997) 445.
9. E.G. Lanza, M.V. Andrés, F. Catara et al., Nucl. Plys. **A654** (1999) 792c.
10. M.V. Andrés, F. Catara, E.G. Lanza et al. Phys. Rev. **C65** (2001) 014608
11. E.G. Lanza, Proceeding of the *Eighth International Spring Seminar on Nuclear Physics, Paestum, Italy ,May 23 to 27, 2004.*
12. M. Hage-Hassan and M. Lambert, Nucl. Phys. **A 188** (1972) 545.
13. C.H. Dasso, L. Fortunato, E.G. Lanza and A. Vitturi, Nucl. Phys. **A724** (2003) 85.
14. S. Szilner, L. Corradi F. Haas, G. Pollarolo et al. Eur. Phys. J. **A21** (2004) 87.
15. S. Szilner, L. Corradi F. Haas, G. Pollarolo et al. in preparation.

16. L. Corradi, A.M. Stefanini, C. Lin, S. Beghini, G. Montagnoli, F. Scarlassara, G. Pollarolo and A. Winther, Phys. Rev. **C59**, 261 (1999).
17. A. Winther, Nucl. Phys. A572 (1994) 191
18. A. Winther, Nucl. Phys. A594 (1995) 203
19. Program GRAZING: htpp:/www.to.infn.it/~nanni/grazing
20. R.A. Broglia, O. Hansen and C. Riedel, *Advances in Nuclear Physics*, edited by M. Baranger and E. Vogt, Plenum, New York, 1973, Vol. 6, p.287.
21. M. Igarashi, K. Kubo and K. Tagi, *Phys. Rep.* **199**, 1 (1991).
22. E. Caurier, K. Langanke, G. Martinez-Pinedo,F. Nowacki and P. Vogel, Phys. Lett **B 240** (2001).

SPONTANEOUS SYMMETRY BREAKING AND RESPONSE FUNCTIONS IN NEUTRON MATTER

M. MARTINI

Dipartimento di Fisica Teorica dell'Università di Torino and Istituto Nazionale di Fisica Nucleare, Sezione di Torino, via P.Giuria 1, I-10125 Torino, Italy

We study the spin response function of an infinite homogeneous system of neutrons interacting through a simple spin-spin Heisenberg force in a non-relativistic context. For a ferromagnetic interaction the spin response along the direction of the spontaneous magnetization displays, for not too large momentum transfers, two distinct peaks. The response along the direction orthogonal to the spontaneous magnetization displays a collective mode to be identified with a Goldstone boson of type II. It is shown that the Goldstone boson contributes to the saturation of the energy-weighted sum rule for $\approx 25\%$ when the system becomes fully magnetized.

1. Introduction

The issue addressed in this work relates to the nature of the collective modes of an infinite homogeneous system of neutrons spontaneously magnetized. In spite of the simplicity of the interaction we consider, which has no pretense of being realistic, it appears that our research bears significance for the physics of the neutron stars, since it explores the extension of the many-body response theory to the situation associated with a broken vacuum in spin space. The latter strongly influences the neutrino cross section and neutrino mean free paths in the medium [1] and, furthermore, it is directly connected with the magnitude of the magnetic field that neutron stars host. A lot of work has actually been lately done on this issue: interestingly, it appears that simple effective interactions — such as the Skyrme ones — give indeed rise to a phase transition of second kind [2,3], whereas more microscopic many-body approaches — such as the Brueckner-Hartree-Fock formalism [4] or quantum simulations [5] — give no indication of a quantum phase transition. Generally speaking, this striking difference can be related to the different predictions these models give for the particle-hole spin interaction at neutron star densities: attractive in the Skyrme models and repulsive in calculations based on realistic nucleon-nucleon potentials [6]. Unfortunately, at present there are no direct phenomenological constraints on this component of the effective nuclear interaction at densities relevant for the neutron stars.

2. The system's longitudinal response

Let us assume the system to undergo a spontaneous symmetry breaking acquiring a magnetization along the z-axis. We wish to explore the system's response to a spin-dependent, but not spin-flipping, external probe acting in the z direction. For sake of simplicity we confine ourselves to assume a ferromagnetic ($V_1 < 0$), spin-dependent, zero-range interaction among neutrons, namely $V(r) = V_1\vec{\sigma}_1 \cdot \vec{\sigma}_2\delta(r)$, clearly constant in momentum space.

We shall compute the response to the probe in the RPA-HF framework both in a normal and in a broken vacuum. To this end it is first necessary to set up the longitudinal HF anomalous polarization propagator in the broken vacuum $\Pi_{zz}^{\mathrm{HF},b}$. This is easily achieved starting from the anomalous single-particle propagator introduced and widely used in Refs. [7,8].

Setting $K \equiv (k_0, \vec{k})$ and $Q \equiv (\omega, \vec{q})$ one gets

$$\begin{aligned}\Pi_{zz}^{\mathrm{HF},b}(Q) &= -i\int \frac{d^4K}{(2\pi)^4}[G_{++}(K)G_{++}(K+Q) + G_{--}(K)G_{--}(K+Q)] \\ &= \Pi_{++}^{\mathrm{HF}}(Q) + \Pi_{--}^{\mathrm{HF}}(Q), \qquad (1)\end{aligned}$$

being $G_{++}(K)$ the HF single-particle propagator in the broken vacuum for a spin up neutron, $G_{--}(K)$ the one for a spin down neutron and

$$\Pi_{\pm\pm}^{\mathrm{HF}}(Q) = \int \frac{d\vec{k}}{(2\pi)^3}\left[\frac{\theta(|\vec{k}+\vec{q}| - k_F^{\pm})\theta(k_F^{\pm} - k)}{\omega + \omega_{\vec{k}} - \omega_{\vec{k}+\vec{q}} + i\eta} - \frac{\theta(|\vec{k}+\vec{q}| - k_F^{\pm})\theta(k_F^{\pm} - k)}{\omega + \omega_{\vec{k}+\vec{q}} - \omega_{\vec{k}} - i\eta}\right] \qquad (2)$$

Note that the HF expressions for Π_{++} and Π_{--} are identical to the free ones in the case of a zero-range interaction. Moreover both their real and imaginary part can be easily computed analytically: one clearly obtains the familiar results for a symmetric vacuum with k_F replaced by k_F^+ and k_F^-, respectively. Concerning the global response region, it is actually made up by two response domains: one associated with k_F^+, where the particles with spin up respond to the external probe, and the other associated with k_F^-, where the particles with spin down respond to the external probe (for the figure see Ref.[8]).

Turning to the propagator in RPA-HF, one gets [8]

$$\Pi_{zz}^{\text{RPA-HF},b} = \frac{\Pi_{++}^{\text{HF}} + \Pi_{--}^{\text{HF}} + 2\Pi_{++}^{\text{HF}}\Pi_{--}^{\text{HF}}(V_{od} - V_d)}{1 - V_d(\Pi_{++}^{\text{HF}} + \Pi_{--}^{\text{HF}}) + \Pi_{++}^{\text{HF}}\Pi_{--}^{\text{HF}}(V_d^2 - V_{od}^2)}. \tag{3}$$

In the above V_d and V_{od} correspond to the diagonal and off-diagonal particle-hole matrix elements of our interaction in spin space. In ring approximation they read $V_d^{\text{ring}} = V_{od}^{\text{ring}} = V_1$, while in RPA $V_d^{\text{RPA}} = 0$ and $V_{od}^{\text{RPA}} = 3V_1$. Equation (3) entails a striking consequence, namely that for a fully broken vacuum (a fully magnetized system, for example in the positive z-direction) no RPA collective mode exists for a zero-range force. Indeed, in this case, since $k_F^- = 0$ then $\Pi_{--}^0 = 0$ and also V_d=0. The situation is clearly illustrated in Fig. 1 where the system's response along the z-axis is shown at a modest value of the momentum transfer, namely $q = 5$ MeV/c (for $q = 50$ and 500 MeV/c see Ref.[8]), and the evolution of the system's response with the amount of breaking of the vacuum is also displayed. Accordingly the responses associated to three pairs of values of k_F^+ and k_F^- are shown: since the density of the system is fixed, these are related by $\rho = \frac{k_F^3}{3\pi^2} = \frac{(k_F^+)^3}{6\pi^2} + \frac{(k_F^-)^3}{6\pi^2}$. We further observe that each choice of (k_F^+,k_F^-) corresponds to a value of the strength of the interaction $V_1 = -\frac{\pi^2}{m}\frac{(k_F^+)^2-(k_F^-)^2}{(k_F^+)^3-(k_F^-)^3}$ obtained from the equilibrium condition $\omega_{k_F^+}^+ = \omega_{k_F^-}^-$, being $\omega_k^\pm = \frac{k^2}{2m} - \frac{V_1 k_F^{\mp 3}}{2\pi^2}$ the single-particle energies for a zero-range force [a]. In panel A of the figure the enhancement and softening of the response in the symmetric vacuum ($k_F^+ = 338.13$ MeV/c), due to the attractive ferromagnetic interaction, is clearly apparent. As one moves towards an increasingly broken vacuum and for not too large momenta one sees the appearance of a second peak in the response at high energy until, for a totally broken vacuum, the collectivity completely disappears in accord with the argument given above and the free response is recovered.

In order to understand the frequency behavior of the response at $q = 5$ MeV/c it helps to keep in mind that

a) the HF (free) response in the broken, but not fully so, vacuum already displays two maxima, when the Pauli principle is active (namely for not too large q);

b) the RPA-HF framework conserves the energy-weighted sum rule even when the vacuum is broken.

[a]We remind the reader that, although our interaction is of pure exchange character, an Hartree term arises in the self-energy owing to the broken vacuum (see Ref. [7]).

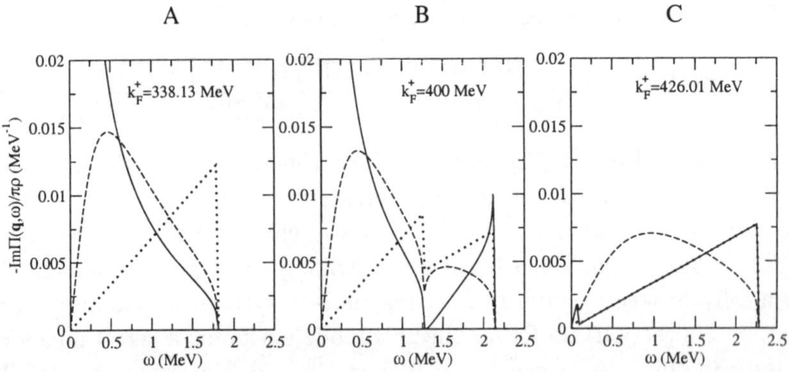

Figure 1. The response of an infinite neutron's system to a z-aligned probe for $q = 5$ MeV/c. Panel (A) refers to the case of a symmetric vacuum ($k_F = k_F^+ = k_F^-$), panel (B) to a partially aligned broken vacuum and panel (C) to a totally broken, fully aligned vacuum. Dotted line: HF (free) response, dashed line: ring approximation, solid line: RPA-HF.

3. The system's transverse response

In this Section we explore the system's response to a probe aligned in the direction orthogonal to the axis along which the spontaneous magnetization of the system occurs. For definitiveness we choose the probe to act in the x-direction. According to the general theory in a non-relativistic context [9], we expect here Goldstone modes to show up. Their number should not be less than the number of the broken generators of the continuous symmetry, provided that the Goldstone bosons of type II are counted twice. In the case we are presently investigating, the number of the broken generators is provided by the dimensions of the coset $O(3)/O(2)$, where $O(3)$ is the rotation group in three dimensions. This group leaves invariant the Hamiltonian of our system of interacting neutrons, whereas $O(2)$ is the rotation group in two dimensions and represents the surviving symmetry after the spontaneous breaking has occurred. Hence in our case two are the generators broken. Accordingly this situation is compatible either with the existence of two Goldstone bosons of type I — characterized by a dispersion relation linear in the momentum — or with the existence of one type II Goldstone boson — which has a dispersion relation quadratic in the momentum.
As we shall see, the latter is actually the occurring case for our system.

In searching for these Goldstone bosons we first ask: where do they live? To answer this question we need to consider the transverse HF polarization propagator

$$\Pi^{\mathrm{HF},b}_{xx/yy}(Q) = -i\int \frac{d^4K}{(2\pi)^4}[G_{++}(K)G_{--}(K+Q)+G_{--}(K)G_{++}(K+Q)]$$
$$= \Pi^{\mathrm{HF}}_{-+}(Q)+\Pi^{\mathrm{HF}}_{+-}(Q), \tag{4}$$

where we have found it convenient to introduce the quantities Π^{HF}_{-+} and Π^{HF}_{+-}, whose vertices embody the spin operators $\sigma_{\pm} = \frac{1}{2}(\sigma_x \pm i\sigma_y)$. In the HF approximation the expressions for Π_{-+} and Π_{+-} are easily deduced starting from the single particle propagators, already employed in deducing the response to a longitudinal external probe. One gets

$$\Pi^{\mathrm{HF}}_{\mp\pm}(Q) = \int \frac{d\vec{k}}{(2\pi)^3}\left[\frac{\theta(|\vec{k}+\vec{q}|-k_F^{\mp})\theta(k_F^{\pm}-k)}{\omega+\omega^{\pm}_{\vec{k}}-\omega^{\mp}_{\vec{k}+\vec{q}}+i\eta} - \frac{\theta(k_F^{\mp}-|\vec{k}+\vec{q}|)\theta(k-k_F^{\pm})}{\omega+\omega^{\pm}_{\vec{k}}-\omega^{\mp}_{\vec{k}+\vec{q}}-i\eta}\right] \tag{5}$$

From the above formula, the response region of the infinite, homogeneous neutron's system in the (ω, q) plane to a spin-flipping probe $(\sigma_{\pm})$ is deduced by searching for the region where, e. g., $\Pi^{\mathrm{HF}}_{-+}(Q)$ develops an imaginary part. We obtain a response region (displayed in Fig. 2 panel A) related to Π^{HF}_{-+} (Π^{HF}_{+-}) which is shifted with respect to the symmetric case upward (downward) by an amount $\Delta\omega = -\frac{V_1}{2\pi^2}\left[(k_F^+)^3-(k_F^-)^3\right]$ directly reflecting the size of the spontaneous breaking of the vacuum. Concerning the response function, it is remarkable that at variance with the symmetric vacuum case, now, for $\omega > 0$, it is also contributed to by the second piece on the right hand side of Eq. (5), the more so, the smaller q is. For the expressions for the real and the imaginary part of Π^{HF}_{-+} and Π^{HF}_{+-} see Ref.[8] and for similar results in the context of asymmetric nuclear matter see Refs.[10,11].

We turn now to discuss the RPA equations. In our case of a zero-range interaction one finds

$$\Pi^{\text{RPA-HF}}_{xx}(\vec{q},\omega) = \frac{\Pi^{\mathrm{HF}}_{-+}(\vec{q},\omega)+\Pi^{\mathrm{HF}}_{+-}(\vec{q},\omega)-6V_1\Pi^{\mathrm{HF}}_{-+}(\vec{q},\omega)\Pi^{\mathrm{HF}}_{+-}(\vec{q},\omega)}{[1-3V_1\Pi^{\mathrm{HF}}_{-+}(\vec{q},\omega)][1-3V_1\Pi^{\mathrm{HF}}_{+-}(\vec{q},\omega)]}. \tag{6}$$

To obtain the dispersion relation of the Goldstone bosons we search for the poles (if any) of the Eq.(6) for positive real ω. Of the two factors appearing in the denominator only the first one (since we have chosen $k_F^+ > k_F^-$) vanishes for just one real and positive value of ω at a given q. In Fig. 2 we display the solution of the equation $[1-3V_1\mathrm{Re}\Pi^{\mathrm{HF}}_{-+}(\vec{q},\omega)] = 0$, which we expect to yield the dispersion relation of the Goldstone boson, should the RPA be a trustworthy theory for our many-body system.

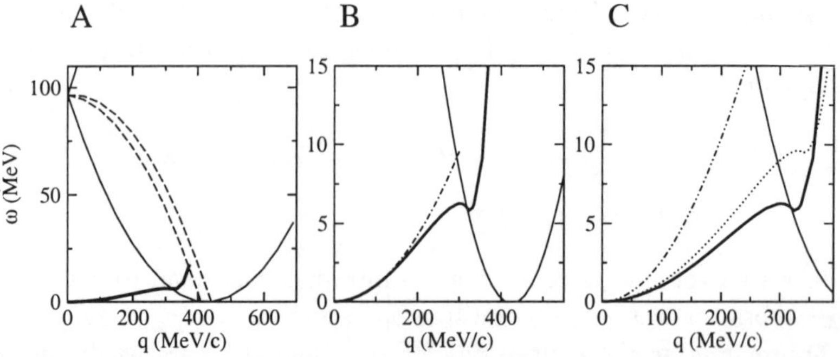

Figure 2. The dispersion relation of the Goldstone boson for $k_F^+ = 426.01$ MeV/c (heavy solid lines), which corresponds to $V_1 = -189.25$ MeV fm^3. Also displayed are the response regions. In panel A one can appreciate how tiny the energy of the Goldstone boson is; in panel B, which enlarges panel A, one can assess the domain of validity of the parabolic dispersion relation of the Goldstone mode (dot-dashed line); in panel C the Goldstone mode is displayed for three different values of the interaction strength, namely $V_1 = -189.25$ (solid), -200 (dot) and -300 MeV fm^3 (dot-dot-dash).

This turns out indeed to be the case, since for small q and ω this solution can be analytically expressed through the expansion of $\mathrm{Re}\Pi_{-+}^{\mathrm{HF}}(\vec{q},\omega)$ getting a parabolic dispersion relation, valid for small values of q. From Fig. 2 (panels B and C) it appears that the parabolic dispersion relation actually remains valid over a substantial range of momenta. Thus, the solution of the above mentioned equation truly corresponds to a type II Goldstone boson as it should. Physically it can be viewed as a twisting of the local spin orientation as the collective wave passes through the system. Furthermore, and remarkably, it turns out that for $V_1 > V_{1,c}^{\mathrm{upper}}$ (the critical value signaling the onset into the system of a total magnetization) the Goldstone mode continues to exist with a dispersion relation that is parabolic over a range of momenta becoming larger as V_1 increases. For $V_{1,c}^{\mathrm{lower}} \leq V_1 \leq V_{1,c}^{\mathrm{upper}}$ (being $V_{1,c}^{\mathrm{lower}}$ the critical value signaling an incipient ferromagnetism) the Goldstone mode displays instead an anomalous behavior: in fact, in this range of couplings, in correspondence to a specific momentum, the collective mode is characterized by a vanishing group velocity.

Concerning the continuum RPA-HF response of the system to an x-aligned probe, whose behavior for different momenta and vacua can be found in Ref.[8], it turns out that

i) for a symmetric vacuum the energy-weighted sum rule S_1 is clearly obeyed;

ii) when the vacuum is broken S_1 is still fulfilled, but the particle-hole con-

tinuum is more depleted;
iii) in accord with ii) the more the vacuum is broken, the stronger the Goldstone boson becomes. This item will be quantitatively addressed in the next section in the sum rule framework.

4. The moments of the response function

In this Section we investigate the non-energy-weighted (S_0) and the energy-weighted (S_1) sum rules, exploring their behavior when a spontaneous symmetry breaking occurs in the vacuum.

Concerning S_1, it is well-known that it is given by the following expression

$$S_1(q) = \int_0^\infty \mathrm{d}\omega\, \omega \frac{-\mathrm{Im}\Pi(\vec{q},\omega)}{\pi\rho} = \frac{1}{2}\langle 0|[\widehat{O},[\hat{H},\widehat{O}]]|0\rangle. \qquad (7)$$

For the density response of a non-interacting gas of fermions of mass m the above is indeed fulfilled and yields $S_1^{\mathrm{free}}(q) = \frac{q^2}{2m}$.
In general, however, Eq. (7) is violated by most of the many-body frameworks, the remarkable exception being the RPA-HF theory [12] . We have indeed verified it by computing numerically with very good accuracy the left hand side of Eq. (7) and by working out the HF expectation value of the double commutator in the same equation — which of course yields $q^2/2m$, if one employs our interaction. Remarkably, as already anticipated, even when the vacuum is broken S_1 keeps the above value, as it can be inferred from the results reported in Table 1. In this instance, at variance with the situation where the probe acts in the direction of the spontaneous magnetization, when the spin-flipping probe is directed orthogonally to the latter, S_1 is contributed to not only by the particle-hole continuum, but by the collective Goldstone mode as well. Actually, this contribution grows with the amount of the symmetry breaking in the vacuum, which, in turn, grows with the strength of the force V_1. For $V_1 = V_{1,c}^{\mathrm{upper}}$, namely when the vacuum is fully aligned in spin-space, the Goldstone mode accounts for roughly 25% of the energy-weighted sum rule.

Concerning S_0 in a symmetric vacuum one observe that the well-known result holding for a non-interacting system of fermions is conserved in the HF theory, but not in RPA or RPA-HF. In the presence of the broken vacuum the HF value of S_0 differs from the value of the symmetric case. Furthermore when the vacuum is broken one finds that the impact of the Pauli correlations on S_0 is lowered with respect to the symmetric case and decreases as the amount of the symmetry breaking grows, as it is apparent from Table 1. In particular, for a fully broken vacuum, S_0 is just 1 for any q — that is the value occurring in the symmetric vacuum for $q \geq 2k_F$ — both when the system is explored in the longitudinal or in the transverse direction by a spin-dependent probe. In other words for a fully broken vacuum the system's constituents no longer feel the Pauli principle, as it should be expected. The reduced influence of the Pauli principle, when the system is only partially aligned, with respect to the situation occurring in the symmetric vacuum, can be exploited to investigate (using a spin-flipping probe) how the collectivity of the Goldstone mode is affected by the degree of spontaneous symmetry breaking of the vacuum. In fact, in Table 1 one sees that the less effective the Pauli correlations are, the more collective the Goldstone boson is.

Table 1. The non-energy-weighted and energy-weighted sum rules at $q = 5$ MeV/c, corresponding to $q^2/2m = 0.0133120$ MeV. In the columns associated with the RPA-HF theory, the first entry represents the contribution to S_0 and S_1, respectively, arising from the particle-hole continuum; the second entry the one arising from the collective Goldstone mode.

k_F^+	S_0^{HF}	$S_0^{\text{RPA-HF}}$	S_1^{HF}	$S_1^{\text{RPA-HF}}$
338.130028	0.0111	(0.0819+0)	0.013312	(0.013312+0)
400	0.65549	(0.000228+0.65527)	36.248	(0.012594+0.000718)
426.01	1.000	(0.000111+0.99979)	96.510	(0.010664+0.002686)

References

1. A. Drago, *this volume*
2. J. Rikovska Stone, J. C. Miller, R. Koncewicz, P. D. Stevenson and M. R. Strayer, *Phys. Rev. C* **68** (2003) 034324.
3. A. A. Isayev and J. Yang, *Phys. Rev. C* **69** (2004) 025801.
4. I. Vidaña, A. Polls and A. Ramos, *Phys. Rev. C* **65** (2002) 035804.
5. A. Sarsa, S. Fantoni, K. E. Schmidt and F. Pederiva, *Phys. Rev. C* **68** (2003) 024308; F. Pederiva, *this volume*
6. S. Reddy, M. Prakash, J. M. Lattimer and J. A. Pons, *Phys. Rev. C* **59** (1999) 2888.
7. A. Beraudo, A. De Pace, M. Martini and A. Molinari, *Ann. Phys. (N.Y.)* **311** (2004) 81.
8. A. Beraudo, A. De Pace, M. Martini and A. Molinari, nucl-th/0409039.
9. H. B. Nielsen and S. Chada, *Nucl. Phys. B* **105** (1976) 445.
10. W.M. Alberico, A. Drago and C. Villavecchia, *Nucl. Phys. A* **505** (1989)309.
11. K. Takayanagi and T. Cheon, *Phys. Lett. B* **294** (1992) 14.
12. D. J. Thouless, *Nucl. Phys.* **22** (1961) 78.

ISOSPIN DYNAMICS IN FRAGMENTATION REACTIONS AT FERMI ENERGIES *

R.LIONTI, V.BARAN,[†] M.COLONNA, M. DI TORO

Laboratori Nazionali del Sud INFN, Phys.Astron. Dept. Catania University Via S. Sofia 44, I-95123 Catania, Italy

E-mail: lionti@lns.infn.it

In this work we have studied the neutrons and protons dynamical behavior in two fragmentation reaction: ^{58}Fe+^{58}Fe (charge asymmetric, N/Z = 1.23) and ^{58}Ni+^{58}Ni (charge symmetric, N/Z = 1.07). We note that isospin dynamic processes take place also in the symmetric system ^{58}Ni+^{58}Ni, that produce more asymmetric fragments and residual nuclei. This is a consequence of the pre-equilibrium phase: we observe a competition between pre-equilibrium evaporation and the phenomenon of the ***isospin-migration***, which is a consequence of the EOS (nuclear equation of state) symmetry term. We have simulated the collision with two different EOS: asy-stiff and asy-soft. Some difference has been noticed, especially about the fragment charge asymmetry. A check of isospin effects has also been done trying to correlate fragment asymmetry with dynamical quantities at the freeze-out time.

1. Introduction

Collisions between charge asymmetric heavy ions (made possible by the recent radioactive beam developments) are the only way to understand the structure of nuclear-*EOS* (Equation of State) isospin term. Isospin also influences the reaction mechanisms, e.g. fragmentation process (the object of this paper), leading to important effects on fragment composition. First of all we will discuss two different kinds of processes that can form fragments: spinodal decomposition and neck fragmentation. We will put in evidence differences between them thanks to the *EOS* isospin term that play a different role in the two reactions; in particular, it introduces a difference from one mechanism to the other due to the low density behaviour

*This work is supported by the PRIN program of MUIR.

[†]On leave from HH-NIPNE and University of Bucharest, Romania.

of the chemical potential for protons and neutrons. In the following pages we will discuss the reactions ^{58}Fe + ^{58}Fe and ^{58}Ni + ^{58}Ni at 47 MeV per nucleon bean energy. We will analyze the charge composition of fragments and residual nuclei with the help of a pre-equilibrium phase study. We will see that isospin dynamics can be derived from the chemical potential behaviour that lead to fluctuations in iso-vectorial density $\rho_i = \rho_n - \rho_p$ [1]. Finally we will study the reactions with an $asy - soft$ EOS [2], looking to explain the difference with the $asy - stiff$ case. In the following, we will refer to an $asy - stiff$ EOS when we consider a potential symmetry term that is always linearly increasing with nuclear density. We will refer to an $asy - soft$ EOS when the potential symmetry term increases up to a saturation around normal density, and then eventually decrease. All the reactions have been simulated by solving the microscopic transport equation BNV (Boltzmann-Nordheim-Vlasov) following a test-particle evolution on a lattice [3 4 5].

2. Isospin-Migration and Fragmentation Reactions

To understand how neutrons and protons move, we must consider the dependance from density of the chemical potential: $\mu_q = \partial\epsilon(\rho_q, \rho_{q'})/\partial\rho_q$, $q = n, p$, ϵ being the density energy. We recall that this quantity contains all contributions to energy per particle (kinetic, potential and simmetrical). In Fig.1 1 is reported the density dependence of the n, p chemical potentials below normal density, where we expect the fragment formation takes place [6], for a system with asymmetry $I = (N - Z)/A = 0.2$ for the two choices of the iso-EOS.

In this figure we can recognize two different regions in which neutrons and protons show a different behavior. The first one appears for $\rho < 0.08$ fm^{-3} in which both neutrons and protons, moving from higher to lower chemical potential, spread toward higher density regions. The second one extends from 0.08 fm^{-3} to an approximate equilibrium density of 0.15 fm^{-3}; here protons move towards a higher density region while neutrons move in opposite direction. This phenomenon is called *isospin migration* [1].

The importance of the different chemical potential behavior in the two regions is evident when we study fragmentation reaction that can appear following two different channels: spinodal decomposition [7 8] and neck fragmentation [6]. In fact we can recognize these two different mechanisms by observing fragment asymmetries (Fig. 2) arising from the reactions ^{124}Sn + ^{124}Sn at 50 $AMeV$ with an impact parameter $b = 2fm$ (central

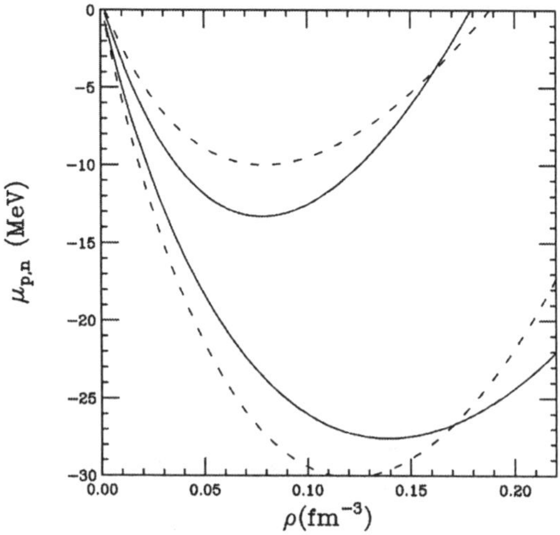

Figure 1. Density dependence of the chemical potential for neutrons (upper curves) and protons (lower curves) for an asy-stiff (solid lines) and asy-soft (dashed lines) *EOS* with asymmetry parameter I=0.2 [1].

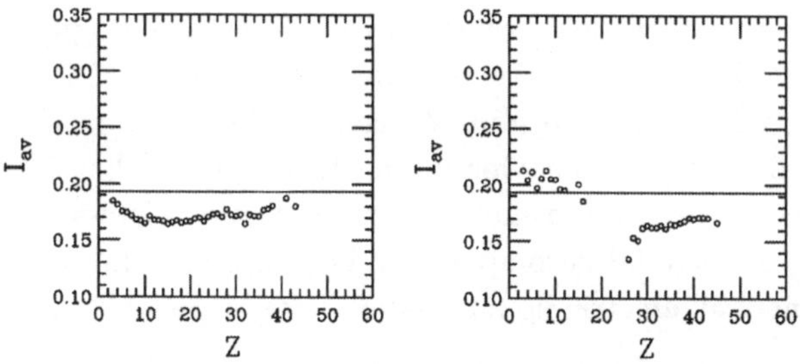

Figure 2. Average asymmetry parameter vs. Z for nuclei from the reactions ^{124}Sn + ^{124}Sn at 50 $AMeV$. There are illustrated two type of collions: central ($b = 2fm$) on the left and semi-peripheral ($b = 6fm$) on the right. Solid lines represent intial asymmetry parameter [1].

collision) and $b = 6fm$ (semi-peripheral collision) [1]: we note how, at the final freeze-out time, intermediate mass fragments (IMF) with Z < 15, arising from the reaction with $b = 6fm$, are more neutron-rich.

We can understand these differences by looking at the region where fragments form [1,6]. Fragments that arise from central collisions form in

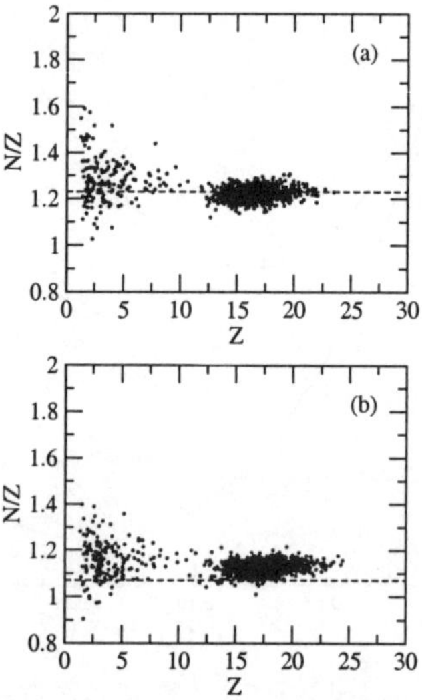

Figure 3. Asymmetry vs. charge of each nuclei arising from the simulation of the reaction ^{58}Fe + ^{58}Fe (a) and ^{58}Ni + ^{58}Ni (b) with an $asy - stiff\ EOS$. Horizontal dashed lines are the initial asymmetries of the colliding systems.

a very dilute region, while fragments arising from neck-fragmentation form in a region at the interface with "spectators", i.e. at densities not very different to the value at equilibrium. We can conclude, thanks to the different n, p chemical potential behavior in the two regions, that isospin migration appears for neck-fragments, and more charge asymmetric fragments form, while, in spinodal decomposition, both protons and neutrons diffuse towards a more symmetric liquid phase.

3. ^{58}Fe + ^{58}Fe and ^{58}Ni + ^{58}Ni at 47 AMeV, b =4 fm

3.1. *The Asy-stiff Case*

Now we study the isospin dynamics in the reactions ^{58}Fe + ^{58}Fe (charge asymmetric $N/Z = 1.23$) and ^{58}Ni + ^{58}Ni (charge symmetric $N/Z = 1.07$). Both reactions have been simulated at a beam energy of 47 $AMeV$ (where some recent data are existing [9]), with an impact parameter $b = 4fm$ (semi-central) and with an $asy - stiff$ EOS: 40% of the events have produced a fragment(*ternary events*). Fig. 3 reports the N/Z ratio of each fragment vs. the charge Z at the freeze-out time.

Residual nuclei (large Z range) show a different behaviour: we note, in fact, that representative points of Fe reaction are along the dashed line that represents initial system asymmetry, while for Ni reaction points are under that line. The IMF behaviour is similar for the two reactions: representative points, for both reactions, lie above the dashed line, although for reaction ^{58}Ni + ^{58}Ni this is a little more evident.

For a better comprehension of Fig. 3, we have to consider what happens during the pre-equilibrium phase: asymmetry of the di-nuclear neutron-rich system changes from 1.23 (initial value) to 1.22 (at t = 100 fm/c, instant in which fragments start to form) since 14 neutrons and 11 protons evaporate, while the di-nuclear neutron-poor system changes from 1.07 to 1.12 as a consequence of a larger proton evaporation (it loses 13 protons against 12 neutrons) becoming an asymmetric system. We can conclude: i) In the neutron rich reaction, neutron evaporation in pre-equilibrium is counterbalanced by the neck neutron enrichment, caused by the isospin-migration; ii) In the neutron-poor collision, however, the fragment asymmetry derives from the proton emission during pre-equilibrium and the neutron enrichment of the neck.

For residual nuclei, it is interesting to study their charge composition distinguishing between binary and ternary events (Table 1).

Table 1. Asymmetry evolution of the residual nuclei arising from binary and ternary events.

	$t = 0$	$t = 100 fm/c$	$t = 200 fm/c$
$^{58}Fe + ^{58}Fe$	1.23	1.22	1.23 *binary*
			1.19 *ternary*
$^{58}Ni + ^{58}Ni$	1.07	1.12	1.17 *binary*
			1.125 *ternary*

For both reactions, we note that the N/Z ratio of residual nuclei from ternary events is lower than the value for binary events, since for the second isospin-migration doesn't appear. This the isospin dynamics effect is rather evident from the comparison with the asymmetry at the time corresponding the the end of the pre-equilibrium phase ($t = 100 fm/c$ in the Table). The Fe + Fe system changes from 1.22 to 1.19; for the Ni + Ni reaction this difference is not as evident (from 1.12 to 1.125) since isospin-migration opposes proton evaporation. All this shows a clear evidence of a dependence from the reaction mechanism of the charge equilibration process (isospin diffusion).

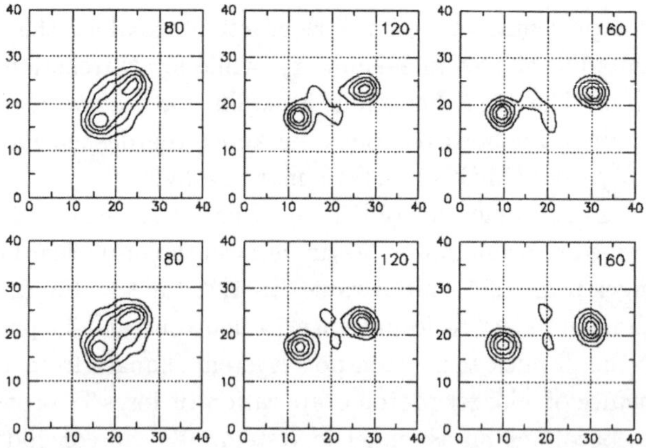

Figure 4. Density contour plots on the reaction plane for two different events of the reaction ^{58}Fe + ^{58}Fe at 47 AMeV.

3.2. *The Dynamics of Reactions*

In Fig. 4, the density contour plot on the reaction plane is presented for two different events coming from the reaction ^{58}Fe + ^{58}Fe with the same charateristics of the reaction described in the previous section ($b = 6fm$, $47\ AMeV$).

The first row shows an event in which a fragment forms with a delay and in a space region correlated to a residual target-like nucleus, unlike the event in the second row, which shows a rapid fragment formation in a region that is not correlated to any spectator remnants. We expect that the first event will form a more asymmetric fragment than the second one. In the latter case, isospin-migration movements of nucleons are quenched by the shorter interaction time; moreover fragments in the first scenario are more subject to the driving force of the residual nucleus. All this suggests to look for some correlation between asymmetry and direction of outgoing fragments. This is shown in Fig. 5, reporting how the average IMF asymmetry varies from 0 to 180° in bin of 20°.

We note how the distribution is flat, in particular for neutron-poor reaction. We would expect less asymmetric fragments to go to 90°, even if this effect is not very visible. In fact statistical errors are different in the various bins. The horizontal lines correpond to the initial asymmetry of the colliding system. We note that the $neck - IMF$ always present a neutron enrichment, *even in the case of a n-poor system.* The latter paradox is due

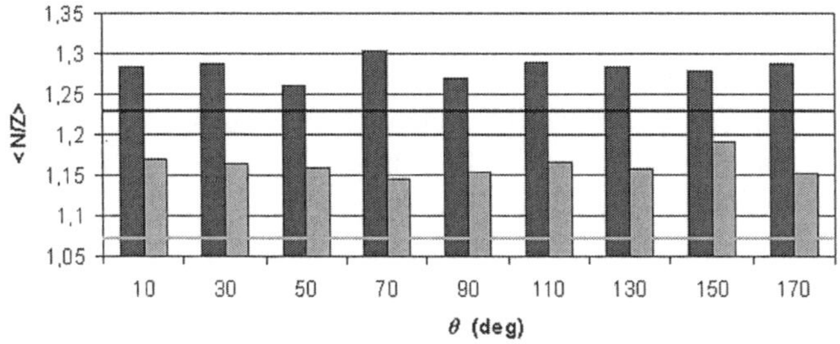

Figure 5. Average asymmetry vs. angle of outgoing fragments for the reaction ^{58}Fe + ^{58}Fe (black histograms) and ^{58}Ni + ^{58}Ni (grey histograms). Zero angle represents a fragment parallel to the beam direction.

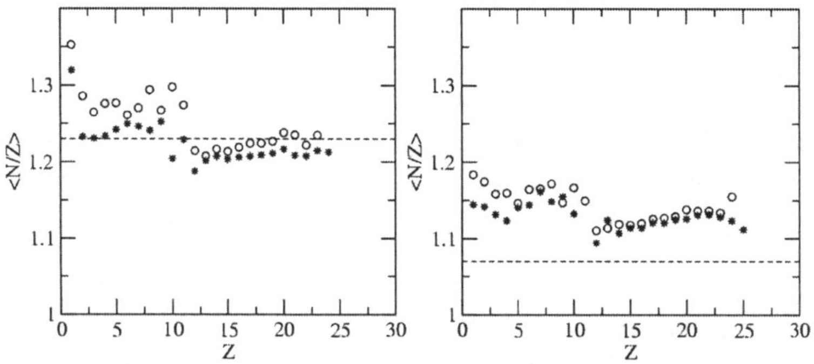

Figure 6. Average asymmetry vs. Z for nuclei from the reactions ^{58}Fe + ^{58}Fe (left panel) and ^{58}Ni + ^{58}Ni (right panel) for an $asy-stiff$ (circles) and $asy-soft$ (triangles) EOS. Horizontal dashed lines represent initial asymmetry of colliding systems [1].

to the pre-equilibrium isospin dynamics. Some evidence has been found in recent data on ^{58}Ni induced fragmentation [10,11,12].

4. The Asy-soft Case

In Fig. 6, the average asymmetry is shown vs. the charge Z of the products arising from the reactions ^{58}Fe + ^{58}Fe and ^{58}Ni + ^{58}Ni at 47 MeV per nucleons, $b = 4fm$, with an $asy - stiff$ and $asy - soft\ EOS$.

We note that both fragments and residual nuclei are more symmetric in the asy-soft case. This difference can be explained with the different behavior of the chemical potential in the two EOS, see Fig. 1. For an asy-soft EOS, proton chemical potential varies not so much in the region where the fragments form (from 0.08 to 0.15 fm^{-3}), while for neutrons there's a significant slope (however smaller than asy-stiff case). Therefore neutron enrichment of the neck will still cause an increase of the fragment asymmetry, but this increase will be smaller than the asy-stiff case, where protons can even migrate out of the neck region.

5. Conclusions

In this paper, we have verified that isospin dynamic processes concern also the symmetric system ^{58}Ni+^{58}Ni that produce very asymmetric fragments and residual nuclei. We have shown how this behavior is a consequence of the pre-equilibrium phase: Fe + Fe system evaporate neutrons, while Ni + Ni evaporate protons. To understand nuclear asymmetries we have compared evaporation during pre-equilibrium to the phenomenon of the isospin-migration, that is connected to the symmetry term of EOS.

It must be stressed that asymmetry effects in the asymmetric system ^{58}Fe + ^{58}Fe are not trivial: from a chemical point of view, the asymmetry spatial distribution of colliding nuclei is uniform, therefore we don't expect concentration gradients which induce asymmetry variations. These variations are instead caused by density gradients during the reaction dynamics since the symmetry term of EOS introduces a different behavior of the chemical potential for neutrons and protons with respect to density. So, when collisions happen, spatial distribution of the isoscalar density induces variation in the iso-vectorial density too.

The reaction analysis with an asy-soft EOS has shown a dependence of the isospin dynamic by the EOS symmetry term. In fact we have noted that $asy - soft$ simulations produce more symmetric IMFs as a consequence of the chemical potential behavior.

Isospin effects verification has also been done by trying to correlate fragment asymmetry with dynamical quantity at the freeze-out instant (like the outgoing angle on the reaction-plane) leading to negligible effects for these systems (with impact parameter b = 4 fm).

Experimentally, the ^{58}Fe + ^{58}Fe and ^{58}Ni + ^{58}Ni reactions have been studied [9] [10] and asymmetric effects have been noted even in the second system. It can be interesting to correlate dynamical quantity (like angles or fragment velocities) with respect to quantities which depend by the EOS isospin term; in this way we can investigate on the structure of EOS and we can also distinguish the reaction mechanism.

References

1. V. Baran *et al.*, *Nucl. Phys.* **A 703**, 603-632 (2002).
2. M. Di Toro *et al.*, *Prog. Part. Nucl. Phys.* **42**, 125 (1999).
3. G:F: Bertsch and S. Das Gupta, *Phys.Rep.* **160**, 189 (1988).
4. A. Guarnera, *TWINGO code*, Ph. D. Thesis, Univ. Caen, July 1996.
5. V. Greco, A. Guarnera, M. Colonna and M. DiToro, *Phys.Rev.* **C59**, 810 (1999).
6. V. Baran, M.Colonna, M.Di Toro, *Nucl. Phys.* **A 730**, 329-354 (2004).
7. H. Heisenberg, C.J. Pethick and D.J. Ravenhall, *Phys. Rev. Lett.* **61**, 818 (1988).
8. M. Colonna and Ph. Chomaz, *Phys. Rev.* **C49**, 1908 (1994).
9. D. V. Shetty *et al.*, *Phys. Rev.* **C68**, 021602 (R)(2003).
10. P. Milazzo *et al.*, *Phys. Lett.* **B509**, 204 (2001).
11. L.Gingros et al., *Phys. Rev.* **C65**, 061604 (2002).
12. R.Moustbchir et al., *Nucl. Phys.* **A 739**, 15 (2004).

ON THE LORENTZ STRUCTURE OF THE SYMMETRY ENERGY

T. GAITANOS, M. COLONNA, M. DI TORO

Laboratori Nazionali del Sud INFN, Phys. and Astron. Dept. Catania University
Via S. Sofia 44, I-95123 Catania, Italy
E-mail: theodor@lns.infn.it

We investigate the Lorentz structure of the symmetry energy at supra-normal densities in intermediate energy nucleus-nucleus collisions of asymmetric nuclei. We present several possibilities offered by such dynamical processes to set constraints on the still unknown and very controversial high density behavior of the symmetry energy.

1. Introduction

Intermediate energy collisions of heavy asymmetric nuclei offer the unique opportunity to access the properties of hadronic matter under extreme conditions of density, temperature and isospin. Such studies are crucial in understanding many astrophysical phenomena (supernovae explosions, neutron starts, etc.). During the last three decades many attempts have been done to explore the high density behavior of hadronic matter, i.e. the Equation of State (EoS) beyond ground state baryon density and/or at finite temperatures beyond the critical one. A nice and extensive overview can be found in Refs. [1].

However, so far Asymmetric Nuclear Matter (ANM) has been only poorly studied for supra-normal densities. Finite nuclei studies predict values for the symmetry energy at saturation in the order of $30 - 35\ MeV$, but for densities far away from saturation one has to rely on extrapolations. On the other hand, in Heavy Ion Collisions (HIC) highly compressed matter can be formed for short time scales, thus the study of such a dynamical process can provide useful information on the high density symmetry energy. Recently theoretical studies on the high density symmetry energy have been started by investigating HIC of asymmetric systems [2,3] and they have been motivated by the planning of new experimental heavy ion facilities with neutron rich radioactive beams.

The aim of this report is to explore the properties of ANM within a relativistic mean field theory in different nuclear systems, i.e. finite nuclei and HIC, with the particular interest on understanding the high density behavior of the symmetry energy in terms of its Lorentz properties.

2. Asymmetric Nuclear Matter in RMF Theory

The description of ANM within a covariant theory is based on the Hartree- or Relativistic Mean Field (RMF) approach of the Quantumhadrodynmics [4]. Within RMF the baryons are given by quantum Dirac spinors Ψ obeying the Dirac equation and the mesons (iso-scalar, scalar σ, iso-scalar, vector ω, iso-vector, scalar δ and iso-vector, vector ρ) are described by their classical field equations [4]. The interaction of the baryons with the classical mesonic fields is characterized through the baryon-meson vertices or coupling functions $g_{\sigma,\omega,\rho,\delta}$ which can be simple constants (Walecka-type models) or explicitely density dependent (Density Dependent Hadronic (DDH) theory), see below.

We focus now on the iso-vector part of the EoS, i.e. the symmetry energy. The symmetry energy E_{sym} is defined from the expansion of the energy per nucleon $E(\rho_B, \alpha)$ in terms of the asymmetry parameter $\alpha = (\rho_n - \rho_p)/(\rho_n + \rho_p)$. From the energy-momentum tensor one obtains the generall form as ($f_i \equiv g_i/m_i^2$, $i = \sigma,\ \omega,\ \rho,\ \delta$) [5]

$$E_{sym} = \frac{1}{6}\frac{k_F^2}{E_F} + \frac{1}{2}\left[f_\rho - f_\delta \left(\frac{m^*}{E_F}\right)^2 \right] \rho \quad . \tag{1}$$

Another important quantity is the effective Dirac mass which depends on isospin in the presence of the iso-vector, scalar δ meson

$$m_i^* = M - f_\sigma(\rho_{sn} + \rho_{sp}) \pm f_\delta(\rho_{sp} - \rho_{sn}) \quad \text{(- proton, + neutron)} \quad . \tag{2}$$

The parameters of the models discussed now have been fixed to nuclear matter saturation properties [6].

From Eq. (1) it is seen that the introduction of the iso-vector, scalar δ channel influences the density dependence of the symmetry energy: in order to reproduce the fixed bulk asymmetry parameter $a_4 = 30.5\ MeV$ one has to increase the ρ-meson coupling g_ρ. On the other hand, the Lorentz decomposition of the potential part of E_{sym} in terms of a vector ρ and a scalar δ field affects the density dependence of the symmetry energy at high densities due to the suppression of the scalar density ($\rho_s \approx \frac{m^*}{E_F^*}\rho$). This will

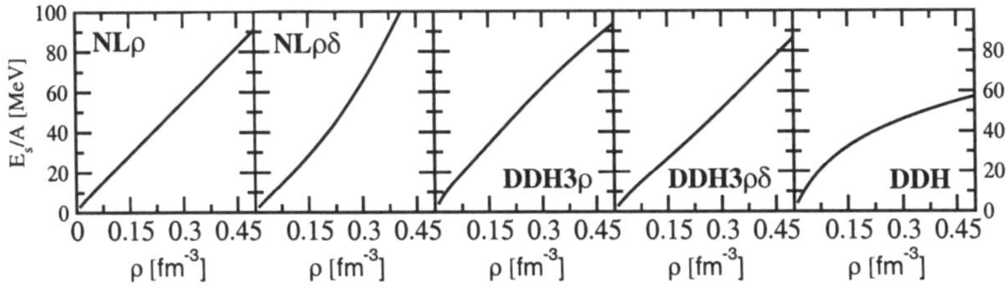

Figure 1. Density dependence of the symmetry energy, Eq. (1), for different models: $(NL\rho, NL\rho\delta)$ non-linear Walecka model including only the ρ-meson and both, the ρ and δ mesons for the iso-vector EOS, respectively. $(DDH\rho, DDH3\rho, DDH\rho\delta)$ Same as in the NL cases for the iso-vector EOS, but within the Density Dependent Hadronic (DDH) mean field theory where all the baryon-meson couplings are explicitely density dependent (taken from [6]).

lead to a stiffer symmetry energy at supra-normal densities because of the stronger ρ-meson coupling when the δ field is taken into acount in this description.

However, the situation turns out to be more complex when considering microscopic models within Dirac-Brueckner-Hartee-Fock (DBHF) theory [7]. The Density Dependent Hadronic (DDH) field theory [7] is based on DBHF. In DDH the baryon-meson vertices are explicitely density dependent with a general decrease of the iso-scalar coupling functions $(g_{\sigma,\omega}(\rho))$ with respect to the baryon density ρ. Such a behavior is consistent with realistic DBHF calculations of symmetric NM where no parameters need to be adjusted [8]. ANM is only poorly investigated within the DBHF theory. In Refs. [8,10] it was shown that the ρ meson coupling strongly decreases with baryon density, but the δ meson coupling, on the other hand, increases for densities above saturation.

The whole picture is summarized in Fig. 1, where the density dependence of E_{sym} within RMF is displayed. We used the non-linear Walecka model (NL) in two different treatments for the iso-vector channel: (a) only with the iso-vector, vector ρ meson $(NL\rho)$ and (b) with both, the iso-vector, vector ρ and iso-vector, scalar δ mesons $(NL\rho\delta)$. The same procedure was applied within the DDH theory by fixing the parameters of the iso-vector channel $(DDH3\rho$ and $DDH3\rho\delta)$ to the density dependence of the iso-vector coupling functions of the parameter free DBHF model [8]. Finally, in the $DDH\rho$ model, which contains only the ρ meson for the description of the iso-vector EOS, the parameters were fixed to finite nuclei properties [9].

We see that the iso-vector, scalar δ channel has important contributions to E_{sym} for densities above saturation due to the relativistic effects as

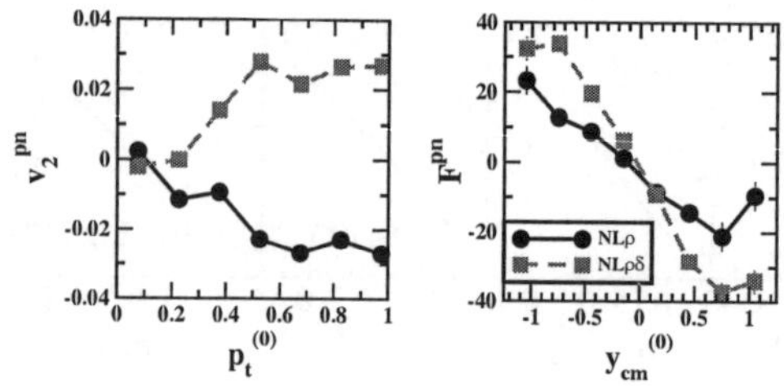

Figure 2. (Right) Isospin flow F^{pn} as function of the rapidity $y_{cm} = \frac{1}{2}\frac{1+\beta_z}{1-\beta_z}$ (β_z being the component of the velocity along the beam direction) and (left) elliptic flow $v_2^{pn} = (p_x^2 - p_y^2)/p_t$ as function of the normalized transverse momentum $p_t^{(0)} = p_t/(p_t^{proj}/A)$. These quantities are calculated from the difference between the proton and neutron flows (indicated with the abbreviation pn). Calculations with the $NL\rho$ (circles), $NL\rho\delta$ (squares) models for a semi-central ($b = 6\ fm$) $Sn + Sn$ are shown (taken from [3]).

discussed above. However, in DDH theory the contribution of the δ meson to the high density symmetry energy is different. Only the comparison between $DDH\rho$ and $DDH3\rho\delta$ leads to the same contribution on E_{sym} as the corresponding one between $NL\rho$ and $NL\rho\delta$. This is due to the fact that in the DDH models the iso-vector couplings has an additional density dependence which also contributes to the density behavior of E_{sym}, apart from the relativistic effects which are always present.

It is important to realize that the relativistic effects, i.e. the suppresion of the iso-vector, scalar δ channel for high densities and the effective mass splitting (Eq. (2)) between protons and neutrons, lead to a natural momentum dependence of the iso-vector EOS even if the baryon-meson vertices do not explicitely depend on energy. This important feature is not included in phenomenological non-relativistic studies [2], where a momentum dependence can be introduced in addition, however, with more parameters to be fixed.

We have applied our models of Fig. 1 both to the static case of finite nuclei and the dynamical one of HIC [6]. In finite nuclei only moderate effects arising from the δ meson were found due to the fact that the symmetry energy shows a similar density dependence for all the models considered for densities at and below saturation. In the next section we will study the more interesting dynamical case where highly compressed ANM can be formed for short time scales.

3. Heavy ion collisions at GSI energies: The key signals

In HIC at SIS energies ($0.1-2\,AGeV$) the highly compressed matter mainly consists of protons, neutrons, intermediate mass fragments and pions. By choosing collisions of asymmetric nucleus like ^{197}Au or $^{132,124,112}Sn$ one can hope to *see* dense ANM at least for some short time scales from which one could select sensitive signals related to E_{sym} at supra-normal densities. The analysis of HIC with the models discussed above was performed within the relativistic transport equation of a Boltzmann-type (RBUU) which describes the dynamical evolution of a 1-particle phase space distribution function under the influence of a mean field (depending on the EOS) and binary collisions (see Ref. [11]). In the following we discuss some of the most important observables which could set constraints on the symmetry energy at high densities.

(1) Collective isospin flows An important observable in HIC is the collective flow due to its high sensitivity on the pressure gradients, i.e. on the degree of the stiffness of the EOS at high densities. Strong collective flow is related to a more repulsive mean field, i.e. to a stiffer EOS. There are different components of collective flow: (a) directed in-plane flow which describes the dynamics into the reaction plane and can be described by the mean transverse in-plane flow $F = < p_x(y) >$ as function of the rapidity y and (b) elliptic flow which describes the dynamics perpendicularly to the reaction plane. The later observable is the most important one due to its earlier formation during the high density phase. It can be extracted from a Fourier analysis of azimuthal distributions as the second Fourier coefficient v_2 [12].

Fig. 2 shows the rapidity transverse momentum dependencies of the isospin flow in- (right) and out- (left) of-the reaction plane, respectively. A stronger collective flow is seen with the calculations including the δ meson in the iso-vector channel. This effect becomes very pronounced for the elliptic flow v_2 of high energetic ($p_t^{(0)} \geq 0.4$) particles due to the fact that those particles are emitted earlier during the formation of the high density asymmetric matter [13].

We can understand the observed effects by referring to Eq. (1). The ρ meson has a repulsive vector character, whereas the δ meson exhibits an attractive scalar one. This Lorentz decomposition is more dominant in the dynamical situation due to relativistic effects: the ρ meson linearly increases with the Lorentz γ factor, whereas the δ meson is not affected by such dynamical effects since the scalar density is a Lorentz scalar. Thus,

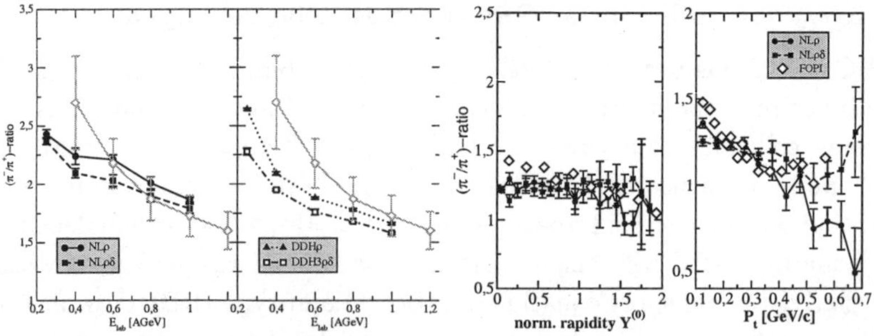

Figure 3. Left: energy dependence of the (π^-/π^+)-ratio for central ($b < 2\ fm$) $Au+Au$ reactions. Calculations with $NL(\rho, \rho\delta)$ and $DDH(\rho, \rho\delta)$ are shown as indicated. Right: rapidity (y^0) and transverse momentum ($p_t^{(0)}$) dependence of the (π^-/π^+)-ratio for central ($b < 1.5\ fm$) $Ru+Ru$ reactions with $NL\rho$ and $NL\rho\delta$ ($y^{(0)}$ and $p_t^{(0)}$ are normalized to the corresponding quantities of the projectile per nucleon). The open diamonds shown in all the figures are FOPI data taken from [14,15] (the figure is taken from [6]).

the stiffness of the symmetry energy is effectively enhanced when including the δ meson in these descriptions which yields more repulsion for neutrons than for protons with the net effect of a stronger "differential" (isospin) collective dynamics in the $NL\rho\delta$ case.

(2) Particle production

Particle production at GSI energies is also directly related to the dynamics of the earlier high density stage of a heavy ion collision. The most dominant inelastic channels are the production of the lowest mass resonances $\Delta(1232)$ and $N^*(1440)$. They are mainly produced during the high density phase in the first nucleon-nucleon collisions and they decay into pions ($\pi^{\pm,0}$). Furthermore, strange particles like kaons are created together with hyperons ($Y = \Lambda, \Sigma$) due to strangeness conservation through baryon-baryon ($BB \longrightarrow BYK^+$ with $B = p, n, \Delta$) and π-baryon ($\pi B \longrightarrow YK^+$) collisions.

Fig. 3 shows the energy, rapidity and transverse momentum dependence of the (π^-/π^+)-ratio. This ratio is reduced on the average in the models which contain the δ meson in the iso-vector EOS, only for high energetic pions ($p_t^{(0)} \geq 0.35$) the trend is seen to be opposite. The observed isospin effects mainly originate from (a) the different density dependence of the symmetry energy and (b) the effective mass splitting ($m_n^* < m_p^*$): (a) due to the stiffer character of E_{sym} neutrons are emitted earlier than protons making the high density phase more proton rich. On the other hand, π^- particles are essentially produced via negative charged resonances Δ^-, for

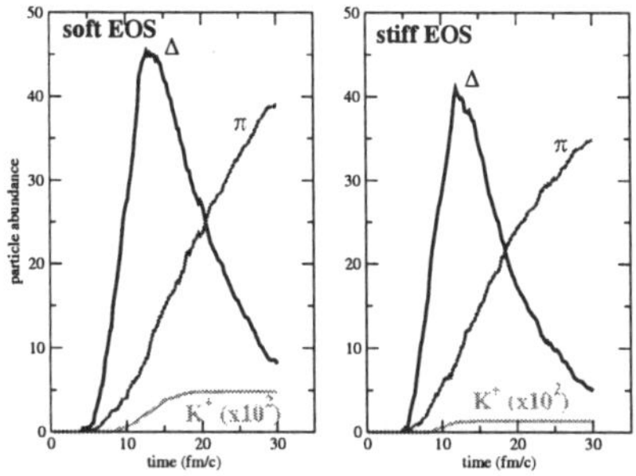

Figure 4. Time evolution of Δ resonances (Δ), pions (π) and kaons (K^+) for central ($b = 0\ fm$) $Au + Au$ reactions at 1 $AGeV$ beam energy. Calculations with a soft ($NL2$ with a compressibility of 200 MeV) and stiff ($NL3$ with a compressibility of 380 MeV) EOS within the non-linear Walecka model are shown (taken from [16]).

example trhough the process $nn \longrightarrow p\Delta^-$, which then decay into π^-. Thus due to the earlier neutron emission one observes a reduction of the (π^-/π^+)-ratio when the δ field is included in the description. This interpretation is also valid for the more complicated cases of the $DDH(\rho, \rho\delta)$ models. (b) The effective mass splitting leads additionaly to threshold effects since in the $(NL, DDH)\rho\delta$ cases less kinetic energy $\sqrt{s^*} = m_n^* + p^{*2}$ is available for resonance production due to the decrease of m_n^*.

However, the comparison with very preliminary FOPI data does not yet support any definitive conlusion. One reason could be that pions interact strongly with the hadronic enviroment due to absorption effects in secondary collisions and the Coulomb interaction. Furthermore, with increasing beam energy these secondary effects increase (more energy available). Therefore, pion production takes place over all the collision process after compression reducing the high density isospin effect. It may be also more useful to select particles directly emitted from the high density region. This can be done by choosing pions with high transverse momenta p_t [17], since in other studies [13] it was found that baryons are emitted the earlier, the higher their transverse momentum is. This is seen in Fig. 3, where the differences between $NL\rho$ and $NL\rho\delta$ turn out to be more important for high transverse momenta $p_t^{(0)}$. In particular, for low $p_t^{(0)} < 0.35$ the (π^-/π^+)-ratio is reduced with the $NL\rho\delta$ model, in consistency with the previous discussion. Since the multiplicity is maximal at this $p_t^{(0)}$ region, on average one obtains a reduction of the (π^-/π^+)-ratio with the $NL\rho\delta$ model. However, for high energetic particles the situation is different. The reason for the increase of the (π^-/π^+)-ratio for $p_t^{(0)} >> 0.35$ arises from a combination of isospin and Coulomb effects, as detailed discussed in Ref. [6].

The kaon production turns out to be a *better candidate* (Fig. 4), since they are produced directly during the high density phase without any secondary effects like the pions. The kaon yield strongly depends on the EOS, in contrary to the pions, see 4. Thus, one will expect to set more stringent constraints on the high density symmetry energy from kaon production since there are a lot of experimental studies. Such a progress is under investigation.

(2) Isospin transparency in the mixed Ru(Zr)+Zr(Ru) systems: Another interesting aspect in HIC's is the isospin transparency which has experimentaly been extensively studied [18]. The idea is to use collisions between equal mass nuclei $A = 96$, but different isotopes (Ru and Zr) which can be taken as projectile and target by making use of all four combinations $Ru(Zr) + Ru(Zr)$ and $Ru(Zr) + Zr(Ru)$. The following imbalance ratio of differential rapidity distributions for the mixed reactions $Ru(Zr) + Zr(Ru)$, $R(y^{(0)}) = N^{RuZr}(y^{(0)})/N^{ZrRu}(y^{(0)})$, was considered, where $N^i(y^{(0)})$ is the particle yield inside the detector acceptance at a given rapidity for $Ru + Zr$, $Zr + Ru$ with $i = RuZr$, $ZrRu$. The observable R can be particularly determined for different particle species, like protons, neutrons, light fragments such as t and ^3He and produced particles such as pions ($\pi^{0,\pm}$), kaons, etc. The observable R chracterizes different stopping scenarios. E.g. in the proton case $R(p)$ rises (positive slope) for partial transparency, falls (negative slope) for full local stopping and is flat when total isospin mixing is achieved in the collision. Therefore, $R(p)$ can be regarded as a sensitive observable with respect to isospin diffusion, i.e. to properties of the symmetry term.

Fig. 5 shows the rapidity dependence of R for different particles and energies. With the $NL\rho\delta$ model R decreases for protons and increases for neutrons at rapidities near target one. At mid rapidity $R \approx 1$ means full isospin mixing, as expected. In an ideal case of full transparency R should approach the initial value of $R(p) = Z^{Zr}/Z^{Ru} = 40/44 = 0.91$ and $R(n) = N^{Zr}/N^{Ru} = 56/52 = 1.077$ for protons and neutrons at target rapidity, respectively. We see that this is approximately the case when the δ meson is taken into account in the iso-vector channel. This effect is obvious since in the $NL\rho\delta$ model the neutrons experience a more repulsive iso-vector mean field at high densities than the protons leading to less degree of stopping. This isospin effect is moderate at low, but more essential at higher beam energy due to the higher compression in the later case.

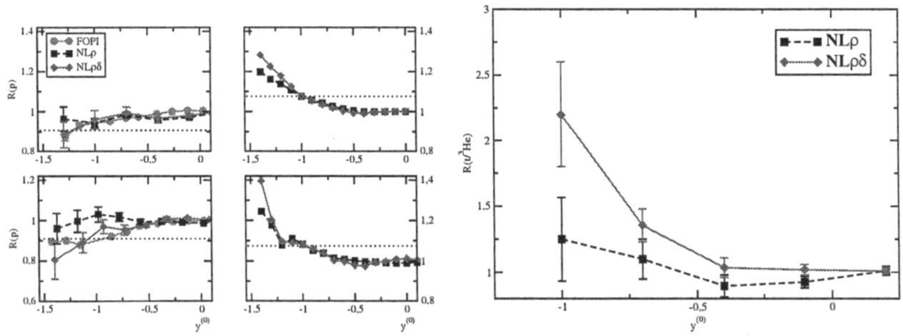

Figure 5. Left: rapidity ($y^{(0)}$) dependence of the imbalance ratio for protons $R(p)$ (top and bottom on the left) and neutrons $R(n)$ (top and bottom on the right) for central ($b = 1.5\ fm$) mixed reactions $Ru(Zr) + Zr(Ru)$ at 0.4 $AGeV$ (top) and 1.528 $AGeV$ (bottom) beam energy. Right: The same but for the ratio of tritons (t) to ^3He at 0.4 $AGeV$ beam energy. Calculations with $NL\rho$ (squares) and $NL\rho\delta$ (circles) are shown and compared with FOPI data [15] as indicated (the figure is taken from [19]).

It is very important to stress the opposite behavior of R as function of rapidity between protons and neutrons which will result to an essential difference between $NL\rho$ and $NL\rho\delta$ models for the same observable R, in particular, of the ratio of t to ^3He fragments as it is, ideed, the case. Our finding for the imbalance ratio R of $R(t/^3He)$ is in full agreement with a transparency scenario which, in particular, becomes more pronounced if the δ meson is taken in these descriptions into acount. The comparison with FOPI data seems to support a stiffer symmetry energy for high densities, i.e. the importance of the δ meson in the description of asymmetric nuclear matter. Corresponding experimental data for $R(t/^3He)$ would give a more precise conclusion.

4. Final remarks

We have analyzed the relativistic features of the iso-vector part of the equation of state by means of a covariant description of symmetric and asymmetric nuclear matter. Nuclear matter studies indicate that the stiffness of the symmetry energy is mainly dominated by the introduction of a iso-vector, scalar δ meson which significantly changes the Lorentz structure of the iso-vector part of the mean field potential at high densities.

In dynamical situations of heavy ion collisions the high density part of the symmetry energy has been studied in terms of different observables which may be directly linked to the density dependence of the symmetry energy. Observables which are related to the earlier high density phase of the process show the strongest effects arising from the different treatment of the iso-vector EOS. The collective isospin flow, the transverse momentum dependence of the (π^-/π^+)-ratio and the imbalance ratio of clusters seem to be very good candidates for studying isospin effects. In fact the kaon production might by the best observable for such investigations. First preliminary resluts strongly support this interpretation [20].

An this level of investigation we conclude that the symmetry energy should exhibit a stiff behavior at supra-normal densities which can be achieved by the introduction of an additional degree of freedom (iso-vector, scalar δ channel). The comparison with microscopic DBHF models supports our findings. Furthermore, more heavy ion data with radioactive beams are needed to make final definite statement.

References

1. W. Reisdorf, H.G. Ritter, *Annu. Rev. Nucl. Part. Sci.* **47**, 663 (1997); N. Herrmann et al., *Annu. Rev. Nucl. Part. Sci.* **49**, 581 (1999).
2. Bao-An Li, *Phys. Rev.* **C67**,017601 (2003).
3. V. Greco et al., *Phys. Lett.* **B562**, 215 (2003).
4. J.D. Walecka, *Ann. Phys. (N.Y.)* **83**, 497 (1974).
5. B. Liu et al., *Phys. Rev.* **C65**, 045201 (2002).
6. T. Gaitanos et al., *Nucl. Phys.* **A732**, 24 (2004).
7. C. Fuchs, H. Lenske, H.H. Wolter, *Phys. Rev.* **C52**, 3043 (1995).
8. F. de Jong, H. Lenske, *Phys. Rev.* **C57**, 3099 (1998).
9. S. Typel, H.H. Wolter, *Nucl. Phys.* **A656**, 331 (1999).
10. E.N.E. van Dalen, C. Fuchs, A. Faessler, nucl-th/0407070.
11. W. Botermans, R. Malfliet, *Phys. Rep.* **198**, 115 (1990).
12. A. Andronic, W. Reisdorf, N. Hermann et al. (FOPI collaboration), *Phys. Rev.* **C66**, 034907 (2003).
13. T. Gaitanos et al., *Eur. Phys. J.* **A12**, 421 (2001).
14. W. Reisdorf (FOPI collaboration), private communication of very preliminary data.
15. B. Hong (FOPI collaboration), GSI-Report 2002.
16. T. Gaitanos, M. Di Toro, G. Ferini, M. Colonna, H.H. Wolter, in: Proccedings XLII International Winter Meeting on Nuclear Physics, Bormio, Italy, 2004, nucl-th/0402041.
17. V.S. Uma Maheswari et al., *Phys. Rev.* **C57**, 922 (1998).
18. W. Reisdorf (FOPI collaboration), *Acta Phys. Polon.* **B33**, 107 (2002); F. Rami et al. (FOPI collaboration), *Phys. Rev. Lett.* **84**, 1120 (2000).
19. T. Gaitanos, M. Di Toro, M. Colonna, H.H. Wolter, *Phys. Lett.* **B595**, 209 (2004).
20. G. Ferini, T. Gaitanos, M. Di Toro, M. Colonna, in preparation.

COMPOUND AND QUASI-COMPOUND STATES IN THE LOW ENERGY SCATTERING OF NEUTRONS AND PROTONS BY THE ^{12}C NUCLEUS

G. PISENT, L.CANTON

Dipartimento di Fisica dell'Università di Padova, and Istituto Nazionale di Fisica Nucleare, sezione di Padova, via Marzolo 8, Padova 35131, Italia

J. P. SVENNE

Department of Physics and Astronomy, University of Manitoba, and Winnipeg Institute for Theoretical Physics, Winnipeg, Manitoba, Canada R3T 2N2

K. AMOS, S. KARATAGLIDIS

School of Physics, University of Melbourne, Victoria 3010, Australia

D. VAN DER KNIJFF

Advanced Research Computing, Information Division, University of Melbourne, Victoria 3010, Australia

In a recent paper [1], a multichannel algebraic scattering (MCAS) theory for nucleons scattering from a nucleus was specified in detail and applied to the analysis of the $n -^{12} C$ low energy scattering.

Here we extend calculations from the ^{13}C to the ^{13}N system, and carry out a comparative analysis of the spectroscopy involved.

It will be shown that, in the $n -^{12} C$ process, the spectrum of resonances up to about 6 MeV (in the Lab system) is almost completely described by a mechanism of excitation of the first 2^+ level of ^{12}C, at energy $\epsilon_2 = 4.4389\ MeV$. The spectrum shows a sequence of compound resonances, generated by the $\frac{1}{2}^-$, $\frac{1}{2}^+$, $\frac{5}{2}^+$ bound states in ^{13}C. The situation is very similar in the $p -^{12} C$ process, with one overall energy shift due to the Coulomb interaction. Because of this shift in energy, many of the compound resonances become quasi-compound ones, and the phenomenology involved becomes more and more interesting. The main idea of the paper is the following. When β tends to zero the compound and quasi compound states tend to be pure states. In the compound resonances the width tends to zero and the resonance energy tends to the energy of the single particle bound state plus the core excitation ϵ_2. In the quasi compound ones, the width tends to the natural width of the single particle resonance and the energy tends to the energy of the single particle resonance plus the core excitation ϵ_2 [2]. It is therefore very interesting to analyze the behavior of the phenomenology contained in the model, as β varies continuously from the physical value to zero, with the double purpose of checking the above outlined rules in a significant physical case, and to describe the spectroscopy of ^{13}C and ^{13}N on the ground of the above outlined schema.

1. The ^{13}C system

Calculations on the ^{13}C system have been carried out with the same parameters of ref [1], reported in table 1 for easy reference.

Table 1. $n - ^{12}C$ Potential Parameters.

parity	central	orbit-orbit	spin-orbit	spin-spin
−	-49.1437	4.5588	7.3836	-4.7700
+	-47.5627	0.6098	9.1760	-0.0520

The other parameters are:

$$r_0 = 1.35\ fm\ ;\ a_0 = 0.65\ fm\ ;\ \beta = -0.52. \qquad (1)$$

Couplings of the input channel with the ground state, the 2^+ ($\epsilon_2 = 4.4389\ MeV$), and the 0^+ ($\epsilon_0 = 7.6542\ MeV$) excited states of the C^{12} target are considered. The Pauli principle is taken into account throughout, as discussed in [1].

In figure 1, the calculated elastic cross section (energies in the laboratory system) is compared with experiments (the data were obtained from the files of the National Nuclear Data Center, Brookhaven, where source references are given). The sequence of resonances is: $\frac{5}{2}^+$, $\frac{7}{2}^+$, $\frac{1}{2}^-$, $\frac{3}{2}^+$, $\frac{3}{2}^+$, $\frac{5}{2}^+$, $\frac{9}{2}^+$.

In table 2, the parameters of the states (calculated by means of the sturmian eigenvalues of the problem, as explained in reference [1]), are compared with the experimental values (reference [3]). The energies of the states in the CM system are in MeV, while the widths are expressed in KeV.

The conventional number given to the state in the first column, is used in the discussion throughout.

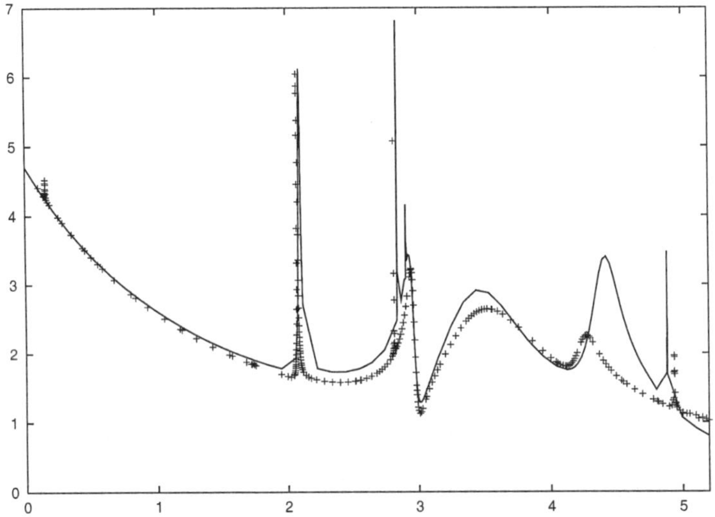

Figure 1. Comparison between experimental (cross) and theoretical (line) elastic cross section.

Let us consider now in particular the states n=3 of Table 2 assuming that it is mainly a single particle bound state. According to the chosen representation of the pure states $\{ J^\pi Ijl \}$ (see [1] for the meaning of the quantum numbers), this amounts to say that it is dominated by the component $\{ \frac{1}{2}^+, 0, \frac{1}{2}, 0 \}$.

Now the expected mechanism under weakly coupling conditions is the following: as the neutron is bound to the $^{12}C(0^+)$ ground state core with the above quantum numbers, giving rise to a bound state of energy $E \simeq -2\ MeV$, so the neutron impinging with energy $\epsilon_2 + E$, loses ϵ_2 for excitation and is bound in the $^{12}C^*(2^+)$ core, giving rise to a doublet of compound resonances $J^\pi = \frac{3}{2}^+$ and $J^\pi = \frac{5}{2}^+$. In fact these are the only states that contain one component of the type $\{J^+, I, j = \frac{1}{2}, l = 0\}$ (considering only couplings with the $^{12}C(0^+)$ excited state core). The possible candidates are the states n=6,10. In fact form Table 2 we see that $E_{th}(n=6) - E_{th}(n=3) = 4.8\ MeV$ and $E_{th}(n=10) - E_{th}(n=3) = 4.0\ MeV$, not far from $\epsilon_2 = 4.4\ MeV$.

A general search of all possible couplings may be carried out by starting from the physical situation $\beta = -0.52$, and switching off gradually the coupling constant, tracking continuously each state up to $\beta = 0$ [a]. By putting $\beta = 0$ and $V_{ss} = 0$, the differences of the energies of all states coupled together is exactly equal to the energy of ϵ_2 or ϵ_0 [b].

By means of these techniques we have analyzed the whole spectrum, arriving to the interpretation which is schematically outlined in the figure 2.

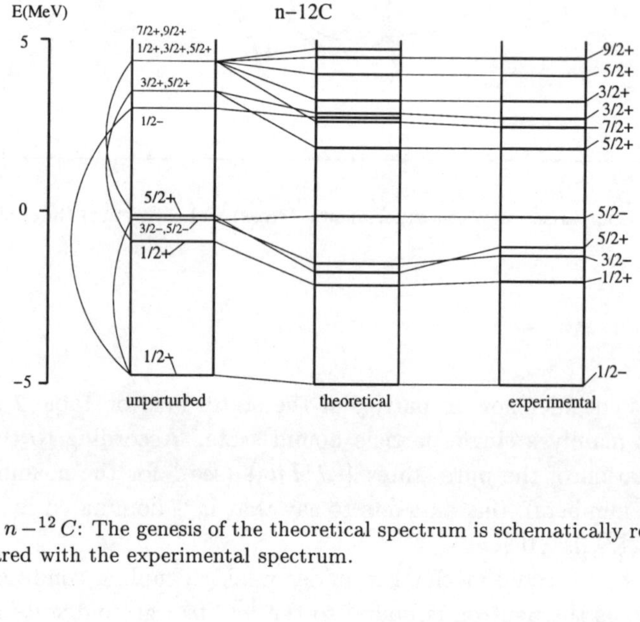

Figure 2. $n - {}^{12}C$: The genesis of the theoretical spectrum is schematically represented and compared with the experimental spectrum.

The ${}^{13}C$ system supports 3 (single particle) bound states, namely $\frac{1}{2}^-, \frac{1}{2}^+, \frac{5}{2}^+$ (see the first box in the figure 2, representing the unperturbed spectrum). The coupling of the incoming neutron with the considered excited core levels, gives rise to meta stable states, whose presumable unperturbed ($\beta = 0$) configurations are represented by thin lines, connected with

[a] Of course when β is exactly null, the cross section loses any trace of compound and quasi compound resonances, but nevertheless the program is able to follow the tracks of the Sturmian eigenvalues and find the unperturbed position of resonances, whose width is zero in the compound case.

[b] With V_{ss} the differences of energies are only approximately null, because it is the only not central potential whose diagonal matrix elements depend on the total momentum J.

the "father" by curve lines (the distance between "father" and "sons" [c] is ϵ_2 or ϵ_0 in the two possible cases); these unperturbed resonances are supposed to be infinitely narrow ($\Gamma \to 0$), and fully degenerate (neglecting in this representation the small effect of the spin-spin interaction). The finite deformation ($\beta = -052$) splits (and enlarges) the resonances, as shown in the second box. The third box gives the experimental spectrum for comparison.

Then we have analyzed the ^{13}N system, assuming that charge symmetry holds exactly, namely using the parameters of Table 1 and switching on the Coulomb interaction. We have only one new parameter which is the coulomb radius R_c. The best agreement with the experiments has been obtained with the value $R_c = 2.4\ fm$.

The table 3 shows the results of calculations relative to proton scattering, in a form similar to table 3 for neutron scattering.

There is a one to one correspondence between states in ^{13}C and ^{13}C. Table 3 shows one resonance more with respect to Table 2 (n=14). As explained below in the figure 5, this comes from coupling of the $\frac{5}{2}^+$ n=9 single particle resonances with the 0^+ excited core state.

Examples of comparison between theory and experiments are shown in figures 3, 4.

Figure 3, gives a comparison between theory and experiment for the differential cross section behavior, between 1 and 7 MeV (in Lab), for the scattering angle $\theta = 54^o$. White circles are taken from reference [4], while black squares come from reference [5]. The resonance $\frac{3}{2}^-$ (n=5) is very well reproduced. The resonances $\frac{5}{2}^+$ (n=10), $\frac{3}{2}^+$ (n=6), and the broad $\frac{3}{2}^+$ (n=7), are well reproduced in shape, with a small shift in energy which can be shown also in Table 3.

The Figure 4 gives an example of A_y angular distribution at fixed energy. The experimental points are taken from reference [4]. A far from resonances energy (3.5 MeV) has been chosen, and the angular distribution shape is seen to be reproduced very well.

The spectrum of ^{13}N has been analyzed with the same techniques of

---

[c]The terms "father" and "son" used in this context are self explanatory; some considerations on this point are nevertheless worthwhile. The correct definition of the parentage relation is possible in the framework of what we call the "unperturbed" ($\beta = 0$) conditions of the system. Of course, under the "physical" conditions, characterized by a finite β value, this parentage becomes a little fuzzy. Nevertheless we find that each state may be followed continuously from physical coupling to zero coupling, and this proves that the parentage between states is still meaningful even with reference to the true physical states.

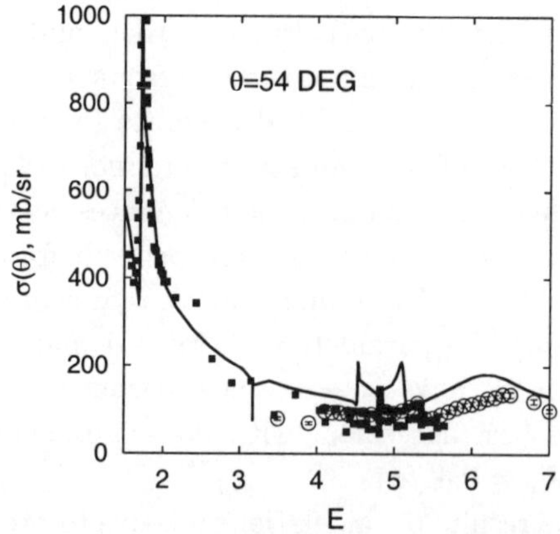

Figure 3. $p - ^{12}C$: Comparison between experimental (points) and theoretical (line) differential cross section for the scattering angle $\theta = 54^o$. Differential cross sections in mb/sr, CM system, energies in MeV, laboratory system.

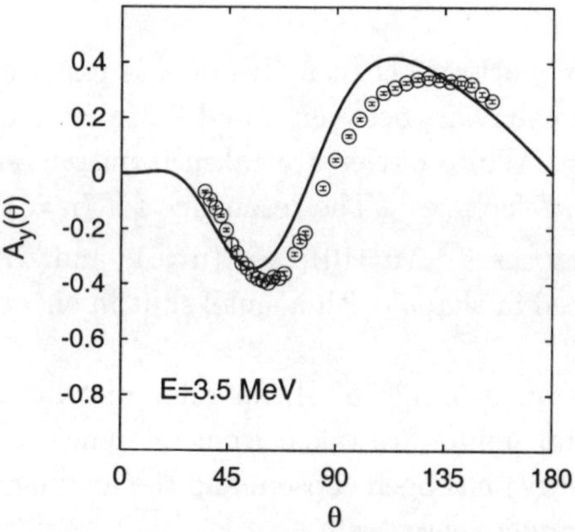

Figure 4. $p - ^{12}C$: Comparison between experimental (points) and theoretical (line) angular distribution of the polarization A_y, at the energy of 3.5 MeV.

^{13}N, and the results are summarize in figure 5, very similar to figure 2.

In conclusion the model reproduces and interprets in a satisfactory and exhaustive way, with a unique set of potential parameters, the phenomenology of both the ^{13}C and ^{13}N systems, in the considered energy range.

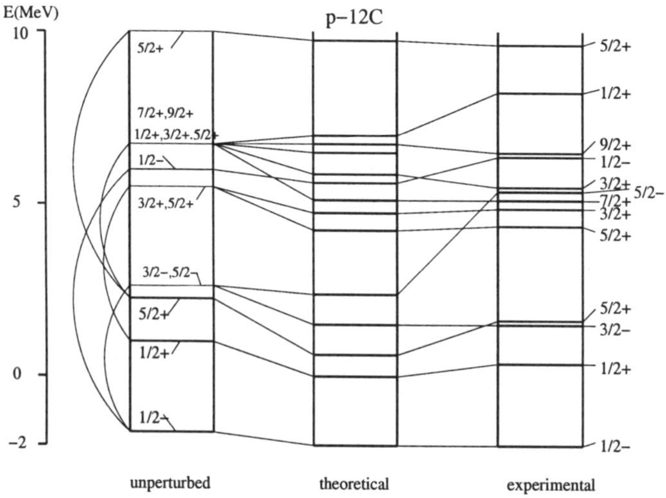

Figure 5. $p - ^{12}C$: The genesis of the theoretical spectrum is schematically represented and compared with the experimental spectrum.

References

1. K.Amos, L. Canton, G. Pisent, J.P. Svenne, and D. van der Knijff, Nucl. Phys. **A728** (2003) 65.
2. G.Pisent, J.P.Svenne, Phys Rev C **51** 3211 (1995)
3. F.Ajzenberg-Selove, Nucl. Phys. **A523** (1991) 1.
4. L.Sydow et al. Nuclear Instruments and Methods in Physics Research **A327** (1993) 441.
5. C.W.Reich, G.C.Phillips and J.L.Russel, Phys Rev **C104** (1956) 143.

Table 2. Comparison between experiment and theory, $n +^{12} C$.

n	J^π	E_{exp}	$\Gamma_{exp}(KeV)$	E_{th}	$\Gamma_{th}(KeV)$
1	$\frac{1}{2}^-$	-4.9463	—	-4.8881	—
2	$\frac{1}{2}^-$	—	—	2.6829	0.332
3	$\frac{1}{2}^+$	-1.8569	—	-2.0718	—
4	$\frac{1}{2}^+$		110	4.6629	555
5	$\frac{3}{2}^-$	-1.2618	—	-1.4783	—
6	$\frac{3}{2}^+$	2.7397	70	2.7309	40.8
7	$\frac{3}{2}^+$	3.2537	1000	3.2447	447
8	$\frac{5}{2}^-$	0.1	—	-0.0338	—
9	$\frac{5}{2}^+$	-1.0925	—	-1.8619	—
10	$\frac{5}{2}^+$	1.9177	6	1.9348	9.65
11	$\frac{5}{2}^+$	3.9314	110	4.0579	126
12	$\frac{7}{2}^+$	2.547	5	2.6220	8.74×10^{-4}
13	$\frac{9}{2}^+$	4.534	5	4.5091	0.745

Table 3. Comparison between experiment and theory, $n +^{12} C$.

n	J^π	E_{exp}	$\Gamma_{exp}(KeV)$	E_{th}	$\Gamma_{th}(KeV)$
1	$\frac{1}{2}^-$	-1.9435	—	-1.9104	—
2	$\frac{1}{2}^-$	6.9745	230	5.6391	17.8
3	$\frac{1}{2}^+$	0.4214	31.7	-0.0158	—
4	$\frac{1}{2}^+$	8.3065	280	6.9911	995
5	$\frac{3}{2}^-$	1.5675	62	1.5793	11.1
6	$\frac{3}{2}^+$	4.9425	115	4.7280	44.5
7	$\frac{3}{2}^+$	5.9565	1500	5.8942	653
8	$\frac{5}{2}^-$	5.4325	75	2.9281	$3.38 \cdot 10^{-3}$
9	$\frac{5}{2}^+$	1.6035	47	0.6379	0.899
10	$\frac{5}{2}^+$	4.4205	11	4.1794	8.87
11	$\frac{5}{2}^+$	—	—	6.5281	726
12	$\frac{7}{2}^+$	5.2115	9	5.1234	1.50
13	$\frac{9}{2}^+$	7.0565	280	6.8341	153
14	$\frac{5}{2}^+$	9.5865	430	9.7895	910

STRUCTURE AND REACTIONS WITH EXOTIC NUCLEI *

ANGELA BONACCORSO †

INFN, Sez. di Pisa,
and Dipartimento di Fisica, Università di Pisa
Largo Pontecorvo 3,
56127 Pisa, Italy
E-mail: bonac@df.unipi.it

The INFN (Italian National Institute for Nuclear Physics) has approved a national theoretical network on "Structure and Reactions with Exotic Nuclei". The project involves the INFN branches of Laboratorio Nazionale del Sud, Padova and Pisa. The aim of the project is to coordinate and homogenize the research already performed in Italy in this field and to strengthen and improve the Italian contribution on the international scenario. Furthermore it aims at creating a solid theoretical structure to support future experimental facilities at the INFN national laboratories such as SPES at LNL and EXCYT at LNS. A review of present and future activities is presented

1. Introduction

Since a few years an increasing number of Italian theoreticians has concentrated his research on the study of exotic nuclei. Such activities have so far been carried out within pre-existing national projects related to a wide spectrum of themes of nuclear dynamics, structure and reactions using many body techniques, shell model, collective modes and semiclassical or fully quantum mechanical approaches to peripheral and central reactions such as transfer and breakup, fusion, elastic scattering via microscopic optical potentials, multifragmentation.

The goal of our project is to start coordinating and homogenizing such efforts to improve our mutual understanding, and to strengthen the Italian contribution on the international scenario. Furthermore our efforts will

---

*within the INFN-PI32 network.

†In collaboration with G. Blanchon and A. Garcia-Camacho, INFN Sez di Pisa; M. Colonna, M. Di Toro, Cao Li Gang, U. Lombardo, J. Rizzo, INFN- LNS; S. Lenzi, P. Lotti, A. Vitturi, INFN Sez. di Padova.

help creating a solid theoretical structure to support future experimental activities at the INFN national laboratories.

In fact, in the last two decades, the use of radioactive beams of rare isotopes in several laboratories around the world (REX-ISOLDE at CERN, GANIL in France, GSI in Germany, CRC, Louvain la Neuve in Belgium, RIKEN in Japan, DUBNA in Russia, Argonne, MSU, Oak Ridge, Notre Dame in USA , etc.) has provided new research directions and an increasing number of researchers all over the world is converging on such subject. The INFN in Italy is also heavily involved in this field. The facility EXCYT and the large acceptance spectrometer called MAGNEX are being completed at Laboratorio Nazionale del Sud. On the other hand the first step of the SPES project at the Laboratorio Nazionale di Legnaro has been approved in the form of a proton driver. Furthermore the INFN is promoting the new European Radioactive Beam Facility (EURISOL). Members of our collaboration are actively participating in NuPECC working groups, in particular in the preparation of "The Physics Case" for EURISOL, whose report is available at http://www.ganil.fr/eurisol/Final_Report/A-Physics-Case-20-Dec-02.pdf, and in general of the NuPECC Long Range Plan.

The relatively new subject of exotic nuclei is of fundamental importance because while all existing theories for the nuclear interaction and the many body nuclear structure have been based on the study of stable nuclei, very little is known about the way in which standard nuclear models work for the description of unstable nuclei with anomalous N/Z ratio. Important questions to answer are for example: the isospin dependence of the effective nuclear interaction, the modification of the traditional shell sequence with possible vanishing of the shell gaps, the persistence of collective features, the properties of nuclear matter at very low density, the form of the EOS for asymmetric nuclear matter. Similarly in the field of nuclear reactions still open questions are the identifications of the most important reaction channels and the clarification of the associated reaction mechanisms. Many of these features are also related to nuclear reactions of astrophysical interest such as those governing the primordial nucleosyntesis.

The proposed research activity will deal with the following aspects: reaction mechanisms and structure information extraction for nuclei close to the driplines, single particle and collective degrees of freedom, dynamical symmetries at the phase transitions, dynamics of heavy nuclei with anomalus N/Z ratios and isospin degrees of freedom, equation of state. The partecipants have complementary competences in the fields of structure and reaction theory. They have common national (LNS,LNL) and international collaborations (ie : IPN, Orsay; GANIL, Caen, France, MSU, USA, etc.). Their present abilities and activities in the above research fields are described in the following.

2. Reaction Mechanisms by the Pisa Group

In Pisa there is a longstanding tradition for studying peripheral reactions such as transfer and breakup, therefore it has been easy and natural to tourn our attention to the study of halo nuclei [1-8].

In recent years we have concentrated on a consistent treatment of nuclear and Coulomb breakup and recoil effects treated to all orders and including interference effects. We have developed a formalism which allows the calculations of energy, momentum and angular distributions for the core and halo particle and absolute cross sections. The possibility of calculating so many observables is almost unique to our model. The dependence on the final state interaction used has been clarified and also the accuracy of the eikonal model compared to fully quantum mechanical theories has been established.

An extension of the method to proton breakup has been recently presented and we plan to apply it to the study of reaction of astrophysical interest such as those involving ^{8}B. Finally a microscopic model for the calculation of the optical potential in the breakup channel has been developed. The method originally used to calculate elastic scattering of halo projectiles on ligth targets is now being extended to heavy targets by the inclusion of recoil effects. Also we are extending our techniques to the calculations of angular correlations.

In the last period we have started to study nuclei unbound against neutron emission, such as ^{10}Li and ^{13}Be. They are the constituents of two neutron halo nuclei (i.e. ^{11}Li and ^{14}Be). The study of their low lying resonance states is of fundamental importance for the understanding of two neutron halo nuclei. The final goal is to clarify the structure of the core-neutron interaction. This is by no means a trivial task as such cores are themselves unstable nuclei (^{9}Li, ^{12}Be) and therefore cannot be used as target in experimental studies. We are at present discussing the differences between the technique of projectile fragmentation and of transfer to the continuum in order to understand whether they would convey the same structure information. This line of research is leading us naturally to study the structure of few-body systems which we plan to undertake in a near future.

3. Reaction Mechanisms and Structure of Rare Isotopes by the Padova Group

The Padova group has similar and complementary lines of research as the Pisa group as far as reaction mechanisms are concerned. However it has a special interest for a somehow lower energy domain where fusion and the coupling to breakup channels are particularly important [9-24]. Besides it is active in studying structure problems such as:

- Study of the pairing correlations in low-density nuclear systems, as in the external part of halo nuclei.

- Microscopic estimate of inelastic excitation to the low-lying continuum dipole strength via microscopic continuum RPA calculations.

- Study of isospin symmetry in low- and high-spin states in medium-mass N=Z nuclei up to ^{100}Sn. Study of the interplay of T=0 and T=1 pairing.

- Study of nuclear structure with algebraic models. This line of research is associated with the use of algebraic models, as the Interacting Boson Model or its variations, to describe different aspects of nuclear spectra. Our traditional approach is based on the use of the concept of boson intrinsic state. In this framework we will study the new symmetries E(5) and X(5) associated with phase transitions and individuate mass region far from stability where such critical points may occur.

- Study of the role of continuum-countinuum coupling in the break-up of weakly-bound nuclei.

4. Isospin Dynamics in Reactions with Exotic Beams at LNS

Two teams are active at the LNS. One is working in the energy range from the Coulomb barrier (Tandem) to the Fermi energies (Superconducting Cyclotron). Our main motivation is to extract physics information on the isovector channel of the nuclear interaction in the medium from dissipative collisions in this energy range using the already available stable exotic ions and in perspective the new radioactive facilties. We have developed very reliable microscopic transport models, in a extended mean field frame, for the simulations of the reaction dynamics in order to check the connection between the tested effective interactions and the experiments, in particular for the isospin degree of freedom [25-39]. This work is of interest for the understanding of the physics behind the reaction mechanisms and for the selection of observables most sensitive to different features of the nuclear interaction. Moreoever we have a more general theoretical activity on the isospin dynamics in nuclear liquid-gas phase transitions. New instabilities have been evidenced with a different "concentration" between the gas and cluster phases, leading to the Isospin Distillation effects recently observed in experiments. A quantitative analysis can give direct information on the density dependence of the symmetry term for dilute asymmetric matter, i.e. around and below saturation. We remind the poor knowledge of the isovector part of the nuclear effective interaction at low densities, which is actually of large interest even for structure calculations of drip line nuclei.

The main results obtained in the last year are related to:

1) Isospin dynamics in low energy dissipative collisions.

2) Isospin in Nuclear fragmentation.

5. Finite Nuclear Systems in Brueckner Theory at LNS

The second team at LNS is interested in relating nuclear properties to elementary interactions between nucleons and to build up an energy density functional starting from a more fundamental level than the present phenomenological energy functionals of non-relativistic mean field or RMF [40-46]. This can be achieved because of the familiarity of the group with the Brueckner theory in infinite nuclear matter including 2-body and 3-body forces. It has been shown that the inclusion of 3-body forces in the Brueckner theory is necessary for obtaining the correct saturation point of nuclear matter and going away from the so-called Coester line. From the results of infinite matter one will construct an energy density functional which can give the same results in nuclear matter and also can be used in finite nuclei. This is quite similar to the energy functional method of atomic physics based on ab initio calculations of the homogeneous electron gas and the local density approximation (LDA). This nuclear energy functional should be trustable away from the stability region since no adjustment will be made to reproduce the properties of stable nuclei, contrarily to phenomenological energy functionals whose extrapolations can be questionable.

The proposed method is a simpler alternative than direct Brueckner calculations of finite systems. It also allows for studies of excitations of nuclei, within RPA-type of calculations built on top of the mean field ground state. This is again in the same spirit as the time-dependent LDA (TDLDA) method which has proved very successful in atomic cluster physics. The main objectives of the project are:

- BHF calculations of asymmetric and polarized matter.

- Construction of the energy functional.
- Ground states of finite nuclei.
- Excitations of finite nuclei.
- Neutron star crust

6. Conclusions

We have presented the main research lines of the new INFN-PI32 theory network on exotic nuclei. They span from low energy reaction theory for elastic scattering and fusion, to intermediate energies studies for breakup and multi-fragmentation for the understanding of the isospin degree of freedom. Structure studies on the pairing problem, on algebraic models and on the Brueckner theory are also actively pursued.

Acknowledgments

We wish to thank Prof. G. Marchesini, head of the national INFN Theory Group, for supporting the institution of the PI32 collaboration.

References

1. J. Margueron, A. Bonaccorso and D. M. Brink, nucl-th/0111011, *Nucl. Phys.* **A703**, 105 (2002).
2. J. Enders, A. Bauer, D. Bazin, A. Bonaccorso...et al, *Phys. Rev.* **C65**, 034318 (2002).
3. A. Bonaccorso and F. Carstoiu, nucl-th/0203018, *Nucl. Phys.* **A706**, 322 (2002).
4. J. Margueron, A. Bonaccorso and D. M. Brink, nucl-th/0303022, *Nucl. Phys.* **A720**, 337 (2003).
5. A. Bonaccorso and D. M. Brink and C. A. Bertulani, nucl-th/0302001, *Phys. Rev.* **C69**, 024615 (2004).
6. G. Blanchon, A. Bonaccorso and N. Vinh Mau, nucl-th/0402050, *Nucl. Phys.* **A739**, 259 (2004).
7. F. Carstoiu, E. Sauvan, N. Orr, A. Bonaccorso, nucl-ex/0406010, *Phys. Rev.* **C70**, 054602 (2004).
8. A. A. Ibraheem, A. Bonaccorso, nucl-th/0411091, *Nucl. Phys.* **A** , (2005) in press.
9. L. Fortunato, A.Vitturi, *J. Phys.* **G 30**, 627 (2004).
10. J. M. Arias, C.E. Alonso, A. Vitturi, J.E. Garcia-Ramos, J. Dukelsky, A. Frank, *Phys. Rev.* **C68**, 041302 (2003).
11. M. Mazzocco, ...A. Vitturi et al., *Eur. Phys. J* . **A18**, 583 (2003).
12. L. Fortunato , A. Vitturi, *J. Phys.* **G29**, 134 (2003).
13. C. H. Dasso, L. Fortunato, E. G. Lanza, A. Vitturi, *Nucl. Phys.* **A724**, 85 (2003).

14. L. Fortunato , A. Vitturi, *Nucl. Phys.* **A722**, 85c (2003).
15. P. Guazzoni,... A. Vitturi et al., *Czech. J. Phys.* **52**, c621, (2002).
16. G. Colo, S. M. Lenzi, E. E. Maqueda, A. Vitturi, *Phys. Rev.* **C67**, 044306 (2003).
17. C. Signorini,... A. Vitturi et al., *Phys. Rev.* **C67**, 044607 (2003).
18. A. Vitturi , *Prog. Theor. Phys. Supp.* **146**, 309 (2002).
19. A. Jungclaus, ... S. M. Lenzi et al., *Eur. Phys. J.* **A20**, 55 (2004).
20. C. Beck, ... S. M. Lenzi et al., *Nucl. Phys.* **A734**, 453 (2004).
21. Z. Podolyak, ... S. M. Lenzi et al., *Eur. Phys. J.* **A17**, 29 (2003).
22. P. Petkov,...S. M. Lenzi et al., *Phys. Rev.* **C67**, 054306 (2003).
23. G. de Angelis, ...S.M. Lenzi et al., *Phys. Lett.* **B 535**, 93 (2002).
24. H. D. Marta, L. F. Canto, R. Donangelo, P. Lotti, *Phys. Rev.* **C66**, 2 (2002).
25. V. Baran, M. Colonna, M. Di Toro, V. Greco, *Phys. Rev. Lett.* **86**, 4492 (2001).
26. M. Colonna, Ph. Chomaz and S. Ayik, *Phys. Rev. Lett.* **88**, 122701 (2002).
27. V. Baran, M. Colonna, M. Di Toro, V. Greco, M. Zielinska-Pfabe' and H. H. Wolter, *Nucl. Phys.* **A703**, 603 (2002).
28. V. Baran, M. Colonna and M. Di Toro, *Nucl. Phys.* **A730**, 329 (2004).
29. T. I. Mikhailova, I. N. Mikhailov, I. V. Molodtsova and M. Di Toro, *Part. Nucl. Lett.* **1**, 13 (2002).
30. P. M. Milazzo, G. Vannini,...M. Colonna..et al., *Nucl. Phys.* **A703**, 466 (2002).
31. D. Santonocito, P. Piattelli, Y. Blumenfeld,..M. Colonna,..et al., *Phys. Rev.* **C66**, 044619 (2002).
32. D. Pierroutsakou, M. Di Toro et al., *Eur. Phys. Journ.* **A16**, 423 (2003).
33. V. Baran, M. Colonna, M. Di Toro, V. Greco, M. Zielinska-Pfabe' and H. H. Wolter, *Phys. Atom. Nucl.* **66**, 1460 (2003).
34. I. N. Mikhailov, C. Briancon, T. I. Mikhailova, I .V. Molodtsova and M. Di Toro, *Phys. Atom. Nucl.* **66**, 1599 (2003).
35. T. X .Liu, X. D. Liu, M. J. .van Goethem, W. G. Lynch,..M. Colonna, M. Di Toro..et al., *Phys. Rev.* **C69**, 014603 (2004).
36. J. Rizzo, M. Colonna, M. Di Toro and V. Greco, *Nucl. Phys.* **A732**, 202 (2004).
37. M. Di Toro, V. Baran, M. Colonna, T. Gaitanos, J. Rizzo and H. H. Wolter, *Prog. Part. Nucl. Phys.* **53**, 81 (2004).
38. E. Geraci...M. Di Toro..et al., *Nucl. Phys.* **A732**, 173 (2004).
39. P. Sapienza, R. Coniglione, M. Colonna et al, *Nucl. Phys.* **A734** , 601 (2004).
40. W. Zuo, A. Lejeune. U .Lombardo and J. F. Mathiot, *Nucl. Phys.* **A706**, 418 (2002).
41. W. Zuo, A. Lejeune. U. Lombardo and J. F. Mathiot, *Eur. Phys. J.* **A14**, 469 (2002).
42. M. Baldo, U. Lombardo, E. E. Saperstein, M. V. Zverev, *Phys. Lett.* **B53**, 17 (2002).
43. W. Zuo, Caiwan Shen, U. Lombardo, *Phys. Rev.* **C67**, 037301 (2003).
44. C. W. Shen, U. Lombardo, N. Van Giai and W. Zuo, *Phys. Rev.* **C68**, 055802 (2003) and *Nucl. Phys.* **A722**, 532 (2003).

45. P. Ring, Z. Y. Ma, N. Van Giai, D. Vretenar, A. Wandelt, and L. G. Cao, *Nucl. Phys.* **A 694**, 249 (2001).
46. Z. Y. Ma, N. A. Wandelt, N. Van Giai, D. Vretenar, P. Ring, and L. G. Cao, *Nucl. Phys.* **A 703**, 222 (2002).

NUCLEAR MATTER PHASE TRANSITION IN INFINITE AND FINITE SYSTEMS

S. TERRANOVA AND A. BONASERA
Laboratori Nazionali del Sud
Via S. Sofia 44
95123 Catania
E-mail: terranova@lns.infn.it; bonasera@lns.infn.it

A new "semiclassical" model of the nuclear matter, composed of u, d colored quarks, is proposed. The approach, named *Constrained Molecular Dynamics* (CoMD) is based on the molecular dynamics simulation of the quarks, which interact through the Richardson's potential, and on a constraint due to Pauli blocking. With a suitable choice of the quark masses, some possible Equation of State (EOS) of the nuclear matter, at temperature equal to zero and finite baryon density, are obtained. These equations of state, not only present some known properties of the nuclear matter, as the *Quark-Gluon Plasma* (QGP) phase transition, but also shown the existence of a new state, the *Exotic Color Clustering* (ECC) state, in which cluster of quarks with the same color are formed. Some new quantities, "indicators" of the phase transition, are introduced: three order parameters, M_{c2}, M_{c3}, M_{c4} defined trough the Gell-Mann matrices λ^a, and the lifetime of the J/Ψ particle. The behavior of the J/Ψ particle is studied also in the "finite" systems, obtained by expanding the corresponding "infinite" systems. It seems that the dynamics and the finite size effects do not wash completely the phase transition occurred in infinite systems, and the J/Ψ particle is still a good signature.

1. Constrained Molecular Dynamic simulation (CoMD)

1.1. *Numerical method*

In the phase space, the dynamics of the quarks is solved classically through the evolution of the distribution function $f(r, p, t)$:

$$\partial_t f + \frac{\mathbf{p}}{E} \cdot \vec{\nabla}_{\mathbf{r}} f - \vec{\nabla}_{\mathbf{r}} U \cdot \vec{\nabla}_{\mathbf{p}} f = 0, \tag{1}$$

where $E = \sqrt{p^2 + m_q^2}$ is the total energy of each quark and m_q is the (u, d) quark mass. The colored quarks interact through the Richardson potential $V(\mathbf{r}, \mathbf{r}_j)$ [1] and $U = U(\mathbf{r}) = \sum_j V(\mathbf{r}, \mathbf{r}_j)$ is the total potential.

Numerically the Eq. (1) is solved by writing the one body distribution function for each particle i through the delta function :

$$f_i(\mathbf{r}, \mathbf{p}, t) = \sum_{\alpha=1}^{Q} \delta(\mathbf{r} - \mathbf{r}_\alpha)\delta(\mathbf{p} - \mathbf{p}_\alpha) \tag{2}$$

$Q = q + \bar{q}$ is the total number of quarks (q) and anti-quarks ($\bar{q}$); here $\bar{q} = 0$.

Inserting the new expression of the distribution function in the exact equation Eq. 1 the *Hamilton equations* are obtained; hence it must solve these equations of motion for the system of interacting quarks.

At the initial condition, the quarks are randomly distributed in a box of side L in coordinate space and in a sphere of radius p_f in the momentum space; p_f is the Fermi momentum estimated in a Fermi gas model. The Fermi momentum is related to the density of the colored quarks ρ_{qc} by the following expression: $\rho_{qc} = \frac{g_q}{3\pi^2} p_f^3$.

$g_q = n_f \times n_c \times n_s$ is the degeneracy number; $n_f = 2$ is the number of flavors (only u, d quarks are considered), $n_c = 3$ is the number of colors, $n_s = 2$ is the number of spins. The Pauli blocking implies that each cell in the phase space of size $h = 2\pi$, can accommodate at most g_q identical quarks.

To simulate an "infinite system", periodic boundary condition are imposed and many events are generated.

The Pauli blocking implies that the *Occupation Average*, which is the probability that a cell in the phase space is occupied, must be less or equal to 1, i.e. ($\bar{f}_i \leq 1$). In order to verify this condition the momenta of the particles are multiplied for a quantity ξ, i.e. $P_i = P_i \times \xi$. In particular if $\bar{f}_i$ is greater than 1, then ξ will be greater than 1; if $\bar{f}_i$ is less than 1, then ξ will be less than 1. This is the: "*Constrained Molecular Dynamics*", (CoMD)[2].

1.2. *The order parameters*

The study of the nuclear matter phase transition from confined hadronic matter to Quark-Gluon Plasma needs some unambiguous quantity able to check if the transition is happened or not. One of this quantity is the order parameter M_c which relates the colors of closest interacting quarks. In particular three different order parameter are defined: M_{c2}, M_{c3}, M_{c4}, the first is related to the color of two closest quark, the second to the color of three closest quarks and the last to the color of four closest quarks. The color matrices (λ^3 and λ^8), allow to define the order parameters,[3]:

$$M_{c2} = -\frac{3}{4} \sum_{a=3,8} \lambda_i^a \lambda_{j(i)}^a \tag{3}$$

where $j(i)$ is the closest quark to quark i. In the same way the other two order parameters can be defined:

$$M_{c3} = -\frac{3}{4}\frac{1}{N} \sum_{i=1}^{N} \sum_{a=3,8} \lambda_j^a \lambda_k^a + \lambda_i^a \lambda_j^a + \lambda_i^a \lambda_k^a \tag{4}$$

and

$$M_{c4} = -\frac{3}{4}\frac{1}{N} \sum_{i=1}^{N} \sum_{a=3,8} \lambda_j^a \lambda_k^a + \lambda_i^a \lambda_j^a + \lambda_j^a \lambda_l^a + \lambda_i^a \lambda_k^a + \lambda_k^a \lambda_l^a + \lambda_i^a \lambda_l^a \tag{5}$$

Here $k(i)$ and $l(i)$ are respectively the second quark and the third quark closest to the quark i.

The order parameters are normalized in this way:

$$\widetilde{M}_{c2} = \frac{2}{3}[M_{c2} + 1]; \ \widetilde{M}_{c3} = \frac{2}{9}[M_{c3} + 3]; \ \widetilde{M}_{c4} = \frac{2}{15}[M_{c4} + 6]. \tag{6}$$

The order parameters can assume different values; each of these values characterizes one different state of the nuclear matter [4]. If the three closest quarks have different colors then $\widetilde{M}_{c2} = 1$ ($M_{c2} = 1/2$), $\widetilde{M}_{c3} = 1$ ($M_{c3} = 3/2$), consequently $\widetilde{M}_{c4} = 1$ ($M_{c24} = 3/2$). This means that quarks are clusterized in isolated white nucleons. This case is recovered in the calculation at small densities, where the system is "locally" invariant for rotation in color space.

If the four closest quarks have the same color $\widetilde{M}_{c2} = \widetilde{M}_{c3} = \widetilde{M}_{c4} = 0$, a new state of the nuclear matter, dubbed *Exotic Color Clustering* materializes. The system is "locally" invariant for rotation in color space.

Finally, if the three closest quarks have two different colors, independently of the color of the two closest quarks, the Quark Gluon Plasma state is reached; hence: $\widetilde{M}_{c2} = \widetilde{M}_{c3} = \frac{2}{3}$. $\widetilde{M}_{c4}$ can assume three different values: $\frac{4}{5}$; 1; $\frac{3}{5}$, according to the colors of the four closest quarks and to the number of pairs of different color which are created. In the Quark-Gluon Plasma case the system is "globally" invariant for rotation in color space.

1.3. *J/Ψ lifetime*

In the high-energy nucleus-nucleus collisions, some J/Ψ particles are produced. Many experimental results seem to suggest that if the Quark-Gluon

Plasma state is formed the multiplicity of the revealed J/Ψ particles reduces. Hence the lifetime of the J/Ψ particle in a quark system can give information on the occurrence or not of the QGP phase transition.

Practically, in the code, when the system of quarks reaches the equilibrium condition, a pair of bound $c\bar{c}$ quarks, with $m_c = m_{\bar{c}} = 1.37GeV$, is embedded in the system. The presence of other light quarks in the system weakens and consequently breaks the bond between $c\bar{c}$ quarks.

The lifetime of the J/Ψ particle in the medium is calculated through its survival probability in the system. The survival probability, (P_{sur}) is calculated by counting the total number of pairs c and $\bar{c}$ that stay bound after they are inserted in the system of u and d quarks. The obtained distribution is fitted by the following expression, similar to the fission [5]:

$$P_{sur}(t) = exp[-(t - t_D)/\tau] \tag{7}$$

where t_D is the delay time, i.e. the time after which the exponential decrease starts happening, and τ is the lifetime of the J/ψ in the system. Hence the survival time (t_{sur}) of the J/ψ particle in the quark matter will be:

$$t_{sur} = t_D + \tau \tag{8}$$

2. Results

When quarks of different colors, are embedded in a dense medium, such as in the nuclear matter, the interaction between them becomes screened (*Debye Screening*). The screening in a quarks system can be obtained "directly", as the potential includes the color charges (Gell-Mann matrices). Moreover, because the considered system is not really an infinite system, as instead the nuclear matter is, the screening is not worth enough to avoid the divergence of the linear potential for $r \to \infty$. A cut-off is introduced: when quark distances are greater than the cut-off, the interaction is equal to zero. The cut-off is a free parameter, which is fixed, with the other free parameters (Λ, and the quark masses). The Λ value is fixed at $0.25GeV$ for all cases.

Of course using a cut-off in the linear term the confinement property of the quarks might be lost. Nevertheless the cut-off is relatively large thus it needs a very large energy to have isolated quarks.

Fig. 1 shows the results obtained by choosing: $m_u = 0.005GeV$, $m_d = 0.01GeV$ and the cut-off equal to $3fm$.

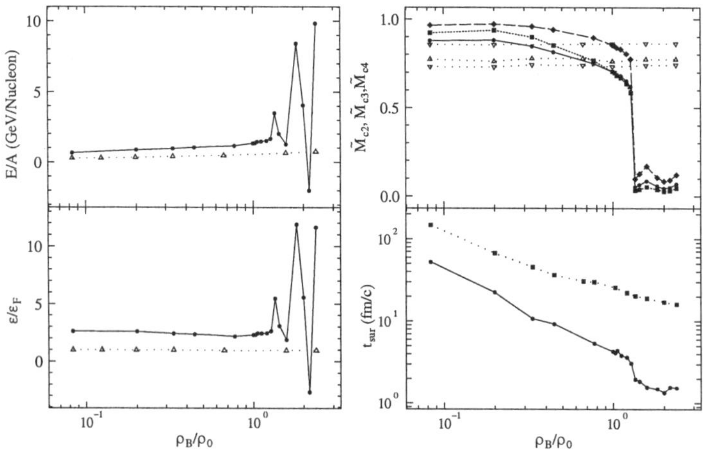

Figure 1. Energy per nucleon (top panel in the left), energy density(bottom panel in the left), normalized order parameters (top panel in the right) and time survival of J/ψ (bottom panel in the right) versus density divided by the normal density ρ_0, for $m_u = 5$ MeV, $m_d = 10$ MeV and cut-off= $3fm$.

The energy per nucleon and the corresponding energy density in units of the ε_F (energy density for a Fermi gas) (panels in the left) have a very irregular behavior already at small density. The three order parameters (top panel in the right) allow to better explain it: for small densities the quarks are condensed in clusters of three different colors, the system is locally white (isolated white nucleon): $\widetilde{M}_{c2}(circles) \simeq \widetilde{M}_{c3}(squares) \simeq \widetilde{M}_{c4}(diamonds) \simeq 1$. At higher densities, the quarks are not in clusters but randomly distributed, $\widetilde{M}_{c2} = \widetilde{M}_{c3} \simeq \frac{2}{3}$ and $\widetilde{M}_{c4} \simeq \frac{4}{5}$. The system reaches the QGP phase at $\rho_B/\rho_0 \sim 1.2$, but it does not remain in this state as it prefers the ECC state, where at least the four closest quarks have the same color. In fact, at density about $1.4 \sim 2.4\rho_0$ $\widetilde{M}_{c2} = \widetilde{M}_{c3} = \widetilde{M}_{c4} \simeq 0$. The system reaches this state through a first order phase transition at about $1.3\rho_0$.

In the energy per nucleon figure, the transition is demonstrated by a discontinuity at the same density. The other discontinuities, at larger densities ($\rho_B/\rho_0 > 1.5$), are probably due to the clusterization of more than four quarks of the same color. The definition of further order parameters, which relate more than four closest quarks, could explain these irregularities.

The calculations have been repeated with the Coulomb term only, (triangles in Fig.1); a constant contribution has been obtained, not only to the energy, but also to the order parameters. This means that the Coulomb term produces a permanent clusterization among quarks which prevents them to reach the ideal QGP state. When the linear term is included, it prevails over the Coulomb term and the system stays in ECC state, $\widetilde{M}_{c2} = \widetilde{M}_{c3} = \widetilde{M}_{c4} = 0$.

· The lifetime t_{sur} of the J/Ψ particle versus baryon density is plotted in the bottom part (right) of the Fig.1, where dotted line represents the time distribution obtained by the code and full line is obtained through Eq. (7). When the density increases, the J/Ψ lifetime decreases because it is more probable that a light quark (u, d) gets in between a $c\bar{c}$ pair and breaks the bond. t_{sur} behaves like an order parameter, in fact it has a jump just where there is the phase transition ($\rho_B/\rho_0 \sim 1.3$) and after there is a saturation of the surviving probability. In the same figure, the behavior of J/Ψ particle in the medium is analyzed by turning off the interaction (squares). Here one observes only a monotonic decrease with the density. The survival time of the J/Ψ in the medium without the interaction is always larger than the survival time calculated with the interaction; in fact, the forces between quarks lead to the Debye screening and break more easily the bonds between particles(c,$\bar{c}$ quarks).

It is clear that the cut-off value changes the transition point as a consequence. In Fig. 3 the reduced order parameters $\widetilde{M}_{c2}, \widetilde{M}_{c3}, \widetilde{M}_{c4}$ versus cut-off are plotted (left panel). The quark masses are $m_u = 0.005GeV$, $m_d = 0.01GeV$ and $\rho_B = \rho_0$.

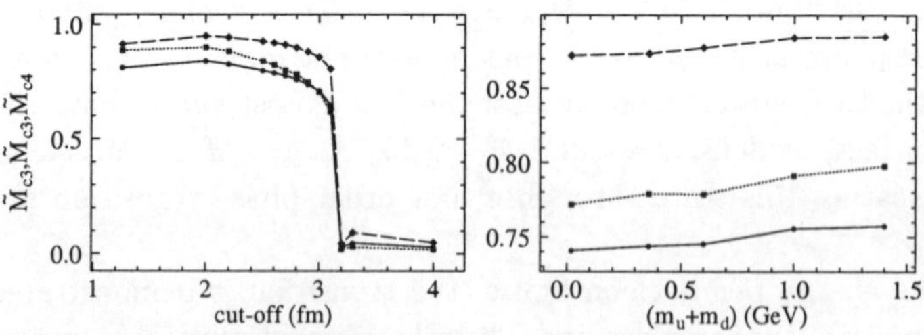

Figure 2. Reduced order parameters vs cut-off (left panel) and vs quark masses (right panel).

The system is in a nucleonic state for small cut-off values, crosses a

Quark-Gluon Plasma region and finally reaches the Exotic Color Clustering state. The phase transition from QGP to ECC state happens at $\rho_B = \rho_0$ if the cut-off values is larger or equal to $3.1 fm$; while in Fig. 1 the transition was at $\rho_B = 1.3\rho_0$. This means that the critical density of the ECC phase transition increases for decreasing cut-off values. Nevertheless, it is interesting to verify if quark system moves to the Exotic Color Clustering state also for higher quark masses. Fig. 3 (right panel) shows the three reduced order parameters versus quark masses when the cut-off is equal to $1.26 fm$. The reduced order parameters are almost constant; hence the cut-off is the main responsible of the nuclear matter phase transition.

In order to study the sensitivity of the results to the input parameters, other calculations have been repeated changing not only the cut-off values, but also the quark masses. The result are presented in [4]. The results shows that the first order phase transition to Exotic Color Clustering can disappear or that also a QGP phase transition can appear.

3. Finite systems

A finite system of quarks can be considered as an infinite system without periodic boundary conditions. When boundary conditions are excluded, the systems expands and cause of the confinement, (kr term in the Richardson potential), the quarks clusterize in nucleons. The expansion is adiabatic, in fact the total energy per particle is preserved.

The dynamic and the finite size effects might wash out completely or hide the phase transition present in the corresponding infinite system. The behavior of the J/Ψ particle in an expanding system can offer many information on the phase transition. While in an infinite quark system the J/Ψ particle sooner or later will dissolve, in a rapidly expanding system the dissolution might not occur.

Fig. 3 shows the results of a finite system. They have been obtained by expanding the infinite systems previously discussed ($m_u = 0.005 GeV, m_d = 0.1 GeV$).

The number of survival J/Ψ (left panel) increases with the density up to $\rho \sim \rho_0$, after it seems to have a jump, at the same density in which a first order phase transition has been obtained for the corresponding infinite system, Fig 1. At low density, the Fermi momentum of the light quarks is small, each J/Ψ particle stays "imprisoned" in a piece of nuclear matter and it will disappear. Instead when the density increases the system of light quarks expands faster and the slow $c\bar{c}$ quarks stay bond. The lifetime

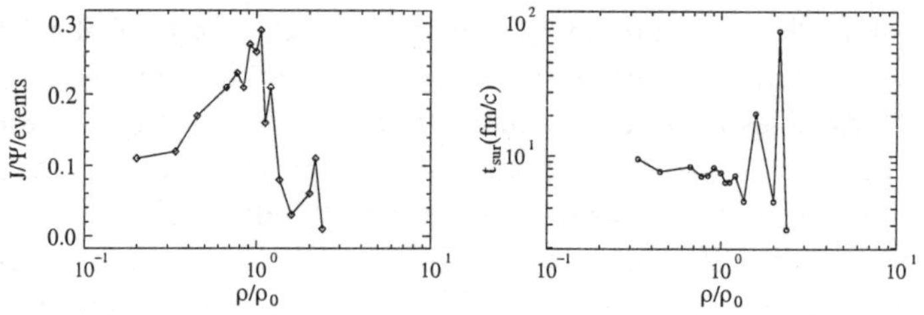

Figure 3. Number of survival J/Ψ (left panel) and J/Ψ lifetime (right panel) vs density.

of J/Ψ versus the baryon density, obtained through the Eq. (7) is shown in the right panel, it has a plateau up to the critical density, and after shows some discontinuities.

4. Conclusions

In conclusion, the Constrained Molecular Dynamics simulation is able to describe some aspects of the nuclear matter system, at zero temperature and finite baryon density. In particular a new state of the nuclear matter appears, the Exotic Color Clustering state, where at least the four closest quarks have the same color. The three reduced order parameters and the J/Ψ lifetime can be considered a "good indicators" of the phase transition. The J/Ψ particle, also in finite system, seems to be a suitable physical observable. Hence, finite size effects and the dynamic of the expansion do not cancel the phase transition.

References

1. J. L. Richardson, *Phys.Lett* **82B** 272 (1979).
2. M. Papa, T. Maruyama, A. Bonasera,*Phys. Rev.* **C64** 024612 (2001).
3. A. Bonasera,*Phys. Rev.* **C62**, 052202(R) (2000).
4. S.Terranova, A. Bonasera, *Phys. Rev.* **C70** 024906 (2004).
5. T. Maruyama, A. Bonasera, M. Papa, S. Chiba, *Eur. Phys.* **J.A 14** 191 (2002).

FUSION ENHANCEMENT BY SCREENING OF BOUND ELECTRONS AT ASTROPHYSICAL ENERGIES

SACHIE KIMURA AND ALDO BONASERA

Laboratorio Nazionale del Sud, INFN, via Santa Sofia, 62, 95123 Catania, Italy
E-mail: kimura@lns.infn.it

We perform molecular dynamics simulations of screening by bound target electrons in low energy nuclear reactions. Quantum effects corresponding to the Pauli and Heisenberg principle are enforced by constraints. We show that the enhancement of the average cross section and of its variance is due to the perturbations induced by the electrons. This gives a correlation between the maximum amplitudes of the inter-nuclear oscillational motion and the enhancement factor. It suggests that the chaotic behavior of the electronic motion affects the magnitude of the enhancement factor.

1. Introduction

The relation between the tunneling process and dynamical chaos has been discussed with great interests in recent years. Though the tunneling is completely quantum mechanical phenomenon, it is greatly influenced by classical chaos. In the sense that the the chaos causes the fluctuation of the classical action which essentially determines the tunneling probability. We study the phenomenon by examining the screening effect by bound electrons in the low energy fusion reaction. In the low energy region the experimental cross sections with gas targets show an increasing enhancement with decreasing bombarding energy with respect to the values obtained by extrapolating from the data at high energies [1]. Many studies attempted to attribute the enhancement of the reaction rate to the screening effects by bound target electrons. In this context one often estimates the screening potential as a constant decrease of the barrier height in the tunneling region through a fit to the data. A puzzle has been that the screening potential obtained by this procedure exceeds the value of the so called adiabatic limit, which is given by the difference of the binding energies of the united atoms and of the target atom and it is theoretically thought to provide the maximum screening potential [2]. Over these several years, the redetermination of the bare cross sections has been proposed theoretically [3] and experimentally [4], using the Trojan Horse Method [5]. The comparison between newly obtained bare cross sections, i.e., astrophysical S-factors, and the cross sections by the direct measurements gives a variety of values for the screening potential. There are already some theoretical studies performed using the time-dependent Hartree-Fock(TDHF) scheme [6,7].

In this paper we examine the subject within the constrained molecular dynamics (CoMD) model [8], even in the very low incident energy region not reached experimentally yet. At such very low energies fluctuations are anticipated to play a substantial role. Such fluctuations are beyond the TDHF scheme. Not only TDHF calculations are, by construction, cylindrically symmetric around the beam axis. Such a limitation is not necessarily true in nature and the mean field dynamics could be not correct especially in presence of large fluctuations. Molecular dynamics contains all possible correlations and fluctuations due to the initial conditions(events). For the purpose of treating quantum-mechanical systems like target atoms and molecules, we use classical equations of motion with constraints to satisfy the Heisenberg uncertainty principle and the Pauli exclusion principle for each event [8]. In extending the study to the lower incident energies, we would like to stress the connection between the motion of bound electrons and chaos. In fact, depending on the dynamics, the behavior of the electron(s) is unstable and influences the relative motion of the projectile and the target. The feature is caused by the nonintegrablility of the 3-body system and it is well known that the tunneling probability can be modified by the existence of chaotic environment. We discuss the enhancement factor of the laboratory cross section in connection with the integrability of the system by looking the inter-nuclear and electronic oscillational motion. More specifically we analyze the frequency shift of the target electron due to the projectile and the small oscillational motion induced by the electron to the relative motion between the target and the projectile. We show that the increase of chaoticity in the electron motion decreases the fusion probability. We mention that the understanding of the fusion dynamics and fluctuations has a great potential for the enhancement of the fusion probability in plasmas for energy production.

The paper is organized as follows. In sect. 2 we determine the enhancement factor f_e and describe the essence of the Constrained molecular dynamics approach briefly. In sect. 3 we apply it to asses the effect of the bound electrons during the nuclear reactions. We discuss also the relation between the amplitudes of the inter-nuclear oscillational motion and the enhancement factor. We summarize the paper in sect. 4.

2. Formalism

2.1. *Enhancement Factor*

We denote the reaction cross section at incident energy in the center of mass E by $\sigma(E)$ and the cross section obtained in absence of electrons by $\sigma_0(E)$. The enhancement factor f_e is defined as

$$f_e \equiv \frac{\sigma(E)}{\sigma_0(E)}. \tag{1}$$

If the effect of the electrons is well represented by the constant shift U_e of the potential barrier, following [9,6], ($U_e \ll E$):

$$f_e \sim \exp\left[\pi\eta(E)\frac{U_e}{E}\right], \tag{2}$$

where $\eta(E)$ is the Sommerfeld parameter [10].

2.2. *Constrained Molecular Dynamics*

We estimate the enhancement factor f_e numerically using molecular dynamics approach;

$$\frac{d\mathbf{r}_i}{dt} = \frac{\mathbf{p}_i c^2}{\mathcal{E}_i}, \qquad \frac{d\mathbf{p}_i}{dt} = -\nabla_{\mathbf{r}} U(\mathbf{r}_i), \tag{3}$$

where ($\mathbf{r}_i$,$\mathbf{p}_i$) are the position, momentum of the particle i at time t. $\mathcal{E}_i = \sqrt{\mathbf{p}_i^2 c^2 + m_i^2 c^4}$, $U(\mathbf{r}_i)$ and m_i are its energy, Coulomb potential and mass, respectively. We set the starting point of the reaction at 10Å inter-nuclear separation. To take into account the quantum mechanical feature of atoms, we put the constraints, i.e., Heisenberg uncertainty principle and Pauli principle for atoms which have more than 2 bound electrons. It is performed numerically by checking $\Delta\mathbf{r}\cdot\Delta\mathbf{p} \sim \hbar$, for Heisenberg principle, and $\Delta\mathbf{r}\cdot\Delta\mathbf{p} \sim 2\pi\hbar(3/4\pi)^{2/3}$, for Pauli blocking. Here $\Delta\mathbf{r} = |\mathbf{r}_i - \mathbf{r}_j|$ and $\Delta\mathbf{p} = |\mathbf{p}_i - \mathbf{p}_j|$. i and j refer to electrons and nuclei. More specifically, to get the atomic ground states, at every time step of the calculation, we calculate $\Delta\mathbf{r}\cdot\Delta\mathbf{p}$ for every pair of particles. If $\Delta\mathbf{r}\cdot\Delta\mathbf{p}$ is smaller(larger) than $\hbar$, in the case of Heisenberg principle, we change $\mathbf{r}_j$ and $\mathbf{p}_j$ slightly, so that $\Delta\mathbf{r}\cdot\Delta\mathbf{p}$ becomes larger(smaller) at the subsequent time step. We repeat this procedure for many time steps until no changes are seen in the energies and mean square radii of the atoms, similarly for the Pauli principle. The approach has been successfully applied to treat fermionic properties of the nucleons in nuclei and the quark system [8]. It can be extended easily in the case of the Heisenberg principle, as stated above. In this way we obtain many initial conditions which occupy different points in the phase space microscopically. Notice that in the D+d case that we investigate here, since one electron is involved, only the Heisenberg principle is enforced for each event. We obtain −13.56 eV as the binding energy of deuterium atom and 0.5327 Å as its mean square radius. These values can be compared with the experimental value of −13.59811 eV [11] and Bohr radius $R_B = 0.529$ Å, respectively [12].

In order to treat the tunneling process, we define the collective coordinates $\mathbf{R}^{coll}$ and the collective momentum $\mathbf{P}^{coll}$ as

$$\mathbf{R}^{coll} \equiv \mathbf{r}_P - \mathbf{r}_T; \qquad \mathbf{P}^{coll} \equiv \mathbf{p}_P - \mathbf{p}_T, \tag{4}$$

where $\mathbf{r}_T, \mathbf{r}_P$ ($\mathbf{p}_T, \mathbf{p}_P$) are the coordinates(momenta) of the target and the projectile nuclei, respectively. When the collective momentum becomes zero, we switch on the collective force, which is determined by $\mathbf{F}_P^{coll} \equiv \dot{\mathbf{P}}^{coll}$ and $\mathbf{F}_T^{coll} \equiv -\dot{\mathbf{P}}^{coll}$, to enter into imaginary time [13]. We follow the time evolution in the tunneling region using the equations,

$$\frac{d\mathbf{r}_{T(P)}^{\Im}}{d\tau} = \frac{\mathbf{p}_{T(P)}^{\Im}}{\mathcal{E}_{T(P)}}; \qquad \frac{d\mathbf{p}_{T(P)}^{\Im}}{d\tau} = -\nabla_{\mathbf{r}} U(\mathbf{r}_{T(P)}^{\Im}) - 2\mathbf{F}_{T(P)}^{coll}, \tag{5}$$

where τ is used for imaginary time to be distinguished from real time t. $\mathbf{r}_{T(P)}^{\Im}$ and $\mathbf{p}_{T(P)}^{\Im}$ are position and momentum of the target (the projectile) during the tunneling process respectively. Adding the collective force corresponds to inverting the potential barrier which becomes attractive in the imaginary times. The penetrability of the barrier is given by [13]

$$\Pi(E) = \left(1 + \exp\left(2\mathcal{A}(E)/\hbar\right)\right)^{-1}, \tag{6}$$

where the action integral $\mathcal{A}(E)$ is

$$\mathcal{A}(E) = \int_{r_b}^{r_a} \mathbf{P}^{coll}\, d\mathbf{R}^{coll}, \tag{7}$$

r_a and r_b are the classical turning points. The internal classical turning point r_b is determined using the sum of the radii of the target and projectile nuclei. Similarly from the simulation without electron, we obtain the penetrability of the bare Coulomb barrier $\Pi_0(E)$.

Since nuclear reaction occurs with small impact parameters on the atomic scale, we consider only head on collisions. The enhancement factor is thus given by eq.(1),

$$f_e = \Pi(E)/\Pi_0(E) \tag{8}$$

for each event in our simulation. Thus we have an ensemble of f_e values at each incident energy.

3. Application to the Electron Screening Problem

Fig. 1 shows the incident energy dependence of the enhancement factor for the D+d reaction. The averaged enhancement factors $\bar{f}_e$ over events in

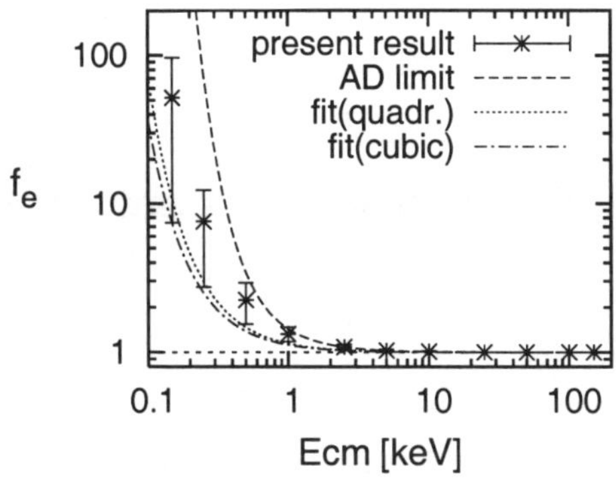

Figure 1. Enhancement factor as a function of incident center-of-mass energy for the D+d reaction.

our simulation are shown with stars and its variance $\Sigma = \left(\bar{f_e^2} - (\bar{f}_e)^2\right)^{1/2}$ with error bars. In the figure we show also several estimations of the enhancement factor by the latest analysis of the experimental data using quadratic(dotted) and cubic(dot-dashed) polynomial fitting [3] with the screening potentials $U_e = 8.7$ eV and 7.3 eV respectively. The dashed curve shows the enhancement factor in the adiabatic limit $f_e^{(AD)}$ for an atomic deuterium target and it is obtained by assuming equally weighted linear combination of the lowest-energy gerade and ungerade wave function for the electron, reflecting the symmetry in the D+d, i.e.,
$f_e^{(AD)} = \frac{1}{2}\left(\exp(\pi\eta(E)\frac{U_e^{(g)}}{E}) + \exp(\pi\eta(E)\frac{U_e^{(u)}}{E})\right)$, where $U_e^{(g)} = 40.7$ eV and $U_e^{(u)} = 0.0$ eV [7,6]. In the low energy region the enhancement factor is very large and exceeds 50. However the averaged enhancement factor does not exceed the adiabatic limit. We performed also a fit of our data using eq. (2)

including very low energy region and obtained $U_e = 15.9 \pm 2.0$ eV. This value, between the sudden and the adiabatic limit, is in good agreement with TDHF calculations[6,7].

Now we look at the oscillational motions of the particle's coordinates as the projection on the z-axis (the reaction axis). We denote the z-component of $\mathbf{r}_T, \mathbf{r}_P$ and $\mathbf{r}_e$ as z_T, z_P and z_e, respectively. Practically, we examine the oscillational motion of the electron around the target $z_{Te} = z_e - z_T$ and the oscillational motion of the inter-nuclear motion, i.e., the motion between the target and the projectile, $z_s = z_T + z_P$, which essentially would be zero due to the symmetry of the system in the absence of the perturbation. In

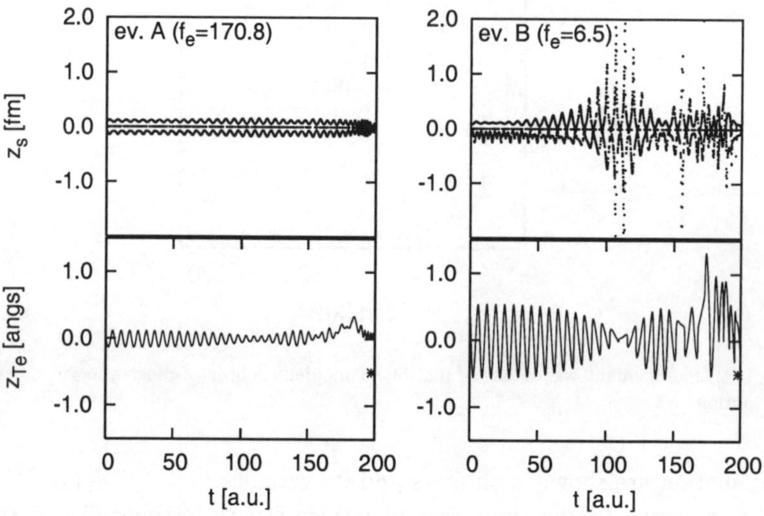

Figure 2. The oscillational motion of the electron around the target (lower panels) and the inter-nuclear motion (upper panels) as a function of time, in atomic unit, for two events, with large f_e(ev. A) and small f_e(ev. B), for the D+d reaction at the incident energy 0.15keV. The inter-nuclear separation is 10Å at $t = 0$.

Fig. 2 these two values are shown for 2 events, which have the enhancement factor $f_e = 170.8$ (ev. A), and $f_e = 6.5$ (ev. B), at the incident energy $E_{cm} = 0.15$ keV. The panels show the z_s, z_{Te} as a function of time. The stars indicate the time at which the system reaches the classical turning point. It is clear that in the case of ev. B the orbit of the electron is much distorted from the unperturbed one than in ev. A. Characteristics of z_s are that (1) its value often becomes zero, as it is expected in the un-perturbed system, and (2) the component of the deviation from zero shows periodical behavior. It is remarkable that the amplitude of the deviation becomes quite large at some points in the case of ev. B which shows the small enhancement factor. Note that in event B one observes clear beats, i.e., resonances. Thus for two events, with the same macroscopic initial conditions, we have a completely different outcome, which is a definite proof of chaos in our 3-body system. We can understand these results in first approximation by considering the motion of the ions to be much slower than the rapidly oscillating motion of the electrons. [14] From the Fig.2 we can deduce the following important fact. If the motion of the electron is initially in the plane perpendicular to the reaction axis, the enhancement factor is large, case A(notice $|z_{Te}| \ll R_B$, i.e., the Bohr radius, at $t \sim 0$). On the other hand if there is a substantial projection of the electron motion, as in case B(the amplitude of $|z_{Te}| \sim R_B$ at $t \sim 0$), on the reaction axis the enhancement factor is relatively small because of the increase of chaoticity. The fact suggests that if one performs experiments at very low bombarding energies with *polarized targets*, the enhancement factor can be controlled by changing the *polarization*. The largest enhancement would be gain with targets polarized perpendicularly to the beam axis. We notice in passing that event A is a case where cylindrically symmetry is approximately satisfied, since the electronic motion remains practically on the $xy-$plane in the target reference frame. This case gives a screening potential U_e =19.5eV closest to the adiabatic limit and to the TDHF result[6,7].

4. Summary

We discussed the penetrability of the Coulomb barrier by using molecular dynamics simulations with constraints and imaginary time. We have shown that both the enhancement factor and its variance increase as the incident energy becomes lower. However we obtained the averaged screening potential smaller than the value in the adiabatic limit. The chaoticity of the electron motion affects the enhancement factor of the cross section. We suggest to perform experiments on fusion at very low energies with polarized targets in order to obtain the large enhancement by the bound electrons.

References

1. A. Krauss, *et al.*, Nucl. Phys. A **467**, 273(1987), S. Engstler *et al.*, Phys. Lett. B **202**, 179 (1988).

2. C. Rolfs, and E. Somorjai, Nucl. Instrum. Meth. B **99**, 297 (1995).
3. F. C. Barker, Nucl. Phys. A **707**, 277 (2002).
4. M. Junker, *et al.* Phys. Rev. C **57**, 2700 (1998).
5. A. Musumarra, *et al.* Phys. Rev. C **64**, 068801 (2001).
6. T. D. Shoppa, S. E. Koonin, K. Langanke, and R. Seki, Phys. Rev. C **48**, 837 (1993).
7. S. Kimura, N. Takigawa, M. Abe, and D.M. Brink, Phys. Rev. C **67**, 022801(R) (2003).
8. M. Papa, T. Maruyama, and A. Bonasera, Phys. Rev. C **64**, 024612 (2001), S. Terranova, and A. Bonasera, Phys. Rev. C **70**, 024906 (2004).
9. H. J. Assenbaum, and K. Langanke and C. Rolfs, Z. Phys. A **327**, 461(1987).
10. D. D. Clayton, *Principles of Stellar Evolution and Nucleosynthesis* (University of Chicago Press, 1983) Chap. 4.
11. J.A. Bearden, and A.F. Burr, Rev. Mod. Phys, **39**, 125(1967).
12. S. Kimura, and A. Bonasera, physics/0409008.
13. A. Bonasera, and V. N. Kondratyev, Phys. Lett. B **339**, 207(1994), T. Maruyama, A. Bonasera, and S. Chiba, Phys. Rev. C **63**, 057601(2001).
14. S. Kimura, and A. Bonasera, Phys. Rev. Lett. (*in press*), nucl-th/0403062.

NUCLEAR PHYSICS WITH ELECTROWEAK PROBES

G. CO'
Dipartimento di Fisica, Università di Lecce
and,
Istituto Nazionale di Fisica Nucleare, sez. di Lecce
Lecce, Italy

The last few years activity of the Italian community concerning nuclear physics with electroweak probes is reviewed. Inclusive quasi-elastic electron-scattering, photon end electron induced one- and two-nucleon emission are considered. The scattering of neutrinos off nuclei in the quasi-elastic region is also discussed.

1. Introduction

In this paper I present the results obtained by the Italian community in the years 2002-2004 in the field of the theoretical study of lepton scattering off medium and heavy nuclei (A>4) [1−39]. These results are the product of numerous collaborations with many foreigner colleagues, their number is about the same of that of the Italian authors.

The range of the problematics covered by the various publications is wide, and I have organized my presentation as follows. First, I shall discuss some general issues concerning the lepton-nucleus interaction. Then I shall present the results of inclusive electron scattering, total photon absorption and those obtained by studying one- and two-nucleon emission processes. Last, I shall be concerned about the application to the neutrino scattering of the nuclear models used to investigate the electron scattering processes.

Since in writing this article I used numerous abbreviations, in order to facilitate the reader, I give their meaning in Table 1.

In the study of the lepton scattering off nuclei it is possible to separate the description of scattering process from that of the nuclear structure. The first, and quite obvious, reason is that projectile and scattered lepton are clearly distinguishable from the hadrons composing the nucleus. In addition, the fact that electroweak processes are well described already at the first-order perturbation theory, helps a lot. In effect, all the calculation I have examined have been done by considering that a single gauge boson is

Table 1. Abbreviations used in the article

FG	Fermi gas
FSI	final state interaction
LDA	local density approximation
LIT	Lorentz inverse transform
LRC	long range correlations
MEC	meson exchange currents
MF	mean field
OB	one body
PWBA	plane wave Born approximation
RFG	relativistic Fermi gas
RPA	Random Phase Approximation
SRC	short range correlations
WS	Woods Saxon

exchanged between the lepton and the target nucleus (see Fig. 1). In addition, also the PWBA has been adopted. With these approximations, the cross section expressions for both electrons [40] and neutrinos [41] scattering processes show a factorization of the leptonic and hadronic variables.

The leptonic vertex is treated within the relativistic theory, since the energies involved are much larger than the leptons masses. On the other hand, the nuclear vertex is usually treated with non-relativistic quantum many-body theory.

2. The electron-nucleus interaction

From now, up to section 4, I shall restrict my discussion to the electromagnetic case. The OB electromagnetic currents are obtained by summing the currents generated by each nucleon. Gauge invariance, i.e. the charge-current conservation law, is not satisfied if only these currents are considered. This indicates the need of including other type of currents, produced by the exchange of mesons between the interacting nucleons and generically called Meson Exchange Currents (MEC).

Gauge invariance indicates the need of MEC, but it does not define them in a unique and unambiguous manner. The various methods used to describe the MEC agree on the fact that the main contributions come from the three diagrams presented in Fig. 2. The most relevant terms are the seagull, diagram (a), the and pionic, diagram (b). They are of the same order of magnitude but they have different sign. They contribute to the

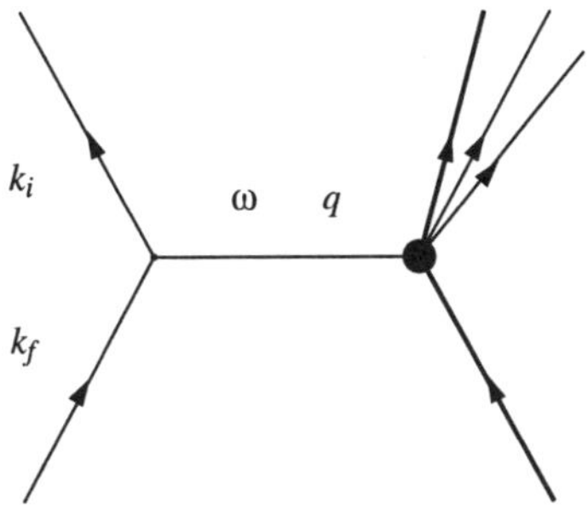

Figure 1. One-boson exchange diagram showing the symbols adopted for the kinematic variables in the scattering process lepton-nucleus. The left vertex represents the lepton, whose four vectors are indicated with $k \equiv (\epsilon, \mathbf{k})$. ω and $\mathbf{q}$ label energy and momentum transfer respectively. The bosons exchanged are the photon, in the case of the electromagnetic interaction, and the Z^o and $W^{\pm}$ in the case of the weak interaction.

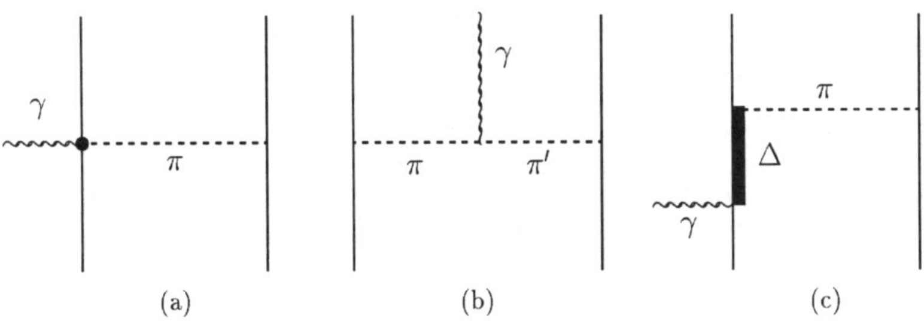

Figure 2. Meson Exchange Diagrams considered in the various calculations. Contact or seagull (a), pionic or pion in flight (b), Δ-current (c).

electromagnetic field of the nucleus only if the exchanged pion is charged. This means that with these two diagrams only proton-neutron pairs are involved.

At energies far from the peak of the nucleonic Δ-resonance the MEC Δ-current terms, diagram (c), are generally smaller than the seagull and pionic ones. The Δ-currents contribute to the electromagnetic field of the nucleus also when the pion exchanged is chargeless. In this case, the two nucleons involved are of the same type. This observation is relevant for the two-nucleon emission processes.

The validity of the non relativistic reductions used to describe the electromagnetic field of the nucleus has been studied by investigating the ideal system of the Fermi gas (FG) [3–9]. The strategy consists in comparing the results obtained for a relativistic Fermi Gas (RFG), which is an exactly solvable model, with those obtained in ordinary non relativistic FG, where various non-relativistic reductions of the currents have been adopted.

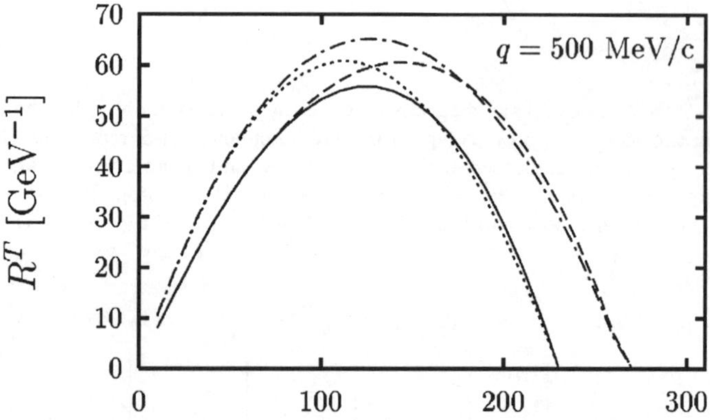

Figure 3. Inclusive transverse response. The quantity on the x axis is the excitation energy in MeV. Full line RFG with OB currents only, dashed line FG with OB currents only, dotted line RFG with OB and MEC, dashed-dotted line FG with OB and MEC.

An example of the results of this investigation [4] is given in Fig. 3 where the RFG results obtained with and without MEC are compared with the analogous results obtained in non relativistic FG. This figure shows the transverse response as a function of the nuclear excitation energy for a given value of the momentum transfer. In these results the effect of the relativity is relatively large. The height of the peak is reduced by about 20%, and also the width of the response is reduced by relativity. On the other hand, the effect of the MEC does not seem to be sensitive to relativity. The shift produced by the MEC on the OB responses is about the same on both RFG and FG results.

The effects shown in Fig. 3, are much weaker in finite nuclei calculations. In Fig. 4 an example of this result is seen. The quantity shown in the figure is the reduced cross section of the ^{16}O(e,e'p)^{15}N reaction as a function of the missing momentum [16]. Also in this case the nuclear wave functions have been described within a MF model. A real WS potential is used to

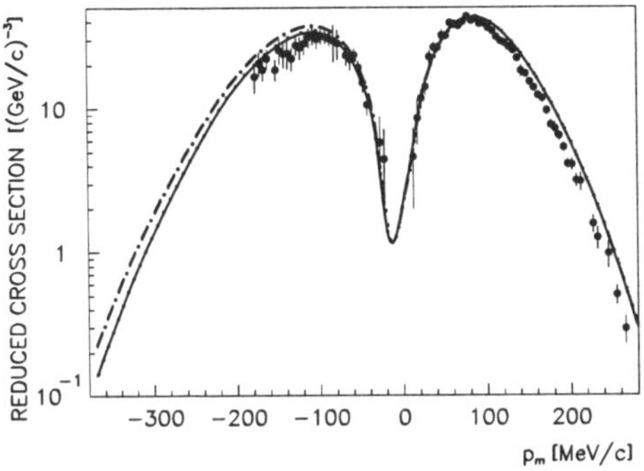

Figure 4. Reduced cross section of $^{16}O(e,e'p)^{15}N$ calculated in the mean field model, as explained in the text, and compared with experimental data [42]. The full line has been calculated by using all the MEC diagrams shown in Fig. 2. The result obtained with OB current only is shown by the dotted line which almost exactly overlaps the full line. The dot dashed line are the results obtained by adding to the OB current the seagull diagram only.

generate the single particle wave functions of the ^{16}O ground state. The parameters have been fixed to reproduce the charge radius and the single particle energies around the Fermi surface. The particle wave function in the nuclear final state, has been obtained by using a complex optical potential whose parameters have been fixed to describe the elastic cross sections of the scattering process between the emitted nucleon and the remaining nucleus with A-1 nucleons.

The result of the calculation where all the MEC diagrams of Fig. 2 are considered is shown by the full line. The dotted line, almost perfectly overlaps the full line, shows the results obtained with the OB currents only. The dashed-dotted line has been obtained by adding to the OB currents only the seagull term. The results of this calculation show that the contributions of the various MEC diagrams cancel each other. In this process the effect of the MEC is so small that they cannot be disentangled by a comparison with the data.

The effects of relativity are also strongly reduced in finite systems [43]. An example of these results is presented in Fig. 5 where the $^{16}O(e,e'p)^{15}N$ reduced cross sections calculated with two different mean field models are compared. The full line shows the result obtained with a relativistic MF, while the dotted line has been obtained with a non relativistic calculation. Here relativity lowers the height of the maximum by about 10%. More relevant is the fact that the shapes of the cross sections are only slightly

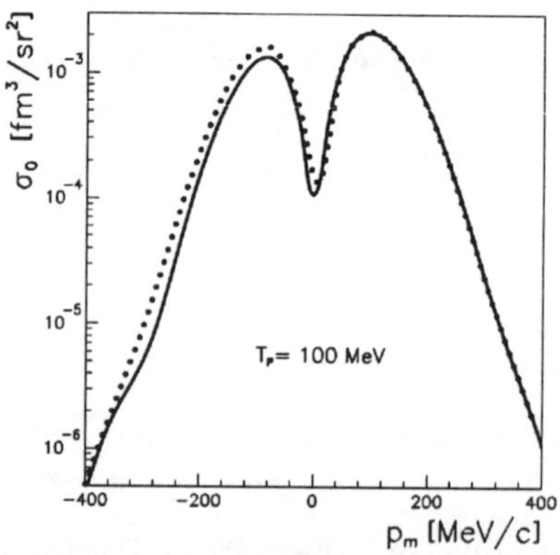

Figure 5. Reduced cross section of $^{16}O(e,e'p)^{15}N$ as a function of the missing momentum. The full line shows the result obtained with a relativistic mean field, the dotted line with a non relativistic one.

modified. Even though Fig. 3 and Fig. 5 show different quantities, it is evident that the global effect of relativity is smaller in finite systems than in FG calculations. A possible explanation of this can be related to the procedures used to define the parameters of the nuclear mean fields in the finite nuclei calculations. In both relativistic and non-relativistic calculations, these parameters are fixed so as to reproduce the same quantities. It is plausible that in this fitting procedure some relativistic effects are effectively included. This is not the case of the FG calculations, where the comparison is done between two ideal systems without any phenomenological parameter to fix.

3. Nuclear structure and electron scattering

In the previous section, I have discussed some source of uncertainties in the description of the reaction mechanism between electron and nucleus. These uncertainties can affect the cross sections on average by a 10%, maximum 20%. The uncertainties insit in the nuclear structure produce much larger effects on the cross sections. As a reminder of this, I would like to briefly recall a set of problems still open which, however, are not at the moment under the attention of the community.

(a) The high momentum data of the elastic scattering cross sections are not well reproduced. As a consequence the theoretical charge distributions are unable to the describe the empirical distributions in the center of the

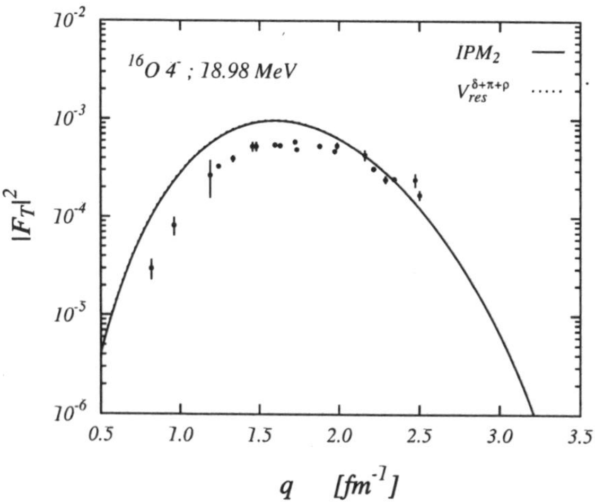

Figure 6. Comparison between (e,e') transition amplitudes calculated with RPA, full line, and MF model, dashed line, with experimental data [45]

nucleus. In spite of some attempts [44] indicating the physical effects responsible for this discrepancy, to the best of my knowledge, there is not a single, fully consistent, calculation able to give a reasonable description of these data.

(b) The (e,e') theoretical cross sections in the discrete excitation usually overestimate the data [46], as shown, for example, in Fig. 6.

(c) The continuum RPA results for total photo-absorption cross sections in the giant resonances region, are able to reproduce the energies of the resonances, but they overestimate the sizes of the cross sections and underestimate their widths, as shown in Fig. 7.

The problems I have just mentioned are due to the limitations of the theoretical models used to calculate the cross sections and the other observables. In medium-heavy nuclei the nuclear excited states are usually described by using the MF model or the RPA. In both these descriptions only one-particle one-hole excitations, and eventually their linear combinations, are considered.

These unsatisfactory results are related to the limitations of the nuclear models adopted and not to the basic assumptions of the theory used to describe atomic nuclei. This information come from the results obtained in the few-body systems ($A \leq 4$) [48]. In these systems the many-body Schrödinger equation describing a set of interacting nucleons, is solved without mak-

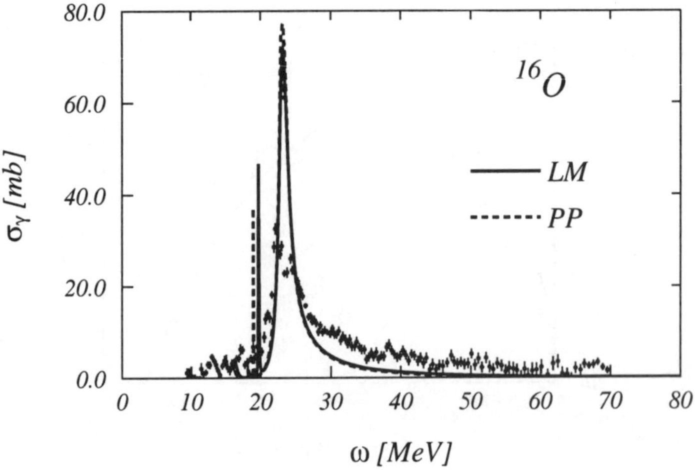

Figure 7. Total photon absorption cross sections calculated within the continuum RPA framework with two different residual interactions compared with the data [47].

ing approximations, and the agreement with electromagnetic experimental data is remarkably superior to what is obtained in heavier systems.

Recently the technique of the Lorentz Inverse Transform (LIT), previously used in few-body systems, has been applied with great success to heavier nuclei, up to A=7 [30,31,32,33]. A more detailed description of the LIT theory is presented elsewhere in these proceeding [49]. Here, I simply want to point out the fact that this technique accounts for all the possible decay channels of the nuclear excited state. On the contrary, the MF model, and also the continuum RPA, consider only the decay in the single nucleon emission channel.

The MF model is unable to describe low-energy data, but it is quite successful in the quasi-elastic region dominated by single-particle dynamics. In the following subsections I present the results obtained in this energy region, considering separately the inclusive processes from those where one, or two, nucleons are emitted and measured in coincidence with the scattered electron.

3.1. *Inclusive scattering: (e,e')*

An example of the agreement between Frascati data [50] on ^{16}O and the results obtained with the relativistic MF model [20] is shown in Fig. 8. The

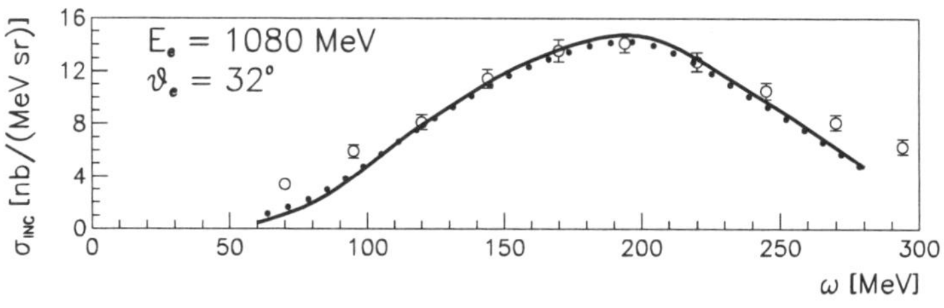

Figure 8. Inclusive electron scattering cross section on ^{16}O nucleus calculated in the framework of the relativistic MF model and compared with the Frascati experimental data [50]. The two lines have been obtained by using two different approximations [20].

same high-quality agreement is obtained for the other measured kinematics. The main point of the calculation is the careful treatment of the Final State Interaction (FSI). As already discussed in Sect. 2, in this MF model, the FSI are taken into account by using a complex optical potential. The imaginary part of the potential removes flux from the elastic channel. In inclusive experiments the total flux should be conserved. What is removed from the elastic channel should go in other decay channels. A more detailed description of the technique used to conserve the flux is presented elsewhere in these proceedings [51].

The same model [20] has been used to study the separated longitudinal and transverse responses of ^{40}Ca and ^{12}C. In ^{40}Ca the MIT data [52] are well reproduced, while the comparison with Saclay data [53,54] suffers the well known failures: the longitudinal responses are overestimated, while the transverse ones are underestimated. The comparison with the separated responses of ^{12}C, measured at various values of the momentum transfer [55], shows a reasonable agreement with the longitudinal responses while the transverse ones are always heavily underestimated. Old calculations of the ^{12}C responses done within a non relativistic framework [56] produce very similar results. This indicates that the effect of the relativity is negligible.

The Frascati (e,e') [50] data have been studied by using a different technique [57]. The basic nuclear model is again the non-relativistic MF. In this case the responses have been calculated with a real potential. The results obtained have been folded with Lorentz functions whose parameters have been fixed to reproduce the energy behavior of the volume integrals of the optical potential. In spite of the technical differences, this approach contains the same physics as that of the Pavia group, and the results obtained are very similar. Also in this case [38] the Frascati data [50] are rather well reproduced. The same kind of agreement is obtained [57] with the MIT ^{40}Ca responses [52]. The comparison with ^{40}Ca and ^{12}C Saclay data, shows the same problems described above [57,58]. The two different techniques used to treat FSI produce very similar results.

3.2. *One-nucleon emission: (e,e'N) and (γ,N)*

The same MF model used to describe the inclusive data has been utilized to study the single-nucleon emission processes induced by electromagnetic probes. The basic ingredients of the models are the two MF potentials. A real potential that describes the ground state of the target nucleus, and a complex optical potential to treat the emitted nucleon wave function.

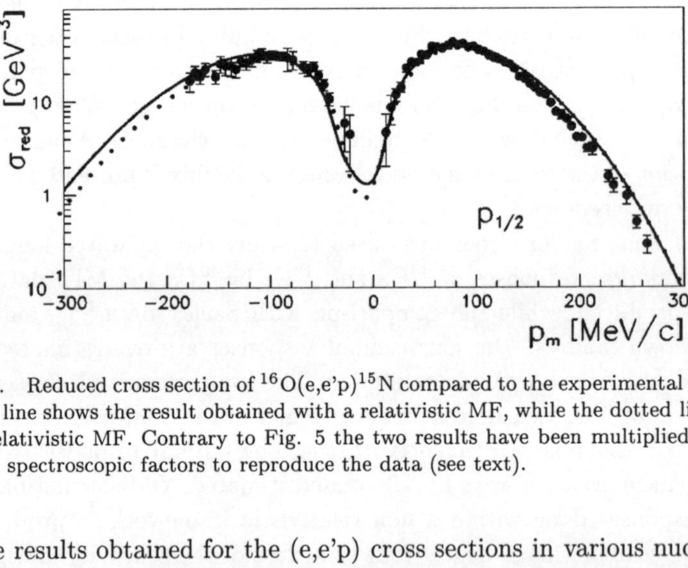

Figure 9. Reduced cross section of ^{16}O(e,e'p)^{15}N compared to the experimental data [42]. The full line shows the result obtained with a relativistic MF, while the dotted line with a non relativistic MF. Contrary to Fig. 5 the two results have been multiplied by two different spectroscopic factors to reproduce the data (see text).

The results obtained for the (e,e'p) cross sections in various nuclei are able to reproduce rather well the behavior of the data after a quenching rescaling factor is applied [40]. This rescaling factor is called spectroscopic factor and it does not depend upon the kinematics of the experiment. This is evident since observables related to the ratio of cross sections are well reproduced without the use of the spectroscopic factor [6]. Furthermore, cross sections measured in very different kinematic conditions are well reproduced by the same spectroscopic factors [24].

The spectroscopic factor is a model dependent quantity, as is deducible, for example, in Fig. 5. In this figure the ^{16}O(e,e'p)^{15}N reduced cross sections calculated [43] with relativistic (full line) and non relativistic (dotted line) MF models are compared with the experimental data [42]. In Fig. 9 the full line has been multiplied by 0.7 while the dotted line by 0.65. This indicates that spectroscopic factors contain some relativistic effects. These effects are not sufficient to explain a large part of the spectroscopic factors. It is necessary to go beyond the MF model, or in other words, to include correlations. The investigation of the effects induced by the correlations has been conducted by using two different approaches.

The basic quantities necessary to calculate the cross sections are the Fourier transforms of the transition densities induced by the current operators $J(\mathbf{r})$:

$$W(\mathbf{q}) = \int d^3r \, e^{i\mathbf{q}\cdot\mathbf{r}} < \Psi_f \mid J(\mathbf{r}) \mid \Psi_0 > \quad . \tag{1}$$

In the approach adopted by the Pavia group these quantities are calculated as:

$$W(\mathbf{q}) = \int d^3r \, e^{i\mathbf{q}\cdot\mathbf{r}} \, \chi(\mathbf{r}) J(\mathbf{r}) \; < \Psi^{A-1}(\mathbf{r}) \mid \Psi_0^A > \tag{2}$$

where $\chi(\mathbf{r})$ is the wave function of the emitted, and detected, nucleon. The important quantity is the overlap function between the wave function describing the ground state of the target nucleus and the wave function describing the state of the nucleus composed by A-1 nucleons. All the complications related to the correlations are contained in the overlap function. The formalism developed by the Pavia group is independent from the methods used to estimate the overlap function. In the MF model the overlap function is the single particle wave function of the nucleon below the Fermi surface.

The approach used in Lecce makes an ansatz on the expression of the nuclear wave function which is supposed to be the product of a symmetrized many-body correlation function F and a Slater determinant.

$$W(\mathbf{q}) = \int d^3r \, e^{i\mathbf{q}\cdot\mathbf{r}} \;\; < \Phi_f \mid F^\dagger J(\mathbf{r}) F \mid \Phi_0 > \tag{3}$$

The Slater determinant $\mid \Phi_0 >$ describing the ground state is composed by all the single particle states below the Fermi surface, while $\mid \Phi_f >$ contains a hole and a particle states. The same correlation function has been used for both ground and excited states. The many-body correlation function is written as a product of two-body correlation functions. A cluster expansion is done, and only the terms containing a linear dependence from the two-body correlation function are retained [59]. This approach is more tuned to investigate the so-called short-range correlations (SRC) due to the hard core repulsion of the nucleon-nucleon potential.

In spite of the differences, the results obtained by the two approaches are very similar. In general, for the considered processes, the effects of the SRC are very small. Certainly the inclusion of the SRC does not reduce sensitively the values of the spectroscopic factors. As an opposite example of the behavior of the correlation, I like to quote the results obtained in ^{40}Ca by using overlap functions produced by the Generator Coordinate Method. In this case the spectroscopic factors are even larger than those of the MF model calculation [17,60].

From the qualitative point of view, the results obtained with the two methods described above, show that the presence of SRC does not modify sensitively the shapes of the (e,e'p) cross sections [13,59]. The differences between cross sections obtained with and without SRC are within the accuracy of the experimental data. An interesting deviation from this general trend is the case of the ^{32}S nucleus [13], which should be worth further investigation.

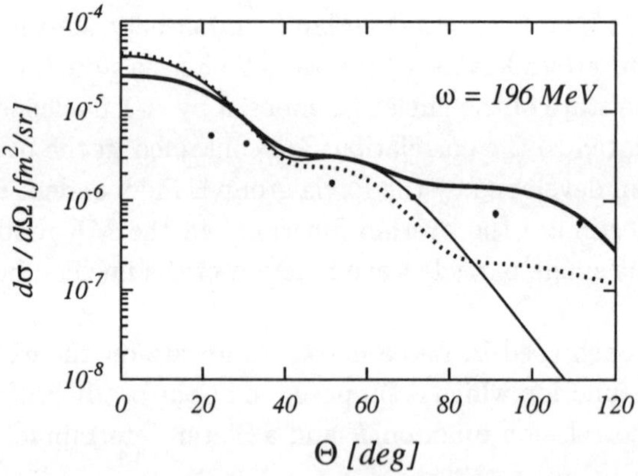

Figure 10. Cross sections of the $^{16}O(\gamma,p)^{15}N$ process as a function of the proton emission angle compared with the experimental data [61]. The thin full line shows the result obtained with OB currents only. The inclusion of SRC produces the dotted line. When also the MEC are included, the thick full line is obtained.

A relatively large effect produced by the SRC has been found in the $(\gamma$,p) reaction on the ^{16}O nucleus [16,34]. In Fig. 10 the cross sections of this process, calculated with OB currents only (thin full line), with the inclusion of SRC (dotted line) and by further adding the MEC (thick full line) are compared with the experimental data [61]. The relative effect produced by the SRC on the OB cross section is quite large. Unfortunately, in this kinematic region, the effects of the MEC are even larger.

3.3. *Two-nucleon emission: (e,e'NN) and (γ,NN)*

In the processes discussed so far, the effects of the SRC correlations have been obscured by the presence of the uncorrelated OB terms, or by the MEC currents. It is possible to eliminate the contribution of the OB terms by considering processes where two nucleons are emitted and detected. In this case the only mechanism competing with the SRC are the MEC. When two-like nucleons are emitted, the only terms of the MEC contributing to the cross section are the Δ-current diagrams, the (c) diagrams of Fig. 2, where the exchanged pion is chargeless.

From the theoretical point of view the description of the two-nucleon emission processes has been treated as a straightforward extension of the single nucleon emission case.

In the Pavia approach Eq.(2) is extended as:

$$W(\mathbf{q}) = \int d^3r\, e^{i\mathbf{q}\cdot\mathbf{r}_{12}}\, \chi_1(\mathbf{r}_1)\chi_2(\mathbf{r}_1)J(\mathbf{r}_{12}) < \Psi^{A-2}(\mathbf{r}_{12}) \mid \Psi_0^A > \quad (4)$$

with the obvious meaning of the symbols. Now the quantity containing the correlations is the two body-overlap function, between the target ground state and the state with A-2 nucleons.

In the Lecce approach the transition density of Eq. (3) is calculated by using a Slater determinant $\mid \Phi_f >$ with two particles in the continuum and, obviously, two holes.

In both approaches the interaction between the two emitted nucleons is not considered. This problem has been investigated by the Pavia group [25,26], by using an approximation. The interaction between the two emitted nucleons has been considered, as has the interaction between each nucleon and the A-2 nucleus. The simultaneous interaction of two nucleons between themselves and with the remaining A-2 nucleus has been neglected. This would be a genuine three-body problem. The results obtained considering the ^{16}O nucleus as a target show that the interaction effect is relevant in (e,e'pp) reaction, but it is negligible in (e,e'pn) reaction. More interesting is the fact that in the (γ,pp) reaction the effect is always negligible.

In order to obtain information on the SRC it is necessary to disentangle the two nucleon emission induced by the correlations from that produced by the MEC. We have already seen that the emission of two-like nucleons eliminate the MEC seagull and pionic diagrams, and also part of the Δ-current terms. It is possible to find kinematic situations where the SRC dominate on the remaining Δ-current terms [12], as is shown in Fig. 11. When the ^{16}O(e,e'pp)^{14}C reaction leads to the ground state of the ^{14}C nucleus, the Δ-currents contribution is much smaller than that of the SRC. The situation is reversed when the nuclear final state is the excited 1^+ state in ^{14}C.

A detailed study of the momentum dependence of the Δ-current contributions [36] shows that they are minimized at small values of the momentum transfer. In photon reactions these contributions are much smaller than those of the SRC for all the ^{14}C final states. From this point of view the two-proton emission induced by real photons with energy far from the Δ resonance peak, is the ideal tool to investigate SRC [36].

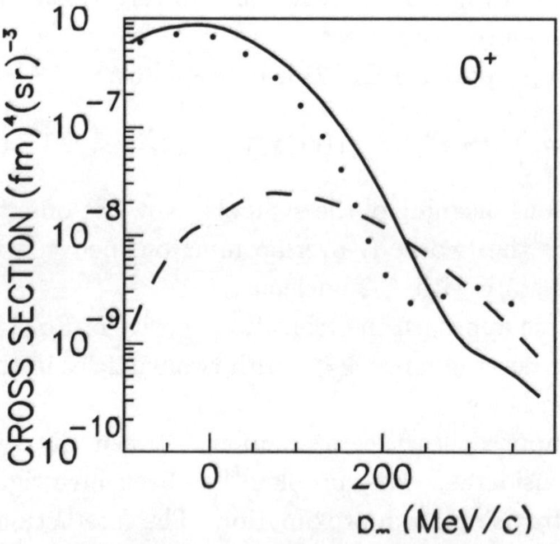

Figure 11. Cross sections of the ^{16}O(e,e'pp)^{14}C process leading to the ground state of the ^{14}C, as a function of the initial momentum of the emitted pair. The dashed line has been calculated by considering the Δ-current only. The dotted line with the SRC only, and the full line with both contributions.

The role of the tensor terms of the SRC is however quenched in the emission of two-like nucleons. The contribution of these terms is significant only in (e,e'pn) processes [12].

4. Neutrino-nucleus interaction

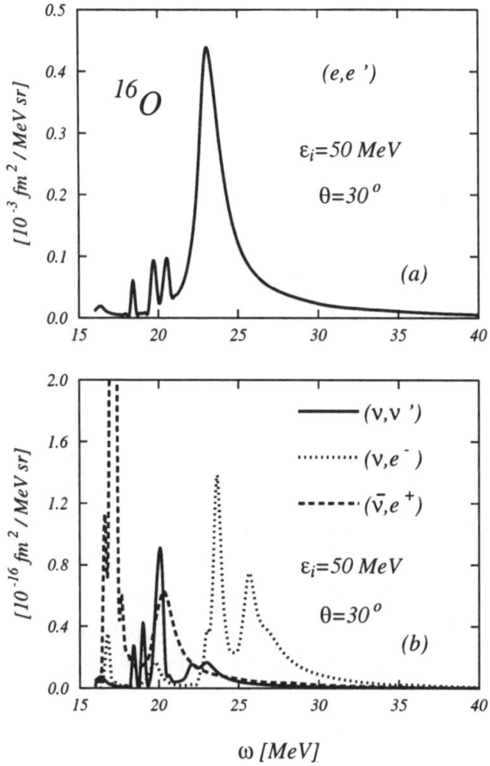

Figure 12. Comparison between electron (above) and neutrino (below) scattering cross sections. The energies of the incoming leptons and the scattering angles are the same in all the reactions considered.

In Fig. 12 the electron scattering cross section is compared with that of neutrino scattering in the same kinematic conditions. All the cross sections have been calculated for the same energy of the projectile and the same scattering angle. The nuclear transitions have been calculated by using the continuum RPA.

The three cross sections have quite a different behavior. This is expected for the charge exchange reactions but it is surprising when electron scattering and neutrino charge conserving neutral current reactions are compared. In this case the particle-hole configuration space describing the nuclear excitation is the same in both processes.

The reason for this difference can be traced by making a multipole decomposition of the cross sections. In the electron scattering case the 1^- excitation is responsible for the 98% of the total cross section. On the contrary, in the (ν,ν') case the 1^- contributes only to the 33% of the cross section, while the main contribution, 58%, is due to the the 2^- multipole [37].

This result is due to the fact that in the neutrino cross section the main contributions are given by the transverse axial vector term of the current operator. This operator excites both natural and unnatural parity states. In electron scattering the main contribution is due to the charge operator exciting natural parity states only. The dominance of the axial vector term is a quite general result. Also the charge exchange reactions are dominated by this term of the current, and this is the dominant term for all the neutrino cross sections also in the quasi-elastic region [62].

As a consequence of this, it is necessary to be careful in relying on the fact that a good description of electron scattering data implies a good description of the neutrino-nucleus cross section. In spite of this warning, electron scattering is still the best guide we have to determine the prediction power of our nuclear models. The extension to the neutrino scattering of the electron scattering formalism is quite straightforward. In these last two years, almost all the Italian groups working in electron scattering have applied their techniques to calculate neutrino-nucleus cross-sections. The calculations have been mainly done in the quasi-elastic region [1,11,21,22,27,28,29,38].

As I have previously pointed out the main correction to the naif MF model is due to the FSI. The FSI calculated with a folding model model [28,38,63], produces a reduction of the total neutrino-nucleus cross section by about 10%. This is due to the fact that the FSI spreads the strength of the response and part of it is moved to excitation energies kinematically prohibited.

The role of the correlations have been estimated to be relatively large [27,28]. The peak of the $^{16}O(\nu,e^-)^{16}F$ cross section for 1 GeV neutrinos is reduced by about 20%. This effect has been estimated by comparing FG results with those obtained by using a correlated spectral function in a local density approximation (LDA). It will be interesting to disentangle the LDA effects from those produced by the correlations.

The neutrino-nucleus scattering in the quasi-elastic regime has been found to be the ideal tool for studying the strangeness content of the nucleon [1]. In this case one has to separate the cross sections where a neutron is emitted from those where a proton is emitted. The calculations done in a FG model show about a 25% difference between the results obtained with and without strangeness in the nucleon form factor [64]. This large effect has been recently confirmed by a finite nucleus calculation with a relativistic MF model [22,51].

5. Conclusions

In these last two years, the activity of the Italian community regarding nuclear physics with electroweak probes has concentrated on the region of the quasi-elastic peak. In this region, the nuclear excitation is dominated by the single-particle dynamics, therefore MF models with OB currents are the starting point of the description of the nuclear excitation. Other effects induced by MEC, FSI, SRC and collective modes, are relatively small, and can be treated as perturbations of the MF results. There is a convergence of the results obtained with different models and techniques, and many of the uncertainties and problems presented and discussed in the past [65,66,67] have been resolved.

The comparison with the experimental data is still problematic. Concerning the inclusive experiments, the Frascati ^{16}O inclusive cross sections are well reproduced [20,38], as well as the ^{40}Ca MIT-Bates separated responses [20,57]. The same models are unable to describe the the Saclay ^{40}Ca and ^{12}C separated responses. I think that at present the major problems are on the experimental side. Experiments to obtain separated responses in ^{16}O are necessary, and the incompatibility between the ^{40}Ca data measured at MIT and at Saclay should be clarified.

The situation regarding one-nucleon emission processes is slightly clearer. The same MF models used to describe inclusive experiments are able to reproduce extremely well the shapes of the cross sections [17]. The open problem is the understanding of the spectroscopic factor needed to obtain quantitative agreement with the data. Contrary to some claims [68,69] it has been shown that the spectroscopic factor is a number [6,24], model dependent, but independent form momentum and energy transfer.

The search for effects induced by SRC has shown that the best way of studying them is the use of two-nucleon emission processes. Since the Δ-current terms become small at low momentum transfer, in the two-nucleon emission processes real photons seem to be a better tool than electrons [16,36]. In this field the various approaches provide similar results, indicating that the theoretical uncertainties are kept under control.

Even though the neutrino scattering is dominated by the axial part of the current, absent in electron scattering, the above mentioned results suggest that the same models could provide a good description of the quasi-elastic neutrino scattering. This is a very important information for the existing and planned experiments which use the nucleus as a detector to investigate the properties of the neutrinos and their sources.

Furthermore neutrino-nucleus experiments in the quasi-elastic region, are perhaps the best tool we have for information on the strange content of the nucleon [1,22]. The conservative estimate of the nuclear structure uncertainties, that resulted to be of the same order of the searched effects, should be updated. In our present understanding of the quasi-elastic excitation this uncertainty is reduced, therefore the strangeness effects could be identified.

Electron scattering off nuclei is a precision tool to control and investigate our understanding of nuclear structure. The evaluation of these cross sections should be used as a benchmark to verify the validity of approaches aimed at predictions in other nuclear physics fields. In these last few years the theories have improved remarkably, and there are signs indicating the possibilities of even greater improvements in the coming years.

It is however disappointing to notice that many of the experimental data are old, incomplete, and quite often not accurate enough to disentangle interesting effects. I join G. Orlandini [33] by pointing out the need of a major experimental program to investigate medium-heavy nuclei with electromagnetic probes.

ACKNOWLEDGMENTS. I thank P. Rotelli for reading, and commenting, the manuscript.

References

1. W.M. Alberico, S.M. Bilenky, C. Maieron *Phys. Rep.* **358**, 227 (2002).
2. W.M. Alberico, S.M. Bilenky, *Phys. Part. Nuclei* **35**, 297 (2004).
3. J.E. Amaro, M.B. Barbaro, J.A. Caballero, T.W. Donnelly, A. Molinari, *Eur. Phys. J.* **A15**, 421 (2002).
4. J.E. Amaro, M.B. Barbaro, J.A. Caballero, T.W. Donnelly, A. Molinari, *Phys. Rep.* **368**, 317 (2002).
5. J.E. Amaro, M.B. Barbaro, J.A. Caballero, T.W. Donnelly, A. Molinari, *Nucl. Phys.* **A697**, 388 (2002).
6. J.E. Amaro, M.B. Barbaro, J.A. Caballero, F.K. Tabatabaei, *Phys. Rev.* **C68**, 014604 (2003).
7. J.E. Amaro, M.B. Barbaro, J.A. Caballero, T.W. Donnelly, A. Molinari, *Nucl. Phys.* **A723**, 181 (2003).

8. J.E. Amaro, M.B. Barbaro, J.A. Caballero, T.W. Donnelly, A. Molinari, I. Sick, arXiv:nucl-th/0409078
9. M.B. Barbaro, J.A. Caballero, T.W. Donnelly, C. Maieron, *Phys. Rev.* **C69**, 035502 (2004).
10. A. De Pace, M. Nardi, W.M. Alberico, T.W. Donnelly, A. Molinari, *Nucl. Phys.* **A726**, 303 (2003).
11. C. Maieron, M.C. Martínez, J.A. Caballero, J.M. Udías, *Phys. Rev.* **C68**, 048501 (2003).
12. C. Barbieri, C. Giusti, F.D. Pacati, W.H. Dickhoff, *Phys. Rev.* **C70**, 014606 (2004).
13. M.K. Gaidarov, K.A. Pavlova, A.N. Antonov, C. Giusti, S.E. Massen, C.C. Moustakidis, K. Spasova, *Phys. Rev.* **C66**, 064308 (2002).
14. C. Giusti, F.D. Pacati, *Nucl. Phys.* **A699**, 57c (2002).
15. C. Giusti, *Eur. Phys. J.* **A17**, 419 (2003).
16. C. Giusti, F.D. Pacati, *Phys. Rev.* **C67**, 044601(2003).
17. M.V. Ivanov, M.K. Gaidarov, A.N. Antonov, C. Giusti, *Nucl. Phys.* **A699**, 336c (2002).
18. A. Meucci, C. Giusti, F.D. Pacati, *Phys. Rev.* **C66**, 034610 (2002).
19. A. Meucci, *Phys. Rev.* **C65**, 044601 (2002).
20. A. Meucci, F. Capuzzi, C. Giusti, F.D. Pacati, *Phys. Rev.* **C67**, 054601 (2003).
21. A. Meucci, C. Giusti, F.D. Pacati, *Nucl. Phys.* **A739**, 277 (2004).
22. A. Meucci, C. Giusti, F.D. Pacati, arXiv:nucl-th/0405004.
23. M. Radici, A. Meucci, W.H. Dickhoff, *Eur. Phys. J.* **A17**, 65 (2003).
24. M. Radici, W.H. Dickhoff, E.R. Stoddard, *Phys. Rev.* **C66**, 014613 (2002).
25. M. Schwamb, S. Boffi, C. Giusti, F.D. Pacati, *Eur. Phys. J.* **A17**, 7 (2003).
26. M. Schwamb, S. Boffi, C. Giusti, F.D. Pacati, *Eur. Phys. J.* **A20**, 233 (2004).
27. O. Benhar, arXiv:nucl-th/0408045.
28. O. Benhar, N. Farina, arXiv:nucl-th/0407106.
29. O. Benhar, arXiv:nucl-th/0307061.
30. S. Bacca, M.A. Marchisio, N. Barnea, W. Leidemann, G. Orlandini, *Phys. Rev. Lett.* **89** 052502 (2002).
31. S. Bacca, H. Arenhoevel, N. Barnea, W. Leidemann, G. Orlandini, arXiv:nucl-th/0406080.
32. S. Bacca, N. Barnea, W. Leidemann, G. Orlandini, *Phys. Rev.* **C69**, 057001 (2004).
33. G. Orlandini, *Nucl. Phys.* **A737**, 210 (2004).
34. M. Anguiano, G. Co', A.M. Lallena, S.R. Mokhtar, *Ann. Phys. (NY)* **296**, 235 (2002).
35. M. Anguiano, G. Co', A.M. Lallena, *J. Phys. G: Nucl. Part.* **29**, 1119 (2003).
36. M. Anguiano, G. Co', A.M. Lallena, arXiv:nucl-th/040841.
37. A. Botrugno, G. Co' arXiv:nucl-th/0409041.
38. G. Co', C. Bleve, I. De Mitri, D. Martello, *Nucl. Phys. B: Proc. Suppl.* **112**, 210 (2002).
39. A.M. Lallena, M. Anguiano, G. Co', arXiv:nucl-th/0407112.
40. S. Boffi, C. Giusti, F.D. Pacati, M. Radici, *Electromagnetic Response of*

Atomic Nuclei, Clarendon, Oxford, 1996.
41. J.D. Walecka, in *Muon Physica*, Academic Press, New York, 1975.
42. M. Leuschner *et al.*, *Phys. Rev.* **C49**, 955 (1994).
43. A. Meucci, C. Giusti, F.D. Pacati, *Phys. Rev.* **C64**, 014604 (2001).
44. M. Anguiano and G. Co', *Jour. Phys. G* **27**,2109 (2001).
45. C.E. Hyde-Wright *et al.*, *Phys. Rev.* **C35**, 880 (1987).
46. S.R. Mokhtar, G. Co', A.M. Lallena, *Phys. Rev.* **C62**, 067304 (2000).
47. J. Ahrens *et al.*, *Nucl. Phys.* **A251**, 479 (1975).
48. A. Kievsky, these proceedings.
49. S. Quaglioni *et al.*, these proceedings.
50. M. Anghinolfi *et al.*, *Nucl. Phys.* **A602**, 405 (1996).
51. A. Meucci, F. Capuzzi, C. Giusti, F.D. Pacati, these proceedings
52. C.F. Williamson et al., *Phys. Rev.* **C56**, 3152 (1997).
53. Z. Meziani et al., *Phys. Rev. Lett.* **52**, 2130(1984).
54. Z. Meziani et al., *Phys. Rev. Lett.* **54**, 1233 (1985).
55. P. Barreau et al., *Nucl. Phys.* **A402**, 515 (1983).
56. F. Capuzzi, C. Giusti, F.D. Pacati, *Nucl. Phys.* **A524**, 681 (1991).
57. J.E. Amaro, G. Co', A.M. Lallena, *Nucl. Phys.* **A578**, 365 (1994).
58. J.E. Amaro, G. Co', A.M. Lallena, *Ann. Phys. (NY)* **221**, 306 (1993).
59. S.R. Mokhtar, M. Anguiano, G. Co', A.M. Lallena, *Ann. Phys. (NY)* **292**, 67 (2001).
60. M.V. Ivanov, M.K. Gaidarov, A.N. Antonov, C. Giusti, *Phys. Rev.* **C64**, 014605 (2001).
61. G.S. Adams *et al.*, *Phys. Rev.* **C38**, 2771 (1988).
62. A. Botrugno, *Ph. D. Thesis*, Università di Lecce, unpublished.
63. C. Bleve *et al. Astr. Phys.* **16** 145 (2001).
64. M. Barbaro, A. De Pace, T.W. Donnelly, A. Molinari, M.J. Musolf, *Phys. Rev.* **C54**, 1954 (1996).
65. F.D. Pacati, *Proc. of the 8th Conference on Problems in Theoretical Nuclear Physics*, Cortona (Italy), World Scientific, (Singapore) 2001.
66. R. Cenni *Proc. of the 9th Conference on Problems in Theoretical Nuclear Physics*, Cortona (Italy), World Scientific, (Singapore) 2003.
67. R. Cenni (Ed.), *Electromagnetic Response Functions of Nuclei*, Nova Science, Huntington (New York), (2001).
68. L. Lapikás, G. van der Steenhoven, L. Frankfurt, M. Strikman, M. Zhalov, *Phys. Rev.* **C61**, 064325 (2000).
69. L. Frankfurt, M. Strikman, M. Zhalov, *Phys. Lett.* **B503**, 71 (2001).

RELATIVISTIC APPROACH TO NEUTRINO-NUCLEUS QUASIELASTIC SCATTERING

ANDREA MEUCCI, CARLOTTA GIUSTI, AND FRANCO DAVIDE PACATI
Dipartimento di Fisica Nucleare e Teorica, Università di Pavia, and Istituto Nazionale di Fisica Nucleare, Sezione di Pavia, I-27100 Pavia, Italy

A relativistic Green's function and a distorted-wave impulse-approximation approach to charged- and neutral-current neutrino-nucleus quasielastic scattering are developed. Results for the neutrino (antineutrino) reactions on ^{16}O and ^{12}C target nuclei are presented and discussed.

1. Introduction

We are here interested in $\nu(\bar{\nu})$-nucleus scattering reactions in the quasielastic (QE) region, where the neutrino interacts with one single nucleon and the mechanisms like RPA or the excitations of the target nucleus become less important. In Sec. 2 a Green's function approach to describe inclusive charged-current (cc) neutrino scattering is developed (1). In such a process, $\nu(\bar{\nu})$'s interact with nuclei via the exchange of weak-vector bosons and charged leptons are produced in the final state. In our model the conservation of flux is preserved and final state interactions are treated consistently with an exclusive reaction. In Sec. 3 a relativistic distorted-wave impulse-approximation (RDWIA) calculation of semi-inclusive neutral-current (nc) ν- and $\bar{\nu}$-nucleus reactions is presented (2). The sensitivity to the strange quark content of the nucleon weak current is discussed.

2. The relativistic Green's function method

The cross section of an inclusive reaction where an incident neutrino or antineutrino interacts with a nucleus and only the outgoing lepton is detected is given by the contraction between the lepton tensor (1) and the hadron tensor, whose components are given by bilinear products of the transition matrix elements of the nuclear weak charged-current operator J^{μ}. The current operator is assumed to be adequately described as the sum of single-nucleon currents, corresponding to the weak charged current

$$j^{\mu} = \left[F_1^{\rm V}\gamma^{\mu} + i\frac{\kappa}{2M}F_2^{\rm V}\sigma^{\mu\nu}q_{\nu} - G_{\rm A}\gamma^{\mu}\gamma^5 + F_{\rm P}q^{\mu}\gamma^5\right]\tau^{\pm} , \qquad (1)$$

where $\tau^{\pm}$ are the isospin operators, κ is the anomalous part of the magnetic moment, $q^{\mu} = (\omega, \boldsymbol{q})$, with $Q^2 = |\boldsymbol{q}|^2 - \omega^2$, is the four-momentum transfer, and $\sigma^{\mu\nu} = (i/2)\left[\gamma^{\mu}, \gamma^{\nu}\right]$. $F_1^{\rm V}$ and $F_2^{\rm V}$ are the isovector Dirac and Pauli nucleon form factors. $G_{\rm A}$ and $F_{\rm P}$ are the axial and induced pseudoscalar form factors, which are usually parametrized as

$$G_{\rm A}(Q^2) = g_{\rm A}\left(1 + Q^2/M_{\rm A}^2\right)^{-2} , \quad F_{\rm P}(Q^2) = 2MG_{\rm A}\left(m_{\pi}^2 + Q^2\right)^{-1} , \ (2)$$

where $g_{\rm A} = 1.267$, m_{π} is the pion mass, and $M_{\rm A} \simeq (1.026 \pm 0.021)$ GeV is the axial mass. Performing the contraction between the lepton and hadron tensors, the inclusive cross section for the QE $\nu(\bar{\nu})$-nucleus scattering can be written as (3)

$$\frac{{\rm d}\sigma}{{\rm d}\varepsilon\,{\rm d}\Omega} = \frac{k\varepsilon G^2}{4\pi^2}\cos^2\vartheta_{\rm c}\Big[v_0 R_{00} + v_{zz}R_{zz} - v_{0z}R_{0z} + v_T R_T \pm v_{xy}R_{xy}\Big], \ (3)$$

where the coefficients v are obtained from the lepton tensor components. All nuclear structure information is contained in the response functions R, which are defined in terms of the hadron tensor components, i.e.,

$$W^{\mu\nu}(\omega, q) = \langle\Psi_0 \mid J^{\nu\dagger}(\boldsymbol{q})\delta(E_{\rm f} - H)J^{\mu}(\boldsymbol{q}) \mid \Psi_0\rangle \ . \qquad (4)$$

Introducing the Green's operator related to the nuclear Hamiltonian H, we have

$$\omega^{\mu\mu} = W^{\mu\mu}(\omega, q) = -\frac{1}{\pi}{\rm Im}\langle\Psi_0 \mid J^{\mu\dagger}(\boldsymbol{q})G(E_{\rm f})J^{\mu}(\boldsymbol{q}) \mid \Psi_0\rangle , \qquad (5)$$

for $\mu = 0, x, y, z$, and simliar expressions for ω^{0z} and ω^{xy} (1).

It was shown in Refs. (1; 4) that the nuclear response in Eq. (4) can be written in terms of the single particle Green's function, $\mathcal{G}(E)$, whose self-energy is the Feshbach's optical potential. A biorthogonal expansion of the full particle-hole Green's operator in terms of the eigenfunctions of the non-Hermitian optical potential $\mathcal{V}$, and of its Hermitian conjugate $\mathcal{V}^{\dagger}$, is performed

$$[\mathcal{E} - T - \mathcal{V}^{\dagger}(E)] \mid \chi_{\mathcal{E}}^{(-)}(E)\rangle = 0 , \quad [\mathcal{E} - T - \mathcal{V}(E)] \mid \tilde{\chi}_{\mathcal{E}}^{(-)}(E)\rangle = 0 \ . \ (6)$$

Note that E and $\mathcal{E}$ are not necessarily the same. The spectral representation is

$$\mathcal{G}(E) = \int_M^{\infty} {\rm d}\mathcal{E} \mid \tilde{\chi}_{\mathcal{E}}^{(-)}(E)\rangle \frac{1}{E - \mathcal{E} + i\eta}\langle\chi_{\mathcal{E}}^{(-)}(E) \mid \ . \qquad (7)$$

The hadron tensor components can be written in an expanded form in terms of the single-particle wave function, $| \varphi_n\rangle$, of the initial state, corresponding to the energy ε_n and whose spectral strength is λ_n as

$$\omega^{\mu\nu}(\omega,q) = -\frac{1}{\pi}\sum_n \mathrm{Im}\left[\int_M^{\infty} \mathrm{d}\mathcal{E}\frac{1}{E_\mathrm{f}-\varepsilon_n-\mathcal{E}+i\eta}T_n^{\mu\nu}(\mathcal{E},E_\mathrm{f}-\varepsilon_n)\right] , \quad (8)$$

where

$$T_n^{\mu\mu}(\mathcal{E},E) = \lambda_n\langle\varphi_n \mid j^{\mu\dagger}(\boldsymbol{q})\sqrt{1-\mathcal{V}'(E)} \mid \tilde{\chi}_{\mathcal{E}}^{(-)}(E)\rangle$$
$$\times \langle\chi_{\mathcal{E}}^{(-)}(E) \mid \sqrt{1-\mathcal{V}'(E)}j^{\mu}(\boldsymbol{q}) \mid \varphi_n\rangle , \quad (9)$$

for $\mu = 0, x, y, z$, and simliar expressions for ω^{0z} and ω^{xy} (1). The factor $\sqrt{1-\mathcal{V}'(E)}$ accounts for interference effects between different channels and allows the replacement of the mean field $\mathcal{V}$ by the phenomenological optical potential $\mathcal{V}_\mathrm{L}$. After calculating the limit for $\eta \to +0$, Eq. 8 reads

$$\omega^{\mu\nu}(\omega,q) = \sum_n \Bigg[\mathrm{Re}T_n^{\mu\nu}(E_\mathrm{f}-\varepsilon_n,E_\mathrm{f}-\varepsilon_n)$$
$$-\frac{1}{\pi}\mathcal{P}\int_M^{\infty} \mathrm{d}\mathcal{E}\frac{1}{E_\mathrm{f}-\varepsilon_n-\mathcal{E}}\mathrm{Im}T_n^{\mu\nu}(\mathcal{E},E_\mathrm{f}-\varepsilon_n)\Bigg] , \quad (10)$$

where $\mathcal{P}$ denotes the principal value of the integral.

Disregarding the square root correction, due to interference effects, The second matrix element in Eq. 9, with the inclusion of $\sqrt{\lambda_n}$ is the transition amplitude for the single-nucleon knockout from a nucleus in the state $| \Psi_0\rangle$ leaving the residual nucleus in the state $| n\rangle$. The attenuation of its strength, mathematically due to the imaginary part of the optical potential, is related to the flux lost towards the channels different from n. In the inclusive response this attenuation must be compensated by a corresponding gain, due to the flux lost, towards the channel n, by the other final states asymptotically originated by the channels different from n. This compensation is performed by the first matrix element in the right hand side of Eq. 9, where the imaginary part of the potential has the effect of increasing the strength. Similar considerations can be made, on the purely mathematical ground, for the integral of Eq. 10, where the amplitudes involved in $T_n^{\mu\nu}$ have no evident physical meaning when $\mathcal{E} \neq E_\mathrm{f}-\varepsilon_n$.

In an usual shell-model calculation the cross section is obtained from the sum, over all the single-particle shell-model states, of the squared absolute value of the transition matrix elements. Therefore, in such a calculation the negative imaginary part of the optical potential produces a loss of flux that is inconsistent with the inclusive process. In the Green's function approach, the flux is conserved, as the components of the hadron tensor are obtained in terms of the product of the two matrix elements in Eq. 9: the loss of flux, produced by the negative imaginary part of the optical potential in χ, is compensated by the gain of flux, produced in the first matrix element by the positive imaginary part of the Hermitian conjugate optical potential in $\tilde{\chi}$. The cross sections and the response functions of the inclusive QE $\nu(\bar{\nu})$-nucleus scattering are calculated from the single-particle expression of the hadron tensor in Eq. 10. After the replacement of the mean field $\mathcal{V}(E)$ by the empirical optical model potential $\mathcal{V}_{\rm L}(E)$, the matrix elements of the nuclear current operator in Eq. 9 are of the same kind as those giving the transition amplitudes of the electron induced nucleon knockout reaction (5) and the same RDWIA treatment can be used (6; 7).

As a study case, we have considered the ^{16}O target nucleus and two different values of the incident neutrino energy $E_\nu = 500$ and 1000 MeV. In order to show up the effect of the optical potential on the inclusive reaction, the results obtained in the present approach are compared with those given by different approximations. In the simplest one the optical potential is neglected and the plane wave approximation is assumed for the final state wave functions $\chi^{(-)}$ and $\tilde{\chi}^{(-)}$. In this plane wave impulse approximation (PWIA) FSI between the outgoing nucleon and the residual nucleus are completely neglected. In another approach the imaginary part of the optical potential is neglected and only the real part is included. This approximation conserves the flux, but it is inconsistent with the exclusive process, where a complex optical potential must be used. Moreover, the use of a real optical potential is unsatisfactory from a theoretical point of view, since the optical potential has to be complex owing to the presence of open channels. In Fig. 1, the differential cross sections of the $^{16}\mathrm{O}(\nu_\mu, \mu^-)$ reaction for $\vartheta_\mu =$ 30 degrees are displayed as a function of the muon kinetic energy T_μ. The behavior of the calculated cross sections is similar for the different energies. The effect of the optical potential increases with T_μ and decreases increasing E_ν. The result of the PWIA is about 20–30% higher at the peak than the one of the Green's function approach. The sum of the exclusive one-nucleon emission cross sections is always much smaller than the complete result. The difference indicates the relevance of inelastic channels and is due to the loss of flux produced by the absorptive imaginary part of the optical potential. In contrast, the cross sections calculated with only the real part of the optical potential are practically the same as the ones obtained with the Green's function approach. Although the use of a complex optical

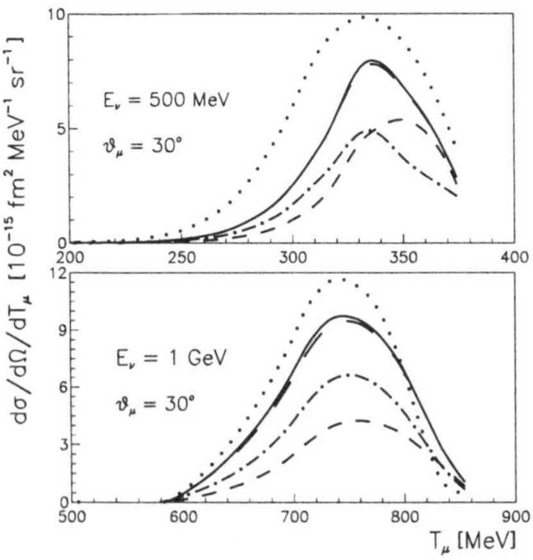

Figure 1. The differential cross sections of the $^{16}O(\nu_\mu, \mu^-)$ reaction for $E_\nu = 500$ and 1000 MeV at $\vartheta_\mu = 30$ degrees. Solid lines represent the result of the Green's function approach, dotted lines give PWIA, long-dashed lines show the result with a real optical potential, and dot-dashed lines the contribution of the integrated exclusive reactions with one-nucleon emission. Short dashed lines give the cross sections of the $^{16}O(\bar{\nu}_\mu, \mu^+)$ reaction calculated with the Green's function approach.

potential is conceptually important from a theoretical point of view, the negligible differences given by the two results mean that the conservation of flux, that is fulfilled in both calculations, is the most important condition in the present situation. In contrast, significant differences are obtained with a real optical potential in the inclusive electron scattering (4). The cross sections for the $^{16}O(\bar{\nu}_\mu, \mu^+)$ reaction are also shown for a comparison. They are always much smaller than the corresponding cross sections with an incident neutrino.

3. The neutral-current semi-inclusive scattering

The differential cross section for the neutral-current $\nu(\bar{\nu})$-nucleus quasielastic scattering is obtained from the contraction between the lepton and hadron tensors, as in Ref. (3). After performing an integration over the solid angle of the final nucleon, we have

$$\frac{\mathrm{d}\sigma}{\mathrm{d}\varepsilon\,\mathrm{d}\Omega\,\mathrm{dT_N}} = \frac{G^2\varepsilon^2}{2\pi^2}\cos^2\frac{\vartheta}{2}\Big[v_0 R_{00} + v_{zz}R_{zz} - v_{0z}R_{0z} + v_T R_T \pm v_{xy}R_{xy}\Big]\frac{|\boldsymbol{p}_\mathrm{N}|E_\mathrm{N}}{(2\pi)^3}\,, \tag{11}$$

where ϑ is the lepton scattering angle and $E_{\rm N}(\boldsymbol{p}_{\rm N})$ the energy (momentum) of the outgoing nucleon. The coefficients v and the responses R are obtained from the lepton and hadron tensor components, respectively (2).

The transition matrix elements are calculated in the first order perturbation theory and in the impulse approximation. Thus, the transition amplitude is assumed to be adequately described as the sum of terms similar to those appearing in the electron scattering case (5; 6). The RDWIA treatment is the same as in Refs. (6; 7). The single-particle current operator related to the weak neutral current is

$$j^\mu = F_1^{\rm V}\gamma^\mu + i\frac{\kappa}{2M}F_2^{\rm V}\sigma^{\mu\nu}q_\nu - G_{\rm A}\gamma^\mu\gamma^5 + F_{\rm P}q^\mu\gamma^5 \ , \qquad (12)$$

The vector form factors $F_i^{\rm V}$ can be expressed in terms of the corresponding electromagnetic form factors for protons ($F_i^{\rm p}$) and neutrons ($F_i^{\rm n}$), plus a possible isoscalar strange-quark contribution ($F_i^{\rm s}$), i.e.,

$$F_i^{\rm V;\, p(n)} = \pm\left\{F_i^p - F_i^n\right\}/2 - 2\sin^2\theta_{\rm W}F_i^{p(n)} - F_i^s/2 \ , \qquad (13)$$

where $+(-)$ stands for proton (neutron) knockout and $\theta_{\rm W}$ is the Weinberg angle ($\sin^2\theta_{\rm W} \simeq 0.2313$). The strange vector form factors are taken as (8)

$$F_2^{\rm s}(Q^2) = \frac{F_2^{\rm s}(0)}{(1+\tau)(1+Q^2/M_{\rm V}^2)^2}, \ F_1^{\rm s}(Q^2) = \frac{F_1^{\rm s}Q^2}{(1+\tau)(1+Q^2/M_{\rm V}^2)^2}, \qquad (14)$$

where $\tau = Q^2/(4M_p^2)$, $F_2^{\rm s}(0) = \mu_{\rm s}$, $F_1^{\rm s} = -\langle r_{\rm s}^2\rangle/6$, and $M_{\rm V} = 0.843$ GeV. The quantity $\mu_{\rm s}$ is the strange magnetic moment and $\langle r_{\rm s}^2\rangle$ the squared "strange radius" of the nucleon. The axial form factor is expressed as

$$G_{\rm A}(Q^2) = \frac{\pm g_{\rm A} - g_{\rm A}^s}{2\left(1+Q^2/M_{\rm A}^2\right)^2} \ , \qquad (15)$$

where $g_{\rm A}^{\rm s}$ describes possible strange-quark contributions.

In order to separate the effects of the strange-quark contribution and of FSI on the cross sections, it was suggested in Refs. (9) to measure the ratio of proton to neutron yields, as this ratio is expected to be less sensitive to distortion effects than the cross sections themselves. Moreover, from the experimental point of view the ratio is less sensitive to the uncertainties in the determination of the incident neutrino flux. In Fig. 2 the ratio for an incident neutrino is displayed as a function of the outgoing nucleon kinetic energy both in RDWIA and RPWIA. The ratio is very sensitive to $g_{\rm A}^{\rm s}$ and exhibits a maximum at $T_N \simeq 0.6\ E_\nu$, which, however, corresponds to values where the cross sections are small. The effect is sensibly reduced for $g_{\rm A}^{\rm s} = -0.10$ with respect to $g_{\rm A}^{\rm s} = -0.19$. An enhancement of the ratio of $\simeq 15\%$ is

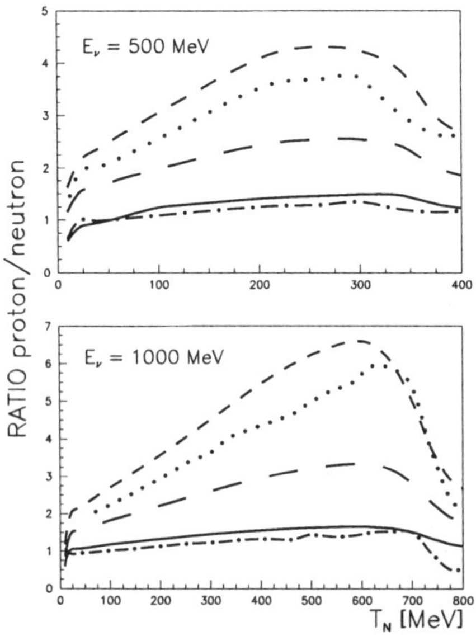

Figure 2. Ratio of proton to neutron total cross sections of the $\nu(\bar{\nu})$ quasielastic scattering on ^{12}C as a function of the incident neutrino (antineutrino) energy. Solid and dashed lines are the results in RDWIA with $g_A^s = 0$ and $g_A^s = -0.19$. Dot-dashed and dotted lines are the same results but for an incident antineutrino. Long-dashed line corresponds to neutrino scattering with $g_A^s = -0.10$.

produced by FSI. This result is due both to Coulomb distortion and to the different coupling of the optical potential with proton and neutron currents. It means that the argument of looking for the strange-quark content in this ratio is strengthened by distortion, but the possibility to fix the exact value of the contribution is affected by the uncertainties due to FSI.

Finally, in Fig. 3 we compare our results with the data of the BNL 734 experiment (10). Experimental results were presented in the form of a flux-averaged differential cross section per momentum transfer squared Q^2. Our results are shown with $g_A^s = -0.19$ and without the strange-quark contribution. We give also the effect of including the strange vector form factors, F_1^s and F_2^s, with $F_1^s = -\langle r_s^2 \rangle/6 = 0.53$ GeV^{-2}, $F_2^s(0) = -0.40$ (11), and the Q^2 dependence given in Eq. 14. The strange-quark contribution produces an enhancement of the cross sections, which makes them slightly higher than the experimental data. The strange weak magnetic contribution decreases the cross section, while the axial and weak electric components give an enhancement.

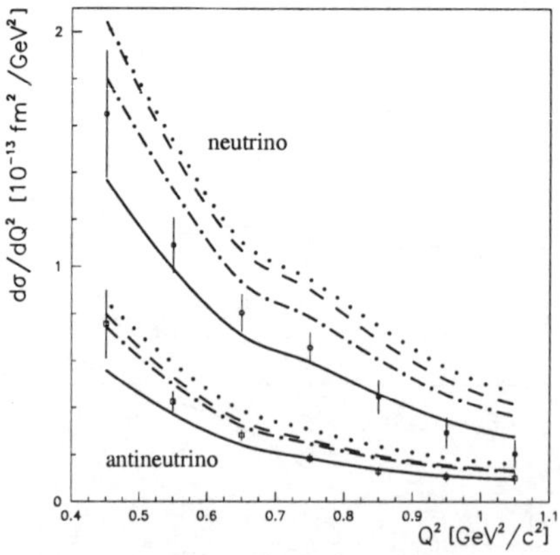

Figure 3. Differential cross sections of the $\nu(\bar{\nu})$ quasielastic scattering, flux-averaged over BNL spectrum (10), as a function of the momentum transfer squared. The four upper curves are for incident neutrino and the four lower ones for incident antineutrino. Solid lines are the results with no strangeness contribution, dashed lines with $g_A^s = -0.19$, dot-dashed lines with $g_A^s = -0.19$ and $F_2^s(0) = -0.40$, dotted lines with $g_A^s = -0.19$, $F_2^s(0) = -0.40$ and $F_1^s = -\langle r_s^2 \rangle/6 = 0.53$ GeV^{-2}. Experimental data from Ref. (10). The errors correspond to statistical and Q^2-dependent systematic errors added in quadrature and do not include Q^2-independent systematic errors.

References

1. A. Meucci, C. Giusti, and F.D. Pacati, *Nucl. Phys.* **A 739**, 277 (2004).
2. A. Meucci, C. Giusti, and F.D. Pacati, *Nucl. Phys.* **A 744**, 307 (2004).
3. J.D. Walecka, in *Muon Physics*, Vol. II, edited by V.H. Hughes and C.S. Wu (Academic Press, New York, 1975), p. 113.
4. A. Meucci, F. Capuzzi, C. Giusti, and F.D. Pacati, *Phys. Rev.* **C 67**, 054601 (2003).
5. S. Boffi, *et al.*, *Electromagnetic Response of Atomic Nuclei*, Oxford Studies in Nuclear Physics, Vol. 20 (Clarendon, Oxford, 1996); S. Boffi, *et al.*, Phys. Rep. **226**, 1 (1993).
6. A. Meucci, C. Giusti, and F.D. Pacati, *Phys. Rev.* **C 64**, 014604 (2001).
7. A. Meucci, C. Giusti, and F.D. Pacati, *Phys. Rev.* **C 64**, 064615 (2001).
8. G.T. Garvey, *et al.*, *Phys. Rev.* **C 48**, 1919 (1993).
9. G.T. Garvey, *et al.*, *Phys. Lett.* **B 289**, (1992) 249; C.J. Horowitz, *et al.*, *Phys. Rev.* **C 48**, 3078 (1993); W.M. Alberico, *et al.*, *Nucl. Phys.* **A 623**, 471 (1997); W.M. Alberico, *et al.*, *Phys. Lett.* **B 438**, 9 (1998).
10. L.A. Ahrens, *et al.*, *Phys. Rev.* **D 35**, 785 (1987).
11. G.T. Garvey, *et al.*, *Phys. Rev.* **C 48**, 761 (1993).

LORENTZ INTEGRAL TRANSFORM METHOD APPLIED TO EXCLUSIVE ELECTROMAGNETIC REACTIONS ON ^{4}He

S. QUAGLIONI, W. LEIDEMANN AND G. ORLANDINI

Dipartimento di Fisica, Università di Trento, and Istituto Nazionale di Fisica Nucleare, Gruppo Collegato di Trento, via Sommarive 14, I-38050 Povo (TN), Italy

N. BARNEA

The Racah Institute of Physics, The Hebrew University, 91904, Jerusalem, Israel

V. D. EFROS

Russian Research Centre "Kurchatov Institute", Kurchatov Square 1, 123182 Moscow, Russia

The Lorentz integral transform method for exclusive processes is applied to the calculation of electromagnetic reactions leading to ^{4}He two–body disintegration. This technique allows to take into account in a rigorous way the final state interaction, without requiring the explicit knowledge of the final state wave function. The obtained results for the cross sections of the $^4\mathrm{He}(\gamma, n)^3\mathrm{He}$ processes as well as the longitudinal response function of the $^4\mathrm{He}(e, e'p)^3\mathrm{H}$ exclusive reaction will be presented.

1. The Lorentz integral transform method for exclusive reactions

For the case of an exclusive perturbation–induced process all the information about the reaction dynamics is contained in the transition matrix element of the perturbation $\widehat{O}$ between the initial (Ψ_0) and final (Ψ_f^-)

states,

$$T_f(E_f) = \left\langle \Psi_f^-(E_f) \left| \widehat{O} \right| \Psi_0 \right\rangle . \tag{1}$$

The calculation of such a matrix element can be carried out with the Lorentz integral transform (LIT) method as outlined in the following.[1,2,3] For simplicity we restrict our discussion to an exclusive reaction leading to a final state with two fragments, a fragment a with n_a nucleons and a fragment b with $n_b = A - n_a$ nucleons, though more complex fragmentations of the initial A-body system can be treated as well. Denoting with H the full nuclear Hamiltonian we have a formal expression for $\Psi^-_{f=a,b}$ in terms of the unperturbed "channel state" $\phi^-_{f=a,b}(E_{f=a,b})$,[4]

$$\left|\Psi^-_{a,b}(E_{a,b})\right\rangle = \widehat{\mathcal{A}} \left|\phi^-_{a,b}(E_{a,b})\right\rangle + \frac{1}{E_{a,b} - i\varepsilon - H} \widehat{\mathcal{A}}\mathcal{V} \left|\phi^-_{a,b}(E_{a,b})\right\rangle , \tag{2}$$

where $\widehat{\mathcal{A}}$ is an antisymmetrization operator. In case that at least one of the fragments is chargeless the channel state $\phi^-_{a,b}(E_{a,b})$ is the product of the internal wave functions of the fragments and of their relative free motion. Correspondingly, $\mathcal{V}$ in Eq. (2) is the sum of all interactions between particles belonging to different fragments. If both fragments are charged, $\phi^-_{a,b}(E_{a,b})$ is chosen to account for the average Coulomb interaction between them, and the plane wave describing their relative motion is replaced by the Coulomb function of the minus type. Correspondingly, $\mathcal{V}$ in Eq. (2) is the sum of all interactions between particles belonging to different fragments after subtraction of the average Coulomb interaction, already considered via the Coulomb function.

When one inserts Eq. (2) into Eq. (1) the transition matrix element becomes the sum of two pieces, a Born term,

$$T^{Born}_{a,b}(E_{a,b}) = \left\langle \phi^-_{a,b}(E_{a,b}) \left| \widehat{\mathcal{A}}\, \widehat{O} \right| \Psi_0 \right\rangle , \tag{3}$$

and an final state interaction (FSI) dependent term,

$$T^{FSI}_{a,b}(E_{a,b}) = \left\langle \phi^-_{a,b}(E_{a,b}) \left| \mathcal{V}\widehat{\mathcal{A}} \frac{1}{E_{a,b} + i\varepsilon - H} \widehat{O} \right| \Psi_0 \right\rangle . \tag{4}$$

While the Born term is rather simple to deal with, the determination of the FSI dependent matrix element is rather complicated. We treat this term within the LIT approach as outlined in the following. Let $\Psi_\nu(E_\nu)$ be the eigenstates of the Hamiltonian labeled by channel quantum numbers ν and normalized as $\langle \Psi_\nu | \Psi_{\nu'} \rangle = \delta(\nu - \nu')$. Using the completeness relation of the set $\Psi_\nu(E_\nu)$ the matrix element $T^{FSI}_{a,b}(E_{a,b})$ can be written in a form which emphasizes its dependence upon the continuum:

$$T^{FSI}_{a,b}(E_{a,b}) = \sum\!\!\!\!\!\!\int_\nu \langle \phi^-_{a,b}(E_{a,b})|\mathcal{V}\widehat{\mathcal{A}}|\Psi_\nu(E_\nu)\rangle \frac{1}{E_{a,b}+i\varepsilon - E_\nu}\langle \Psi_\nu(E_\nu)|\widehat{O}|\Psi_0\rangle$$
$$= -i\pi F_{a,b}(E_{a,b}) + \mathcal{P}\int_{E^-_{th}}^{\infty} \frac{F_{a,b}(E)}{E_{a,b}-E} dE, \tag{5}$$

where $F_{a,b}(E)$ is defined as

$$F_{a,b}(E) = \sum\!\!\!\!\!\!\int_\nu \left\langle \phi^-_{a,b}(E_{a,b}) \left| \mathcal{V}\widehat{\mathcal{A}} \right| \Psi_\nu(E_\nu) \right\rangle \left\langle \Psi_\nu(E_\nu) \left| \widehat{O} \right| \Psi_0 \right\rangle \delta(E-E_\nu), \tag{6}$$

and E_{th} is the lowest excitation energy in the system, i.e. the breakup threshold energy. To calculate $T^{FSI}_{a,b}$ one needs to know the function $F_{a,b}$ for all energy values. The direct calculation of $F_{a,b}$ is of course far too difficult, since one should know all the eigenstates Ψ_ν for the whole eigenvalue spectrum of H. However, an indirect calculation of $F_{a,b}$ is possible applying the LIT method. To this end, one introduces an integral transform of $F_{a,b}$ with a kernel of Lorentzian shape,

$$L\left[F_{a,b}\right](\sigma) = \int_{E^-_{th}}^{\infty} \frac{F_{a,b}(E)}{(E-\sigma_R)^2+\sigma_I^2}\, dE$$
$$= \frac{1}{2i\sigma_I}\int_{E^-_{th}}^{\infty} F_{a,b}(E)\left[\frac{1}{E-\sigma} - \frac{1}{E-\sigma^*}\right] dE$$
$$= \frac{1}{2i\sigma_I}\left\langle \phi^-(E_{a,b}) \left| \mathcal{V}\widehat{\mathcal{A}}\left[\frac{1}{H-\sigma} - \frac{1}{H-\sigma^*}\right]\widehat{O} \right| \Psi_0 \right\rangle, \tag{7}$$

where σ is a complex parameter given by $\sigma = \sigma_R + i\sigma_I$. The final expression of Eq. (7) can be directly evaluated via the Lanczos technique[5] without explicit knowledge of $F_{a,b}$. Actually, in terms of the Lanczos orthonormal basis $\{\varphi_i, i = 0, ..., n\}$, where the starting Lanczos vector is given by

$$|\varphi_o\rangle = \frac{\widehat{O}\,|\Psi_0\rangle}{\sqrt{\left\langle \Psi_0 \left| \widehat{O}^\dagger\widehat{O} \right| \Psi_0 \right\rangle}}, \tag{8}$$

starting from Eq. (7) one can write

$$L[F_{a,b}](\sigma) = \frac{\sqrt{\left\langle \Psi_0 \left| \widehat{O}^\dagger\widehat{O} \right| \Psi_0 \right\rangle}}{2i\sigma_I} \sum_{i=0}^{n} \left\langle \phi^-_{a,b}(E_{a,b}) \left| \mathcal{V}\widehat{\mathcal{A}} \right| \varphi_i \right\rangle$$
$$\times \left\langle \varphi_i \left| \frac{1}{H-\sigma} - \frac{1}{H-\sigma^*} \right| \varphi_o \right\rangle, \tag{9}$$

where matrix elements $\langle\varphi_i|(H-\sigma)^{-1}|\varphi_o\rangle$ can be calculated as continued fractions of the Lanczos coefficients.

After having calculated $L[F_{a,b}](\sigma)$ one obtains the function $F_{a,b}(E)$, and thus $T_{a,b}(E_{a,b})$, via the inversion of the LIT, as described in Ref. 6. The knowledge of $F_{a,b}(E)$ then allows to obtain the FSI part of the transition matrix element via Eq. (5).

2. Two-body photodisintegration of ^{4}He

The technique outlined in Sec. 1 is applied to the calculation of the total cross sections for the processes $^4\text{He}(\gamma,p)^3\text{H}$ and $^4\text{He}(\gamma,n)^3\text{He}$. For these exclusive reactions we investigate the question of the giant dipole peak height, but we also consider higher energies. We would like to emphasize that FSI is taken into account rigorously also in the region beyond the three-body break-up threshold. Here (as well as in the calculation of the inclusive cross section[7,8]) only the transitions induced by the unretarded dipole operator D are taken into account. In terms of the transition matrix elements the total exclusive cross section of the ^{4}He photodisintegration into the two fragments N and *3*, where N refers to the scattered proton (neutron) and *3* refers to the ^{3}H (^{3}He) nucleus, is given by

$$\sigma_{N,3}=4\pi^2\frac{e^2}{\hbar c}k\mu\,\omega_\gamma\int\sum_{M_3,M_N}\left|\left\langle\Psi^-_{N,3}(E_{N,3})\,|D|\,\Psi_\alpha\right\rangle\right|^2 d\Omega_k\ ,\ E_{N,3}=\omega_\gamma+E_\alpha \tag{10}$$

(in the equation above we neglect the very small nuclear recoil energy). With μ and k we denote the reduced mass and the relative momentum of the fragments, respectively, ω_γ is the incident photon energy, Ψ_α is the ground–state wave function, $\Psi^-_{N,3}$ is the final–state continuum wave function of the minus type pertaining to the N,*3* channel,[4] and E_α and $E_{N,3}$ are the energies of the corresponding initial and final states ($E_{N,3}=k^2/2\mu+E_3$, E_3 being the energy of fragment *3*), respectively. The sum goes over projections M_3 and M_N of the fragment angular momenta in the final state.

The ground states of ^{4}He, ^{3}He and ^{3}H as well as the LIT in Eq. (9) are calculated using the correlated hyperspherical harmonics (CHH) expansion method. In order to speed up the convergence, state independent correlations are introduced as in Ref. 9. We use the MTI-III[10] potential and identical CHH expansions for the ground state wave functions of ^{4}He and of the three–nucleon systems as in Ref. 7 and Ref. 11, respectively.

In Fig. 1 we present our results for the $^4\text{He}(\gamma,n)^3\text{He}$ cross section together with experimental data (see Ref. 3). The shaded area around the

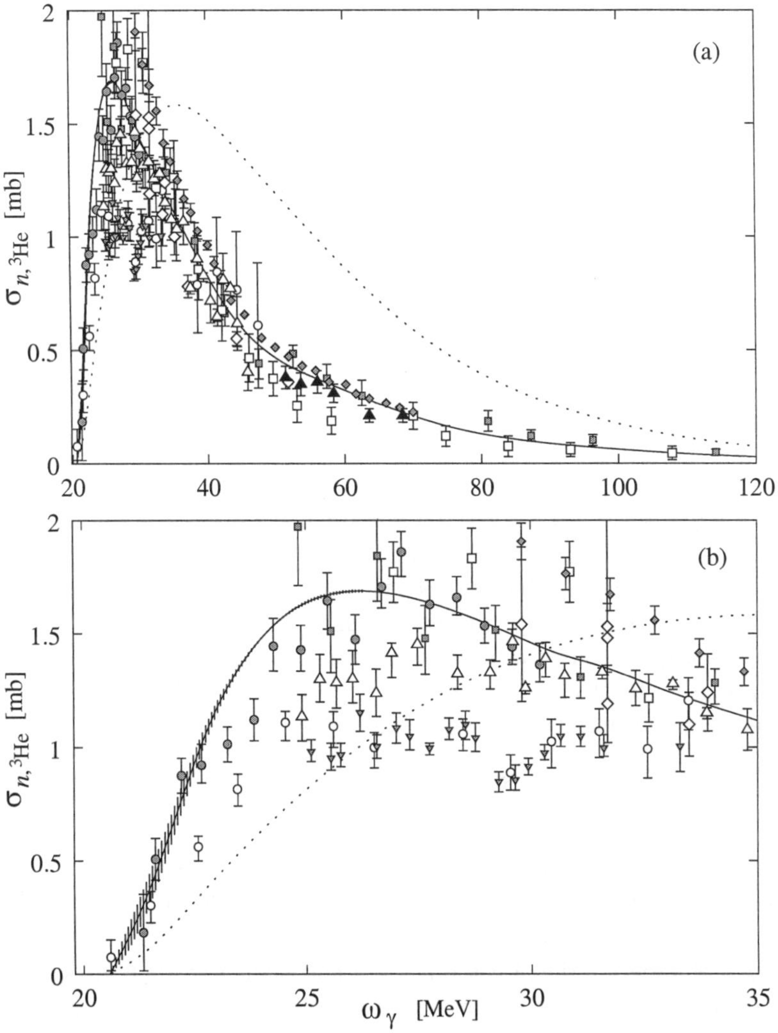

Figure 1. ^{4}He$(\gamma, n)^3$He cross section up to 120 MeV (a) and 35 MeV (b) together with experimental data (see Ref. 3): full result with FSI included (solid curve), Born approximation only (dotted curve).

full result represents an estimate of the uncertainties of our calculation.[3] The difference between the total cross section and its Born approximation shows large effects of FSI.

The experimental results do not show a unique picture. In the energy region beyond 35 MeV there is an overall agreement of the data whereas in the dipole resonance region big discrepancies are present. In the low–energy region our full calculation favours the strongly peaked data. It is seen from the figure that our full calculation agrees quite well with the higher energy experimental cross section.

Very similar results are obtained for the ^{4}He$(\gamma, p)^3$H cross section.[3]

3. The ^{4}He$(e, e'p)^3$H longitudinal response function

The longitudinal response function of the exclusive process ^{4}He$(e, e'p)^3$H,[12]

$$R_L(\mathbf{q},\omega) = \sum_{M_{^3\mathrm{H}},M_p} \left|\left\langle \Psi^-_{p,^3\mathrm{H}}(E_{p,^3\mathrm{H}})\,|\hat\rho(\mathbf{q})|\,\Psi_\alpha\right\rangle\right|^2 \,,\ E_{p,^3\mathrm{H}} = \omega - \frac{|\mathbf{q}|^2}{2M_\alpha} + E_\alpha , \tag{11}$$

is calculated for the first time in a microscopic way, taking fully and consistently into account the interaction both in the initial and final states via the LIT method, as outlined in Sec. 1.

In the equations above we use the same notation as Sec. 2 with the only exceptions of ω and $\mathbf{q}$, which represent the energy and the momentum transfer from the electron to the nucleus, respectively. We take the nuclear charge operator $\widehat{\rho}$ in its non relativistic form:

$$\hat\rho(\mathbf{q}) = \sum_{j=1}^{4} (G_E^p)^2 \frac{1+\tau_j^3}{2} \exp\left(i\mathbf{q}\cdot\mathbf{r}_j\right). \tag{12}$$

Here τ_j^3 denotes the third component of the j-th nucleon isospin and $\mathbf{r}_j$ represents the position of the j-th nucleon with respect to the center of mass of the four-body system.

The reliability of the direct knock-out and plane wave hypothesis (plane wave impulse approximation (PWIA)) and the role of FSI are investigated in the parallel kinematics (outgoing proton momentum $\mathbf{p}_p$ parallel to the momentum transfer $\mathbf{q}$) of Ref. 13 and also in some non parallel ones. The results have been obtained using the CHH expansion method and the semirealistic MTI-III potential as already mentioned in Sec. 2.

In Figure 2 we show the effects of the "exchange terms" (exc.) originating from the proper antisymmetrization of the p–^{3}H final state, as well as those of FSI. As one should expect for parallel kinematics one notices only weak influences of the antisymmetrization for almost all cases, however, at lower energies, such effects can increase up to about 10%. In comparison the role of FSI is much more important, especially at low q, but the

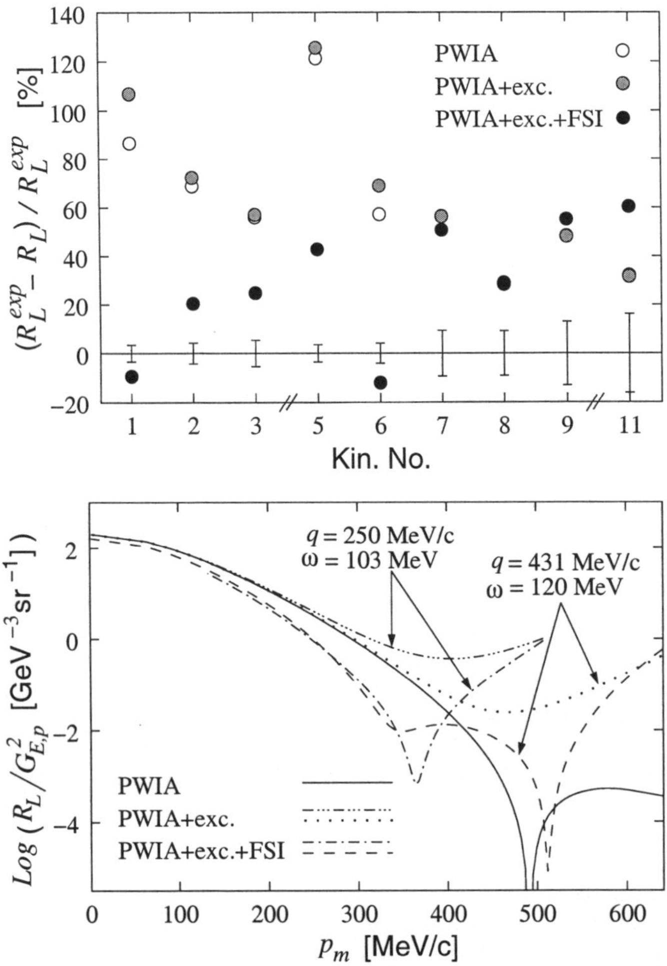

Figure 2. Upper panel: deviation from experiment of the calculated longitudinal response function in parallel kinematics; kinematics (Kin. No.) and experimental data from Ref. 13. Lower panel: longitudinal response function divided by proton electric form factor for two nonparallel kinematics as function of the modulus of the missing momentum ($\mathbf{p}_m = \mathbf{q} - \mathbf{p}_p$).

FSI–effect decreases considerably beyond $q = 500$ MeV/c. An exception is Kinematics 11 with $q = 680$ MeV/c exhibiting a rather high FSI contribution of about 20%. The relatively large effect is probably due to the rather high p_m ($\mathbf{p}_m = \mathbf{q} - \mathbf{p}_p$) and the fact that one is further away from the quasi elastic ridge. For the nonparallel kinematics we find a strong influence of both exchange terms and FSI at high p_m. At low p_m the role of FSI is to lower the response by about 20%. In addition, in the upper panel we show the deviation of the theoretical results for R_L with respect to the experimental data of Ref. 13, extracted via a Rosenbluth separation. We find a fair overall agreement between theory and experiment especially at lower ω and q (Kinematics 1 and 6). For the other cases the theoretical results are always somewhat higher than data. The difference ranges from about 25% for the kinematics closer to the quasi elastic ridge, to about 50 and even 60% for the others ones. Though the calculation is complete regarding the fully consistent treatment of the interaction both in the initial and in the final states it is too premature to draw any conclusion from the discrepancy between theory and experiment. The use of more realistic nuclear force could affect the results and give interesting insight in the role of the different potential terms.

References

1. V. D. Efros, *Yad. Fiz.* **41**, 1498 (1985) [*Sov. J. Nucl. Phys.* **41**, 949 (1985)].
2. A. La Piana and W. Leidemann, *Nucl. Phys.* **A677**, 423 (2000).
3. S. Quaglioni, N. Barnea, V. D. Efros, W. Leidemann, and G. Orlandini, *Phys. Rev.* **C69**, 044002 (2004).
4. M. L. Goldberger and K. W. Watson, *Collision theory*, Wiley, New York, 1964.
5. M. A. Marchisio, N. Barnea, W. Leidemann, and G. Orlandini, *Few-Body Syst.* **33**, 259 (2003).
6. V. D. Efros, W. Leidemann, and G. Orlandini, *Few-Body Syst.* **26**, 251 (1999).
7. N. Barnea, V. D. Efros, W. Leidemann, and G. Orlandini, *Phys. Rev.* **C63**, 057002 (2001).
8. V. D. Efros, W. Leidemann, and G. Orlandini, *Phys. Rev. Lett.* **78**, 4015 (1997).
9. V. D. Efros, W. Leidemann, and G. Orlandini, *Phys. Rev. Lett.* **78**, 432 (1997).
10. R. A. Malfliet and J. Tjon, *Nucl. Phys.* **A127**, 161 (1969).
11. V. D. Efros, W. Leidemann, and G. Orlandini, *Phys. Lett.* **B408**, 1 (1997).
12. D. Drechsel and M. M. Giannini, *Rep. Prog. Phys.* **52**, 1083 (1989).
13. J. E. Ducret *et al.*, *Nucl.Phys.* **A556**, 373 (1993).

WEAK DECAY OF Λ–HYPERNUCLEI

W. M. ALBERICO AND G. GARBARINO
Dipartimento di Fisica Teorica, Università di Torino and INFN, Sezione di Torino, I–10125 Torino, Italy

A. PARREÑO AND A. RAMOS
Departament d'Estructura i Constituents de la Matèria, Universitat de Barcelona, E–08028 Barcelona, Spain

We shortly review the open problems in the physics of the weak decay of Λ–hypernuclei. For many years the main question has been the discrepancy between theory and experiment for the ratio Γ_n/Γ_p between the neutron– ($\Lambda n \to nn$) and proton–induced ($\Lambda p \to np$) decay rates. Recent indications towards a solution of the puzzle are discussed here. The other open question concerns the asymmetric non–mesonic decay of polarized hyperuclei. While theory predicts negative asymmetries, with a moderate dependence on the hypernucleus, the measurements favor negative values for ${}^{12}_{\Lambda}$C but small, positive values for ${}^{5}_{\Lambda}$He. Further theoretical and experimental investigations are needed to clarify this issue.

1. Introduction

A Λ–hypernucleus is a bound system of nucleons and one Λ hyperon. When it becomes stable with respect to electromagnetic and strong processes, the hyperon is in the $1s$ level. Such a hypernucleus can then only decay via a strangeness–changing weak interaction, through the disappearance of the Λ. In the following we introduce the main features of the mesonic and non–mesonic decay modes of the Λ in the nuclear medium.

1.1. *Mesonic decay*

The mesonic decay mode:

$$\begin{aligned} \Lambda \to \pi^- p \quad (\Gamma_{\pi^-}) \\ \pi^0 n \quad (\Gamma_{\pi^0}) \end{aligned}$$

is the main decay channel of a Λ in free space, with a Q–value for a decay at rest $Q_\Lambda \simeq m_\Lambda - m_N - m_\pi \simeq 40$ MeV. Taking into account energy–

momentum conservation, the momentum of the final nucleon turns out to be $p \simeq 100$ MeV. Inside a hypernucleus, the binding energies of the recoil nucleon ($B_N \simeq -8$ MeV) and of the Λ ($B_\Lambda \geq -27$ MeV) tend to decrease Q_Λ [$Q_{\Lambda,\mathrm{bound}} = Q_\Lambda + B_\Lambda - B_N$] and hence p.

As a consequence, in nuclei the Λ mesonic decay is disfavored by the Pauli principle, particularly in heavy systems. It is strictly forbidden in normal infinite nuclear matter (where the Fermi momentum is $k_F^0 \simeq 270$ MeV), while in finite nuclei it can occur because of three important effects:

(1) In nuclei the hyperon has a momentum distribution that allows larger momenta to be available to the final nucleon;
(2) The final pion feels an attraction by the medium such that for fixed momentum $\vec{q}$ it has an energy smaller than the free one [$\omega(\vec{q}) < \sqrt{\vec{q}^2 + m_\pi^2}$], and consequently, due to energy conservation, the final nucleon again has more chance to come out above the Fermi surface;
(3) At the nuclear surface the local Fermi momentum is considerably smaller than k_F^0, and the Pauli blocking is less effective in forbidding the decay.

The mesonic channel provides information on the pion–nucleus optical potential since the mesonic widths Γ_{π^-} and Γ_{π^0} are very sensitive to the pion self–energy in the medium: the latter is enhanced by the attractive P–wave π–nucleus interaction and reduced by the repulsive S–wave one.

1.2. *Non–mesonic decay*

In the nuclear medium the decay can also occur through processes which involve a weak interaction of the Λ with one or more nucleons, resulting in non–mesonic processes of the following type:

$$\Lambda n \to nn \qquad (\Gamma_n) \tag{1}$$

$$\Lambda p \to np \qquad (\Gamma_p) \tag{2}$$

$$\Lambda NN \to nNN \ \ (\Gamma_2)\,. \tag{3}$$

The total weak decay rate of a Λ–hypernucleus is then:

$$\Gamma_{\mathrm{T}} = \Gamma_{\mathrm{M}} + \Gamma_{\mathrm{NM}},$$

where:

$$\Gamma_{\mathrm{M}} = \Gamma_{\pi^-} + \Gamma_{\pi^0}, \quad \Gamma_{\mathrm{NM}} = \Gamma_1 + \Gamma_2, \quad \Gamma_1 = \Gamma_n + \Gamma_p,$$

and the lifetime is $\tau = \hbar/\Gamma_{\mathrm{T}}$. The channel (3) can be interpreted by assuming that the pion emitted by the Λ vertex is absorbed by a pair of nucleons, correlated by the strong interaction. Obviously, the non–mesonic processes can also be mediated by the exchange of more massive mesons than the pion.

The non–mesonic mode is only possible in nuclei and, nowadays, the systematic study of the hypernuclear decay is the only practical way to get information on the weak process $\Lambda N \to NN$ (which provides the first extension of the weak $\Delta S = 0$ $NN \to NN$ interaction to strange baryons), especially on its parity–conserving part, which is masked by the strong interaction in the weak $NN \to NN$ reaction.

The final nucleons in the non–mesonic processes emerge with large momenta: disregarding the Λ and nucleon binding energies and assuming the available energy $Q = m_\Lambda - m_N \simeq 176$ MeV to be equally splitted among the final nucleons, it turns out that $p_N \simeq 420$ MeV for the one–nucleon induced channels [Eqs. (1), (2)] and $p_N \simeq 340$ MeV in the case of the two–nucleon induced mechanism [Eq. (3)]. Therefore, the non–mesonic decay mode is not forbidden by the Pauli principle: on the contrary, the final nucleons have great probability to escape from the nucleus. The non–mesonic mechanism dominates over the mesonic mode for all but the s–shell hypernuclei.

Since the non–mesonic channel is characterized by large momentum transfer, the details of the hypernuclear structure do not have a substantial influence (then providing useful information directly on the hadronic weak interaction). On the other hand, the NN and ΛN short range correlations turn out to be very important.

It is interesting to observe that there is an anticorrelation between mesonic and non–mesonic decay modes such that the total decay rate is quite stable from light to heavy hypernuclei (see Fig. 1). The non–mesonic rate saturates with increasing hypernuclear mass number.

2. The Γ_n/Γ_p puzzle

Up to very recent times, the main challenge of hypernuclear weak decay studies has been to provide a theoretical explanation of the large experimental values for the ratio Γ_n/Γ_p[1,2]. The large uncertainties involved in the extraction of the ratio from data did not allow to reach any definitive conclusion. The data were quite limited and not precise due to the difficulty of detecting the products of the non–mesonic decays, especially the neutrons. However, due to recent theoretical[3,4,5,6,7] and experimental[8] progress, we are now towards a solution of the Γ_n/Γ_p puzzle.

The one–pion–exchange (OPE) approximation provides small Γ_n/Γ_p

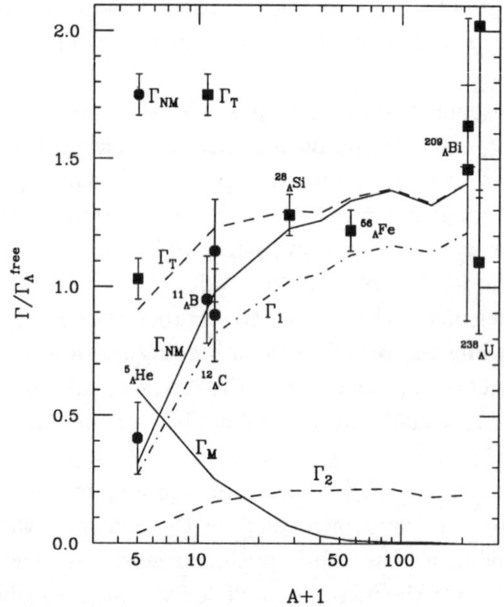

Figure 1. Theoretical predictions and experimental data for the hypernuclear partial decay widths as a function of the nuclear mass number A. (taken from Ref. [1]).

values, in the range $0.05 \div 0.20$, for all considered systems. This is due to the $\Delta I = 1/2$ rule, which fixes the vertex ratio $V_{\Lambda\pi^- p}/V_{\Lambda\pi^0 n} = -\sqrt{2}$ (both in S– and P–wave interactions), and to the particular form of the OPE potential, which has a strong tensor component requiring isospin 0 np pairs in the antisymmetric final state. However, the OPE model has been able to reproduce the total non–mesonic rates observed for light and medium hypernuclei[1,2].

Other interaction mechanisms beyond the OPE might then be responsible for the overestimation of Γ_p and the underestimation of Γ_n. Many attempts have been made up to now in order to solve the Γ_n/Γ_p puzzle. Those which have improved the situation are: the inclusion in the $\Lambda N \to nN$ transition potential of mesons heavier than the pion[4,5,6], the inclusion of interaction terms that explicitly violate the $\Delta I = 1/2$ rule[9] and the description of the short range baryon–baryon interactions in terms of quark degrees of freedom[3], which automatically introduces $\Delta I = 3/2$ contributions.

The analysis of Γ_n/Γ_p is also influenced by two–nucleon induced processes[7], $\Lambda NN \to nNN$, whose importance has been established theoretically. Unfortunately, its experimental identification is rather difficult and it is a challenge for the future. Moreover, since Γ_n/Γ_p turns out to be sensitive to the detailed kinematics of the non–mesonic processes and to the experimental threshold for nucleon detection, a carefull evaluation of the nucleon rescattering inside the residual nucleus has to be performed[7,10].

2.1. *Towards a solution of the puzzle*

Fortunately, recent important developments have contributed to approach the solution of the Γ_n/Γ_p puzzle. The old analyses of the ratio based on the measurement of *single* nucleon energy spectra revealed to be quite indirect and inaccurate methods[7]. The study of nucleon–nucleon *coincidence* measurements permitted a more direct extraction of Γ_n/Γ_p from data.

The experiments KEK–E462 and KEK–E508[8] measured nn and np angular and energy correlations for the decay of ${}^{5}_{\Lambda}$He and ${}^{12}_{\Lambda}$C. In Ref. [7] these data are analyzed on the basis of a model in which the weak decay is described in a finite nucleus framework using a one–meson–exchange $\Lambda N \to nN$ transition potential[5]. The two–nucleon stimulated decay is evaluated within a polarization propagator method and by treating the nuclear finite size effects via a local density approximation scheme. An intranuclear cascade code[10] takes into account the nucleon final state interactions (FSI). From KEK data the following ratios have been determined:

$$\frac{\Gamma_n}{\Gamma_p}\,({}^{5}_{\Lambda}\mathrm{He}) = 0.40 \pm 0.11 \text{ if } \Gamma_2 = 0\,, \quad \frac{\Gamma_n}{\Gamma_p}\,({}^{5}_{\Lambda}\mathrm{He}) = 0.27 \pm 0.11 \text{ if } \frac{\Gamma_2}{\Gamma_1} = 0.2,$$

$$\frac{\Gamma_n}{\Gamma_p}\,({}^{12}_{\Lambda}\mathrm{C}) = 0.38 \pm 0.14 \text{ if } \Gamma_2 = 0\,, \quad \frac{\Gamma_n}{\Gamma_p}\,({}^{12}_{\Lambda}\mathrm{C}) = 0.29 \pm 0.14 \text{ if } \frac{\Gamma_2}{\Gamma_1} = 0.25.$$

These values are in agreement with the pure theoretical predictions of Refs. [4,5,6] but are substantially smaller than those obtained experimentally from previous single nucleon spectra analyses. Actually, all these experimental analyses of single nucleon spectra, supplemented in some cases by intranuclear cascade calculations, derived Γ_n/Γ_p values in disagreement with all existing theoretical predictions.

In our opinion, the achievements of Ref. [7] clearly exhibit the interest of analyses of correlation observables and represent an important progress towards the solution of the Γ_n/Γ_p puzzle. Forthcoming coincidence data could be directly compared with the results discussed in that paper. This will permit to achieve better determinations of Γ_n/Γ_p and to establish the first constraints on the two–nucleon induced decay channel.

3. Non–mesonic decay of polarized hypernuclei: the asymnmetry puzzle

Lambda hypernuclear states can be produced with a sizable amount of polarization. The development of angular distribution measurements of decay particles (photons, pions and protons) from polarized hypernuclei is of crucial importance in order to extract new information on hypernuclear production, structure and decay. A recent and intriguing problem is open in this area: it concerns a strong disagreement between theory and experiment on the asymmetry of the angular emission of non–mesonic decay protons from polarized hypernuclei.

The intensity of protons emitted in $\vec{\Lambda} p \to np$ decays along a direction forming an angle θ with the polarization axis is given by[11]:

$$I(\theta) = I_0 \left(1 + p_\Lambda \, a_\Lambda \, \cos\theta\right) , \tag{4}$$

where p_Λ is the Λ polarization and a_Λ the intrinsic Λ asymmetry parameter. This asymmetry, due to the interference between parity–violating and parity–conserving $\vec{\Lambda} p \to np$ transition amplitudes, is supposed to provide new constraints on the dynamics of the non–mesonic decay. Theory predicts negative a_Λ values, with a very weak dependence on the hypernucleus.

Nucleon FSI acting after the non–mesonic weak decay modify the weak decay intensity of Eq. (4). Experimentally, one has access to a proton intensity:

$$I^{\rm M}(\theta) = I_0^{\rm M}(1 + p_\Lambda \, a_\Lambda^{\rm M} \cos\theta), \tag{5}$$

where the observable asymmetry $a_\Lambda^{\rm M}$ could depend on the hypernucleus and can be obtained by measuring $I^{\rm M}(0^\circ)$ and $I^{\rm M}(180^\circ)$.

Until now, four KEK experiments measured the proton asymmetric emission from polarized Λ–hypernuclei. They favor negative values for $a_\Lambda^{\rm M}({}^{12}_{\Lambda}\vec{\rm C})$ and small, positive values for $a_\Lambda^{\rm M}({}^{5}_{\Lambda}\vec{\rm He})$.

In order to make a direct comparison between theory (which gives predictions for a_Λ) and experiment (which observes $a_\Lambda^{\rm M}$) possible, we have recently estimated the effects of the nucleon FSI in the non–mesonic weak decay of s– and p–shell polarized hypernuclei[12]. We summarize here some results of this investigation, which is the first one evaluating $a_\Lambda^{\rm M}$.

In Table 1 we show predictions for the weak decay and observable proton intensities, $I(\theta)$ and $I^{\rm M}(\theta)$, respectively. As a result of the nucleon rescattering in the nucleus, $|a_\Lambda| \gtrsim |a_\Lambda^{\rm M}|$ for any value of the proton kinetic energy threshold: when $T_p^{\rm th} = 0$, $a_\Lambda/a_\Lambda^{\rm M} \simeq 2$ for ${}^{5}_{\Lambda}\vec{\rm He}$ and $a_\Lambda/a_\Lambda^{\rm M} \simeq 4$ for ${}^{12}_{\Lambda}\vec{\rm C}$; $|a_\Lambda^{\rm M}|$ increases with $T_p^{\rm th}$ and $a_\Lambda/a_\Lambda^{\rm M} \simeq 1$ for $T_p^{\rm th} = 70$ MeV in both cases.

Asymmetries $a_\Lambda^{\rm M}$ rather independent of the hypernucleus are obtained for $T_p^{\rm th} > 30$ MeV. The KEK data quoted in the table refer to a $T_p^{\rm th}$ varying between 30 and 50 MeV: the corresponding predictions of Ref. [12] agree with the ${}^{12}_{\Lambda}\vec{\rm C}$ datum but are inconsistent with the observation for ${}^{5}_{\Lambda}\vec{\rm He}$.

Table 1. Results of Ref. [12] for the proton intensities [Eqs. (4) and (5)] from the non–mesonic weak decay of ${}^{5}_{\Lambda}\vec{\rm He}$ and ${}^{12}_{\Lambda}\vec{\rm C}$.

	${}^{5}_{\Lambda}\vec{\rm He}$ $I_0^{\rm M}$	$a_\Lambda^{\rm M}$	${}^{12}_{\Lambda}\vec{\rm C}$ $I_0^{\rm M}$	$a_\Lambda^{\rm M}$
Without FSI	$I_0 = 0.69$	$a_\Lambda = -0.68$	$I_0 = 0.75$	$a_\Lambda = -0.73$
FSI and $T_p^{\rm th} = 0$	1.27	−0.30	2.78	−0.16
FSI and $T_p^{\rm th} = 30$ MeV	0.77	−0.46	1.05	−0.37
FSI and $T_p^{\rm th} = 50$ MeV	0.59	−0.52	0.65	−0.51
FSI and $T_p^{\rm th} = 70$ MeV	0.39	−0.55	0.38	−0.65
KEK–E462 (preliminary)[13]		0.07 ± 0.08		
KEK–E508 (preliminary)[13]				-0.44 ± 0.32

In conclusion, nucleon FSI turn out to be an important ingredient also when studying the non–mesonic weak decay of polarized hypernuclei, but they cannot explain the present asymmetry data. Further investigations are then required to clarify the issue. On the theoretical side there seems to be no reaction mechanism which may be responsible for positive or vanishing asymmetry values. On the experimental side the present anomalous discrepancy between different data needs to be resolved.

4. Conclusions

The hypernuclear non–mesonic rates have been considered within a variety of phenomenological and microscopic models. In spite of the fact that several calculations were able to reproduce, already at the OPE level, the total non–mesonic width, the values therewith obtained for Γ_n/Γ_p revealed a strong disagreement with the experimental data. Although some recent calculation represented an improvement of the situation, further efforts were required in order to approach a solution for the Γ_n/Γ_p puzzle. From the experimental side, nucleon–nucleon coincidence observables have been measured recently with good statistics. Analyses of these data, complemented with the theoretical estimate of final state interactions, allowed the determination of Γ_n/Γ_p values in agreement with the theoretical expectations. Yet, good statistics coincidence measurements of nn and np emitted pairs are further required. They will also allow one to establish the first constraints on the two–nucleon induced decay width.

While theory predicts negative values for both the intrinsic asymmetry a_Λ and the observable asymmetry $a_\Lambda^{\rm M}$, with a moderate dependence on the hypernucleus, experiments favour negative values for $a_\Lambda^{\rm M}({}^{12}_{\Lambda}\vec{\rm C})$ but small, positive values for $a_\Lambda^{\rm M}({}^{5}_{\Lambda}\vec{\rm He})$. Further investigations are then required to clarify the issue. In particular, improved experiments, establishing with certainty the sign and magnitude of $a_\Lambda^{\rm M}$ for s– and p–shell hypernuclei, are strongly awaited.

Acknowledgments

Work partly supported by EURIDICE HPRN–CT–2002–00311, MIUR 2001024324_007, INFN, DGICYT BFM2002-01868 and Generalitat de Catalunya SGR2001-64.

References

1. W. M. Alberico and G. Garbarino, *Phys. Rep.* **369**, 1 (2002); *International School of Physics Enrico Fermi*, CLVIII Course, *Hadron Physics* Varenna, Italy, nucl-th/0410059 [IOS press, Amsterdam (to be published)]
2. E. Oset and A. Ramos, *Prog. Part. Nucl. Phys.* **41**, 191 (1998).
3. K. Sasaki, T. Inoue and M. Oka, *Nucl. Phys.* **A 669**, 331 (2000); **A 678**, 455(E) (2000); **A 707**, 477 (2002).
4. D. Jido, E. Oset and J. E. Palomar, *Nucl. Phys.* **A 694**, 525 (2001).
5. A. Parreño and A. Ramos, *Phys. Rev.* **C 65**, 015204 (2002); A. Parreño, A. Ramos and C. Bennhold, *Phys. Rev.* **C 56**, 339 (1997).
6. K. Itonaga, T. Ueda and T. Motoba, *Phys. Rev.* **C 65**, 034617 (2002).
7. G. Garbarino, A. Parreño and A. Ramos, *Phys. Rev. Lett.* **91**, 112501 (2003); *Phys. Rev.* **C 69**, 054603 (2004).
8. H. Outa, *International School of Physics Enrico Fermi*, CLVIII Course, *Hadron Physics* Varenna, Italy [IOS press, Amsterdam (to be published)].
9. A. Parreño, A. Ramos, C. Bennhold and K. Maltman, *Phys. Lett.* **B 435**, 1 (1998).
10. A. Ramos, M. J. Vicente–Vacas and E. Oset, *Phys. Rev.* **C 55**, 735 (1997); **C 66**, 039903E (2002).
11. A. Ramos, E. van Meijgaard, C. Bennhold and B. K. Jennings, *Nucl. Phys.* **A 544**, 703 (1992).
12. W. M. Alberico, G. Garbarino, A. Parreño and A. Ramos, nucl-th/0410107.
13. T. Maruta *et al.*, *VIII International Conference on Hypernuclear and Strange Particle Physics* (HYP2003), JLAB, Newport News, Virginia, nucl-ex/0402017 [*Nucl. Phys.* **A** (to be published)].

STUDY OF STRONGLY INTERACTING MATTER (I3HP)*

C. GUARALDO
Laboratori Nazionali di Frascati dell'INFN
C.P. 13 - 00044 Frascati, Italy

The Project entitled: "Study of strongly interacting matter" (acronym: HadronPhysics) promotes the access to nine European Research Infrastructures (RI), and covers seven Networking Activities and twelve Joint Research Activities (JRA). The Project originates from a common initiative of more than 2000 European scientists working in the field of Hadron Physics. Hadron physics deals with the study of strongly interacting particles (hadrons), as the proton and the pion. Hadrons are composed of quarks and gluons. Their interaction is described by Quantum Chromo Dynamics, the theory of the strong force. Hadrons form more complex systems, in particular atomic. Under extreme conditions of pressure and temperature hadrons may lose their identity and dissolve into a new state of matter similar to the primordial matter of the early Universe.

The Networking Activities are related to the organisation of experimental and theoretical collaborative work concerning both ongoing activities at present Research Infrastructures and planned experiments at future facilities. In hadron physics, in fact, the close connection between experimentalists and theoreticians is of paramount importance (therefore, three of the Networking Activities are theoretical ones).

The Joint Research Activities concentrate on technological innovations for present and future experiments at the participating RIs. Applications in material science, medicine and information technology will be pursued vigorously.

The main objective of this Project is to strengthen the European Research Area by promoting access to the leading Hadron Physics RIs in Europe and by improving their performances. This will be achieved by

*For technical reasons the whole manuscript was not available in time. More details about the project I3HP within the 6th Framework Programme of the European Union can be found at the following web site: http://www.infn.it/eu/i3hp.

developing new methods and experimental tools in the JRAs and by Networking Activities, unifying, for the first time, three previously separated communities of researchers: communities who are using leptons, hadrons and high energy heavy ion beams, respectively, for studying hadrons and their properties.

THE PANDA EXPERIMENTAL PROGRAM

P. GIANOTTI
Laboratori Nazionali di Frascati INFN, P.O. Box 13, 00044 Frascati, Italy
E-mail:paola.gianotti@lnf.infn.it

A major upgrade of the GSI accelerator complex, presently running in Darmstadt, has been recently funded by the German Government. This new facility will also include a machine for hadronic physics studies giving an intense, high momentum resolution, antiproton beam, with momenta between 1.5 and 15 GeV/c: the High Energy Storage Ring (HESR). This will allow to exploit a wide physics program, mainly devoted to hadron spctroscopy, by means of a general purpose detector (PANDA). In this talk the main topics that might be addressed, in the next future, by the PANDA experiment will be illustrated.

1. Introduction

With the planned upgrade of the GSI accelerator facility of Darmstadt (FAIR), many aspects of modern experimental physics [1] will be addressed:

- Nuclear structure physics: research with rare isotope beams to study nuclei far from the stability line.
- Nuclear matter physics: study of compressed and dense hadronic matter in nucleus-nucleus collisions.
- Plasma physics: high energy density matter produced using ion and laser beams.
- QED studies: ion-matter interactions in extremely strong electromagnetic fields.
- Antiproton physics: hadron spectroscopy studies to understand the hadron mass spectrum, and the fundamental properties of the strong interaction like the quark confinement, and the chiral symmetry breaking mechanism.

In the past years, experiments with antiprotons have demonstrated to be rich sources of high quality information for hadronic physics. With the new High Energy Storage Ring (HESR) of GSI, the physics of strange and charm quarks will be deeply explored. This is an energy region of transition between the perturbative regime and the low energy domain where phenomenological approaches are used to describe the strong interaction. With a high performance, full solid angle, magnetic spectrometer (the PANDA detector) some crucial points of this scientific field will be analyzed and hopefully clarified.

Quantum Chromo Dynamics (QCD) is nowadays the accepted theory of strong interaction. QCD is believed to be well understood in the high energy range, nevertheless this is no longer true if one goes down to lower energies where the quark-gluon coupling become stronger. Here QCD is governed by non-perturbative phenomena leading to the formation of hadrons. The mechanisms of confinement and of chiral symmetry breaking, which play a crucial role at this energy scale, are not perfectly understood and their implications must be clarified. This is the challenge that both theory and experiments have to struggle. From the experimental point of view, the antiproton beams available at the CERN LEAR machine, and at the antiproton accumulator of Fermilab, have produced a wide set of results in the field up to the charmonium energy. Therefore, the new high energy storage ring for antiprotons at FAIR will allow to continue the systematic studies started by the previous mentioned factories.

2. The antiproton experimental program

Experimentally, hadron's structure can be investigated using different probes (e^-, π, p, $\bar{p}$,...) each one with some specific advantage. Nevertheless, antiproton-proton annihilations copiously produce particles with gluon content, as well as particle-antiparticle pairs, allowing to access any quantum number for the final states. Therefore, HESR antiprotons are an excellent tool to perform spectroscopic studies of ordinary and exotic mesons in the energy range between 3 and 5 GeV. Furthermore, the use of antiprotons can address other open problems like in-medium modifications of hadrons properties, with the aim of checking the effects of chiral symmetry partial restoration on light mesons, and like the study of double hypernuclei and of hyper-atoms. These are unique playgrounds to get information on the hyperon-nucleon and hyperon-hyperon interactions. Finally, as soon as the luminosity of the HESR will reach the value of 10^{32} cm^{-2} s^{-1}, other more challenging topics could be accessed by the PANDA experiment: D-meson decay spectroscopy, the search for CP-violation signals in the charm and strange sector, the extraction of parton distribution functions from the inversed Deeply Virtual Compton scattering process and

Drell-Yan reactions.

In the following sub-sections the main topics of the PANDA scientific program will be more deeply illustrated.

2.1. *Exotic states*

QCD is a non-Abelian gauge theory and therefore the gluons can interact with each other. Therefore, QCD predicts the existence of bound states of gluons (glueballs, G) and other kind of matter in which gluons explicitly contribute to the overall quantum numbers (hybrids,H). Thus, glueballs and hybrids can exhibit quantum numbers that cannot be achieved by $q\bar{q}$ systems. The antiproton annihilation process is rather complicated at the microscopic level, nevertheless it can lead to the production of exotic final states with cross sections of the same order of magnitude of those of ordinary mesons. In $\bar{p}p$ annihilation two different mechanisms can lead to the formation of exotic states (see fig. 1:

- **production:** together with the exotic state G/H a recoil ordinary meson M is produced. Thus, G/H can have even exotic (non-$q\bar{q}$) quantum numbers. The cross sections are of the order of 100 pb.
- **formation:** in this case the exotic state G/H is directly formed in the annihilation process. This state could then have only ordinary quantum numbers. The formation cross sections are of the order of 1 μb.

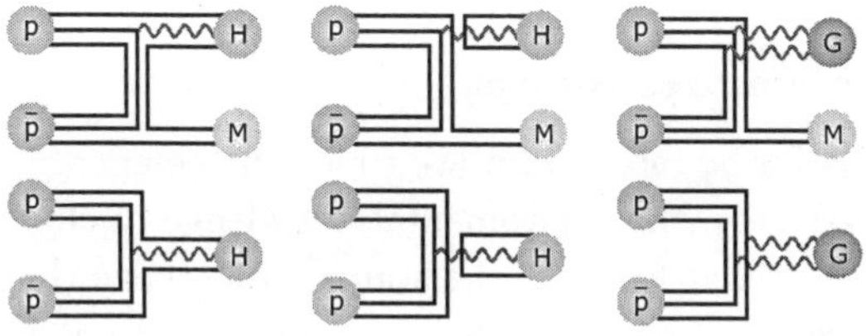

Figure 1. Diagram of production (top) and formation (bottom) mechanisms of exotic states in $\bar{p}p$ annihilation.

The unambiguous identification of exotic states is not just academic; the origin of the mass of the elementary particles is not completely clear. The mere sum of quark masses is not able to reproduce the mass of the known particles; glueballs would be massless without the strong interaction. Therefore, the possibility of studying the whole spectrum of glueballs might help in understanding the mechanism of mass generation by the strong interaction. Some experimental evidence of exotic states already exist [2]; nevertheless none of them is so striking to convince the whole community. A paradigmatic example is the $\eta_L(1440)$ meson comprehensively studied at LEAR by the Obelix collaboration [3]. It is not widely accepted to be the pseudoscalar glueball only because LQCD calculations predict its mass above 2 GeV/c^2 (see fig. 2 from [4]).

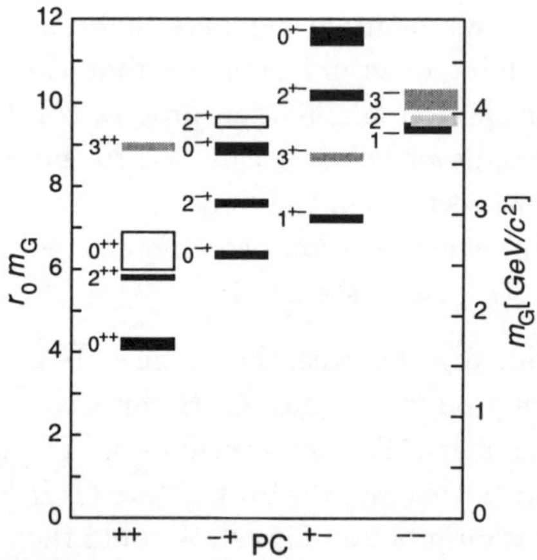

Figure 2. Mass spectrum of glueballs obtained by LQCD calculation from ref. 4

2.2. *Charmonium Spectroscopy*

Potential models for $q\bar{q}$ interaction are tuned in the energy region of charmonium. Nevertheless, the experimental knowledge of charmonium states is far to be complete, at least for the states lying above the $D\bar{D}$ threshold (see tab.1). Here e^+e^- experiments have only measured $R = \sigma(e^+e^- \rightarrow hadrons)/\sigma(e^+e^- \rightarrow \mu^+\mu^-)$ in large energy steps. Recently, the results of the more accurate measurements performed by the BES collaboration [5] do not confirm the existence of the state at 4040, 4160 and 4415 MeV/c^2. The Fermilab experiments E760 and E835 [6] showed that cooled antiproton beams are extremely well suited to perform precise studies of charmonium states. Figure 3 shows the excitation function of χ_{c0} measured by E835 experiment in $\bar{p}p$ annihilation and by the Cristal Ball experiment in e^+e^- process. The better quality of the Fermilab result in undoubtedly. In fact,

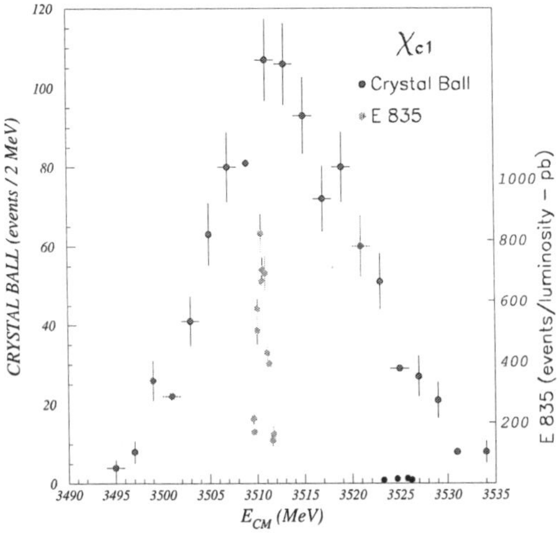

Figure 3. χ_c spectrum measured by Crystal Ball and E835 collaborations.

in $\bar{p}p$ annihilation all the charmonium states can be directly formed and the only parameter limiting the mass resolution is the antiproton momentum uncertainty, that, at the Fermilab antiproton accumulator, was very good ($\Delta p/p \sim 10^{-4}$). On the other hand, in e^+e^- collisions only vector mesons are formed, and the other states can be obtained via radiative decay processes. Thus, the measurements of masses and widths are limited by the photon energy resolution of the detector system.

Table 1. Summary of our knowledge on charmonium states (data from ref. 2)

	Mass	Decay channels studied	Total BR seen (%)	Decay channels with error < 30%
η_c	2979.6 ± 1.2	21	60.3	3
η_c'	3654.0 ± 6.0	4	~ 0	0
J/ψ	$3096.92 \pm .01$	135	44.2	83
ψ'	$3686.09 \pm .03$	62	77.1	29
χ_{c0}	3415.2 ± 0.3	17	12.3	10
χ_{c1}	3510.59 ± 0.1	13	33.6	6
χ_{c2}	3556.26 ± 0.11	19	26.3	10
h_c	-	2	?	0
$\psi(3770)$	3770.0 ± 2.4	2	~ 0	1
$\psi(3836)$	3836 ± 13	1	~ 0	0
$X(3872)$	3872.0 ± 0.6	3	~ 0	0
$\psi(4040)$	4040 ± 10	6	~ 0	1
$\psi(4160)$	4159 ± 20	1	~ 0	0
$\psi(4415)$	4415 ± 6	2	~ 0	0

Nevertheless, E760/E835 experiments could not access the energy region above the $D\bar{D}$ threshold. This is the region where narrow $^1D_2$, $^3D_2$ states and the first radial excitation of h_c and χ_{cJ} are also expected. A first evidence of one of those state came out in the $J/\psi\pi^+\pi^-$ invariant mass [7]. This state, labeled $X(3872)$, does not fit easily in the present $c\bar{c}$ model and other interpretation, like $D^0\bar{D}^{*0}$ molecule, have been proposed.

2.3. *In medium modification of hadron properties*

The investigation of hadron properties modification induced by nuclear matter is presently one of the main research activity of GSI. The aim is to understand the effect of the spontaneous chiral symmetry breaking mechanism on the process of hadron masses generation. This is done by going into an environment, the nucleus, where the chiral symmetry is partially restored. Experimental evidences of mass shift of light mesons have been already seen by studing deeply bound pionic atoms [8], and kaon production in heavy-ion collisions [9]. A high intensity antiproton beam up to 15 Gev/c will allow to extend this research to the charm meson family. For the low lying charmonium states J/ψ, η_c recent calculation [10] indicate, however, small in-medium mass reductions, of the order of 5-10 MeV/c^2, but since the effect is expected to scale with the volume occupied by the $c\bar{c}$ pair, the situation may change for excited charmonium states.

For D meson's family the situation is different: made of a c quark and of a light antiquark, they represent QCD analogue of hydrogen atom. Hence, they provide the unique opportunity of studing the in-medium dynamics of a system with a single light quark. Recent theoretical works predict a mass splitting for D mesons of different entity and sign: a positive value of 50 MeV/c^2 is calculated in Ref. [11] (see fig 4); a negative one of 160 MeV/c^2 is reported in Ref. [12]. Up to now, few experimental information is available on charm propagation in the nuclear medium, and theoretical predictions are strongly model dependent. Therefore, to better understand the behavior of charmed hadrons in nuclear matter, studies devoted to the measurement of J/ψ and D meson production cross-section in $\bar{p}$ annihilation on different nuclear targets might help to test and tune the models. The comparison of the experimental yields obtained in $\bar{p}p$ annihilation and in $\bar{p}$-nuclei reactions is of enormous importance also for the understanding of J/ψ suppression in ultra-relativistic heavy-ion collisions, interpreted as a signal of quark-gluon plasma formation.

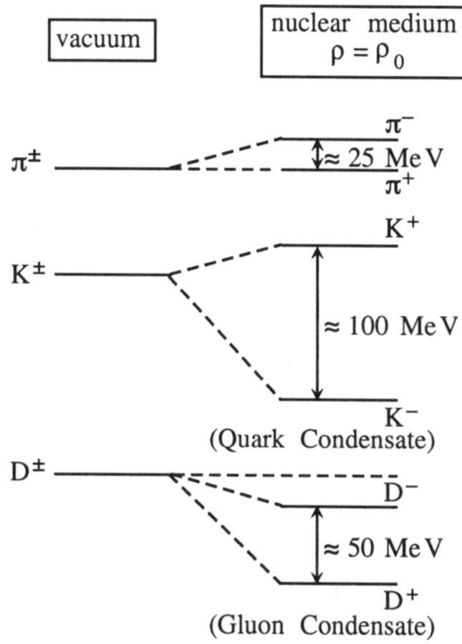

Figure 4. Measured ($\pi^{\pm}$ and $k^{\pm}$) and predicted ($D^{\pm}$) meson's mass shifts in nuclear matter.

2.4. *Hypernuclear physics*

Replacing a u or a d quark with an s one inside a nucleon of the nucleus leads to the formation of a hypernucleus. Hypernuclear physics experiments have a double valence: on one side they allow to study nuclear matter in presence of an explicit s quark, on the other, they offer a unique source of data on hyperon-nucleon interaction not accessible otherwise. Hypernuclear physics is not new, but in spite of its age, it is experiencing a renewed interest thanks to the availability of better experimental conditions that are well suited to clarify some long standing problems. One of these is the precise evaluation of double Λ-hypernuclei's binding energy. Up to now, only 3 double Λ-hypernuclei have been completely identified via their double pion decay, but if they could be produced at a reasonable rate, they could be a unique source of data on hyperon-hyperon interaction allowing to determine the $\Lambda\Lambda$ strong interaction strength. The existence of an S=-2 six-quark state, the H particle [13], is an other challenging topic that can be addressed producing $\Lambda\Lambda$-hypernuclei. Finally hyperatoms, created during the capture process of the hyperon, will provide new information on fundamental properties of hyperons.

Recent hypernuclear physics experiments, carried out at KEK and BNL, have demonstrated how powerful are Ge-detectors for performing high resolution γ-spectroscopy of Λ-hypernuclei. By using Ge-detectors, unprecedented measurements of the spin-orbit component of the hypernon-nucleon interaction have been performed. The idea of the PANDA collaboration is that of using the same experimental technique to study double Λ hypernuclei and Ω atoms. The production of hypernuclei and of hyperatoms at the HESR will go through a two step mechanism: baryons-antibaryons pairs will be produced inside a first nuclear target. Then antibaryon (e.g. a $\bar{\Xi}$) could be used as a trigger for the reaction, while the baryon (in this case Ξ^-) is slowed down and subsequently absorbed in a second active target. When a Ξ^- interacts with a proton, it produces two Λ particles with an energy release of only 28 MeV. Therefore, the probability that these two hyperons remain stuck to the nucleus is high. A new conception Ge-array detector will surround the second target allowing high precision spectroscopy of hypernuclear levels.

3. The GSI Antiproton Accelerator Complex

The central goals of the GSI facility upgrade are a substantial increase of the intensity and of the energy of the ion beams, and to provide energetic antiproton beams. The layout of the existing GSI facility, together with the proposed one, is shown in fig. 5. The 30 GeV protons form the SIS100 will be used to produce antiprotons that after being collected and cooled in two small storage rings (CR and NESR) will be available for users in the HESR (High Energy Storage Ring) with a momentum ranging form 1.5 to 15 Gev/c, corresponding to a c.m. energy up to 5.5 GeV. The beam will be stochastically cooled over the whole momentum range obtaining a momentum spread $\Delta p/p \sim 10^{-4}$. For high precision charmonium spectroscopy measurements, a momentum spread $\Delta p/p \sim 10^{-5}$ will be achieved with an additional electron cooler. The antiproton program is only a part of the whole GSI upgrade program. However, the accelerator complex is designed for parallel operations to optimize the facility usage.

4. The PANDA detector

The HESR will be equipped with an internal target station located in one of the two straight sections of the storage ring. Around the target an almost 4π general purpose detector will be build. $\overline{P}$ANDA (Antiproton

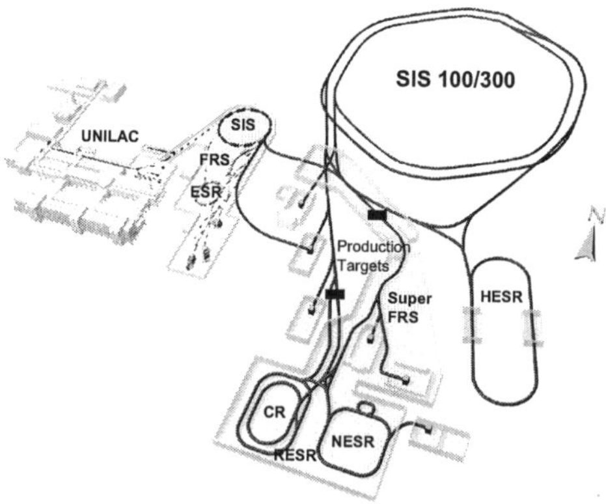

Figure 5. Layout of the FAIR accelerator complex: the present setup (left) will be the injection system of the new (right) facility.

ANnihilation at DArmstadt) intends to continue and extent the scientific program carried out at the LEAR and Fermilab facilities. To achieve these physics aims, the detector should have momentum resolution at the % level, good particle identification on a wide momentum range ($0.1 \sim 3$ GeV/c), possibility to detect secondary vertexes in order to identify K^0_S, Λ, D mesons. Therefore, the internal target will be surrounded by Si-pixel vertex detectors; a charged tracking system, with straw tubes in the barrel region, and mini-drift chambers in the forward one; ring imaging Cherenkov counters will provide the particle identification, and an electromagnetic calorimeter, with $PbWO_4$ crystals read-out by avalanche photo diodes, will be used to detect neutral particles. A superconducting solenoid (2 T), and a dipole magnet in the forward region will provide the magnetic field for the tracking. Finally, muon counters will be placed outside the iron yoke. To close the solid angle in the forward region, a planar spectometer with a hadronic and an electromagnetic calorimeter and a muon detector will be installed. A schematic view of the PANDA detector is shown in fig. 6

5. Summary

After the LEAR shutdown and the end of the Fermilab fix target program, a new challenging project involving antiprotons is officially started in Europe. The characteristics of the new beam, together with the high performance of the detector involved, will determine a step forward in the hadron physics sector, allowing to continue the investigations on gluonic degrees of freedom

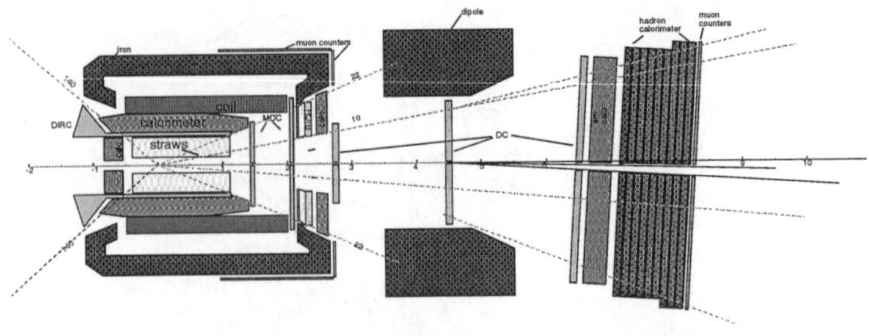

Figure 6. Top view of the PANDA detector.

and quark-antiquark potential.

References

1. "GSI Future Project, Conceptual Design Report", *see www.gsi.de/GSI_Future/cdr.*
2. S. Eidelman *et al.*, Phys. Lett. B **592** (2004) 1.
3. F. Nichitiu *et al.*, Phys. Lett. B **545** (2002) 261 and references there in.
4. C. J. Morningstar and M. Peardon, Physics Rev. D **60** (1999) 034509.
5. J. Z. Bai *et al.*, Phys. Rev. D **57** (1998) 3854.
6. T. A. Armstrong *et al.*, Phys. Rev. D **48** (1993) 3037. M. Ambrogiani *et al.*, Phys. Rev. **D 62** (2000) 052002.
7. S.-K. Choi *et al.*, Phys. Rev. Lett. **91** (2003) 262001. D. Acosta *et al.*, Phys. Rev. Lett. **93** (2004) 072001.
8. K. Suzuki *et al.*, Phys. Rev. Lett. **92** (2004) 072302.
9. F. Laue *et al.*, Phys. Rev. Lett. **82** (1999) 1640; K. Wisniewski *et al.*, Eur. Phys. J. A **9** (2000) 515.
10. F. Klingl *et al.*, Phys. Rev. Lett. **82** (1999) 3396.
11. A. Hayshigaki, Phys. Lett. B **487** (2000) 96.
12. A. Sibiritsev *et al.*, Eur. Phys. J. A **6** (1999) 351.
13. R. L. Jaffe, Phys. Rev. Lett. **38** (1977) 195.

EURONS – THE INTEGRATED INFRASTRUCTURE INITIATIVE OF NUCLEAR-STRUCTURE PHYSICS IN EUROPE WITHIN FP6

A. C. MUELLER[1] K.-D. GROSS[2], D. MÜLLER[2], I. REINHARD[2], AND C. SCHEIDENBERGER[2] FOR THE I3 **EURONS**

[1] *CNRS-IN2P3,* [2] *GSI Darmstadt*

EURONS is an approved Integrated Infrastructure Initiative, which will receive funding from the European Commission Services. The development, the goals, the structure, and the organization of **EURONS** are outlined. All activities, which will be pursued within the four years after the start of the contract, are briefly described.

1. Introduction

EURONS is the Integrated Infrastructure Initiative (I3) of the European nuclear structure scientists in the 6th framework programme (FP6). The project has been approved quite recently by the European Commission Services. This I3 comprises 75 involved institutions from 27 countries reflecting the community at large and within an equal opportunity structure. It consists of a coherent and complementary ensemble of Networking, Transnational Access (TA) and Joint Research Activities (JRA). **EURONS** builds on a successful and rich tradition of interplay between theory and experiment, and universities and large-scale infrastructures and continues collaborative European research actions. The organization of **EURONS**, its present status, and the planned activities within the three branches will be described in this contribution.

2. Development of EURONS

Starting spring of 2002, a combined bottom-up/top-down procedure was launched by addressing all 3500 European nuclear physicists electronically from a list provided by NuPECC. This call for ideas and proposals brought the community together, at the occasion of a meeting devoted to the presentation of novel integrating projects without any prior restriction. A panel of

European Scientists selected by FINUPHY evaluated the submitted ideas, taking into account the requirements of the infrastructure facilities. Thus the selection of activities within **EURONS** has been made by the community at large, in a self-organised and transparent way. This anticipated to some extend the *modus operandi* of the new instruments as introduced by the EU for FP6. The milestones elaborated on the occasion of several FINUPHY meetings were

- the meeting at Trento (28. Oct. 2002), where first ideas for JRAs were presented,
- the meeting of the Writing Committee at Mainz (7. Dec. 2002), where JRAs were selected and first suggestions for budgets of JRA, TA, and Network activities were presented,
- the meeting of the Writing Committee on the occasion of the NuPECC Town Meeting at Darmstadt (31 Jan.2003), where A. C. Mueller was appointed coordinator and it was decided that GSI should act as managing institution,
- the meeting at Catania (27./28. Feb. 2003), where the status was presented and decisions on the assignments to individual and overall budgets were taken,
- and finally the submission to the EC, where the proposal was handed in in person on 14. April 2003.

Evaluation of the proposal took place in spring and early summer of 2003 by a refereeing procedure of the EU. End of August 2003 the Evaluation Summary Report was received stating, that this "...is an outstanding proposal of great European impact". And in the letter of the Research Directorate-General dated 12. May 2004 a maximum financial contribution of 14.056.000 Euro was assigned. After a phase of negotiations lasting until November 2004, the contract has been finalized and signed recently.

3. Organization, role and goals of EURONS

At variance to the preceeding framework programmes FP4 and FP5, where numerous other networking and research activities relevant to nuclear-structure physics were funded by the EU as individual proposals, i. e. with no direct coordination between them, in the present framework programm FP6 the activities are treated and coordinated as a whole. The EC has only one contact person, the co-ordinator. The co-ordinator is supported by the management team and by the Project Coordination Council (PCC, includes activity coordinators and overall management of **EURONS**). An Executive Board of six PCC members will ensure rapid interaction with the **EURONS** coordinator and its managing team on pressing issues for efficient project management, both scientifically and administratively. A General Assembly (GA), constituted by one representative of each participating laboratory will ensure the feedback to the community at large and monitor the overall progress of **EURONS**. The very broad adhesion of the community is considered as being a guarantee for its high degree of integration. It is a fundamental aim of **EURONS**

- to ensure that the European nuclear-structure community concentrates on the most prominent joint research activities, for further improvements and extensions of the infrastructure facilities,
- to promote the most needed R&D, as identified by the community, using as main criterion scientific and technical promise, combined with a rather rapid applicability,
- to focus on activities that are in general relevant to more than one facility,
- to benefit from the R&D potential of the European university groups, often in leading positions.

In practice, these goals will be pursued by the activities described in the following.

4. The three pillars of the I3 EURONS

4.1. *Transnational access:*

The backbone of **EURONS** are its eight infrastructures. This follows the lines of two previous contracts between the European nuclear-structure community and the European Union (EU):

- The Concerted Action "Frontiers In Nuclear physics and Astrophysics" (FINA), which ran between October 1, 1997 and March 31, 2001, under the Fourth Framework Programme (FP4), Training and Mobility of Researchers (TMR), contract number ERBFMBGECT 970087.
- The Infrastructure Cooperation Network "Frontiers In Nuclear PHYsics" (FINUPHY), between October 1, 2000 and September 30, 2004, under FP5, Improving Human Potential (IHP), contract number HPRI-CT-1999-40004.

Both FINA and FINUPHY have led to a culture of cooperation between the European research infrastructures in nuclear physics at the top level. The integration of the feedback from the users has been a second important accomplishment. They have furthermore triggered a common elaboration of the development plans of the different research infrastructures, particularly in the field of radioactive nuclear beams, and of scientific instrumentations around them. The experimental programmes of the European research infrastructures in nuclear physics were thereby fine-tuned to each other as far as possible at that time. All this has been accomplished under the auspices of NuPECC, which assured a global view on the needs for development of nuclear physics within Europe.
For nuclear-structure research and also for inter-/multidisciplinary research exploiting nuclear beams, a large number of excellent, in parts forefront, research infrastructures are available in Europe. They are prominent with regard to accelerator specifications and/or instrumentation and also with respect to the users' interest in being offered access. The fundings for the transnational-access activities (TA) support and enhance the research and training opportunities that should be offered to users of the nuclear-structure community in Europe. Within **EURONS**, the following eight facilities have been selected by the EC for funding in FP6. These are:

- CERN-ISOLDE at Geneva (Europe/Switzerland)
- ECT* at Trento (Europe/Italy)
- GANIL at Caen (France)
- GSI at Darmstadt(Germany)
- KVI at Groningen (The Netherlands)
- JYFL at Jyväskylä (Finland)
- CRC at Louvain-la-Neuve (Belgium)
- LNL at Legnaro (Italy)

At all facilities one can single out unique instrumentation for a wide range of experiments with stable and unstable ion beams. Examples are an high-energy heavy-ion storage-cooler ring at GSI (unique in the world), post-acceleration (ditto: the number of accessible nuclear species from CERN-ISOLDE and the energy range accessible at GANIL), and high-performance spectrometers for particle and gamma-ray detection. ECT* is the only center of its kind in Europe, bringing together theorists and experimentalists. An important goal of **EURONS** will be to identify, realise, and coordinate the improvements and extensions of the instrumentation and the experimental programme of the facilities for increasing the quality and quantity of access.

4.2. *Joint research projects*

The Joint Research Activities of **EURONS** are proposed as a direct consequence of the strategy described in the preceeding section. It is to note, that, besides the particle generation in ion sources, no joint research activities for accelerator R&D have been proposed within **EURONS**. It has been assessed that, for nuclear structure, this is not directly relevant for the I3. The dedicated design studies for the future facilities largely include this issue. The following list gives an overview and a brief description of the joint research activities within **EURONS**.

- **ACTAR**: development of **ac**tive **tar**get detectors for the study of very exotic nuclei using inelastic scattering and nuclear reactions at in-flight facilities which provide beams at low and medium energies. This novel concept will allow to improve the experimental possibilities for kinematically complete experiments with 4π solid angle, and provides full particle identification and improved spatial and angular resolution.
- **AGATA**: the principal objectives of the project are the development, construction, commissioning and evaluation of the first modules for an advanced gamma-ray tracking array, including dedicated digital front-end electronics and a data acquisition system capable of handling the γ-ray tracking procedure in real time.
- **Charge breeding**: the goal is to apply and optimize new charge breeding and cooling techniques to facilities, where RIBs are post-accelerated. The main goal of the advanced charge breeding activity is to improve the present charge breeding schemes by narrowing the charge state distributions, to improve the beam emittances by ion-ion cooling and to shorten the breeding time.
- **DLEP**: this activity for **d**etection of **l**ow **e**nergy **p**articles from exotic β-decays aims at new schemes for low energy particle detection, which can be applied at ISOL and IGISOL facilities. It will investigate a new approach, to simultaneously detect the mass and charge of particles impinging upon a Si-detector, made possible by the advent of fast digitization circuits. In addition we wish to explore the potential of using this digital pulse shape recording with the aim of achieving neutron/gamma discrimination at lower energies in neutron detectors.

- **EXL**: this project, **ex**otic nuclei studied with **l**ight hadronic probes, will exploit the capabilities of light hadron scattering as an essential and indispensable tool for nuclear structure investigations, that yet await full implementation in case of secondary beams of exotic nuclei. The main objective is to capitalize on hadron scattering in inverse kinematics by using novel storage-ring techniques.
- **INTAG**: this Joint Research Activity has as its main objective the **i**mprovement of **n**uclear **tag**ging techniques. It will increase the sensitivity for very weak decay channels, which depends on the possibility how well the detected emission can be tagged. Several scenarios are envisaged with stopped beams or with beams at Coulomb-barrier energies (post-accelerated beams or slowed-down fragmentation products).
- **ISIBHI**: the objective of the project is to improve the performances of ECR ion sources for heavy ions, in order to enhance the accelerator final performances in terms of beam variety, intensity, and quality. R&D will be performed, and new prototypes will be tested. The reliability, reproducibility, stability and easy maintenance of modern ion sources will be optimized.
- **LASER**: the main goal is to develop tools and to perform R&D for the Resonance Ionisation **LASER** Ion Source (RILIS) in order to i) produce pure ground-state and isomeric beams of exotic nuclei and develop the in-source laser spectroscopy of short-lived nuclei, and ii) accumulate, cool, bunch and polarize radioactive ion beams. This will improve the capabilities of ISOL and IGISOL facilities and allow for measurements with unprecedented sensitivity for the most exotic species.
- **RHIB**: RHIB covers experimental reaction studies with rare isotope beams, with emphasis on nuclear structure, dynamics, and astrophysical aspects. The objectives of the RHIB project are to define the demands for, and to develop a versatile experimental setup, which can accomplish the different requirements of the various reaction experiments with high-energy radioactive beams.
- **SAFERIB**: for the increased radioactive inventory of upgraded existing or newly built Europan radioactive beam facilities, several tasks have been identified that need research and development in order to ensure safe operation also for present high-intensity radioactive ion beam facilities, such as characterization of radi-

ation at/from production target, characterization of mechanical, thermal and radiation-correlated material properties for production targets/fission sources and handling devices, optimization of radiation shielding, study on migration of radioactivity, technical solutions for a contamination-free transport, safety studies on potential risk and failure scenarios.

- **TRAPSPEC**: planned are improvements and developments of ion **trap**s, **spec**trometers and related detectors. This project aims at the development of new technologies to upgrade existing ion trap based infrastructures and to develop new instrumentation and advanced detectors to be used in combination with ion traps. These will allow for strongly increased precision and/or efficiency in measurements with the existing ion trap set-ups and, in addition, for a wider application of ion traps in nuclear physics, especially for in-trap spectroscopy.

4.3. *Networking activities*

EURONS will make large use of the possibilities for networking. The managing network MANET is not only organized to fulfil the requirements of the EU for the administrative handling of an I3, but also to include the natural continuation of the FP5 FINUPHY activities. Besides the managing network, eight other networking activities have a particular prospective character with an emphasis on

- fostering future cooperations,
- pooling of resources (including human capital),
- stimulating complementarity and ensuring broad dissemination of results,
- integrating the activities of east-European scientists from candidate countries.

The networks correspond to the subjects identified by the community. They are listed in the following together with a short description of the planned activities.

- **MANET**: the management network of **EURONS**. Coordination and monitoring of all technical, scientific, financial, administrative, contractual, and legal activities of the **EURONS** project. Knowledge and quality management within **EURONS**.

- **CARINA**: **c**hallenges **a**nd **a**dvanced **r**esearch **in** **n**uclear **a**strophysics. Identify key forefront studies and coordinate an inter-disciplinary effort in nuclear-physics theory, stellar modelling and experimental techniques.
- **GAMMAPOOL**: coordinate the use of the resources and for experimental campaigns in various countries owing state-of-the-art equipment for high-resolution gamma-ray spectroscopy (Euroball detectors and ancillary equipment).
- **East-West-Outreach**: the activities of the East-West nuclear physics network aim at the further integration of the nuclear physics community in the north- and south-eastern part of Europe. The network will explore and enhance the perspectives for these regions.
- **NuPECC Mapping Studies**: these mapping studies aim at a survey and the presentation of the nuclear physics activity in Europe: quantify present scale of collaborative efforts, identify the potential of European facilities to contribute in other areas of science, examine training and career progression relevant to the European nuclear skills shortage, provide the community and the funding agencies with a coherent picture and advice on the field.
- **PANS-I3**: **p**ublic **a**wareness of **n**uclear **s**tructure in Europe. Enhance dissemination of information on nuclear research within **EURONS** and transfer the achievements of **EURONS** to the scientific communities, including also those EC-countries without specific large infrastructure facilities and to the public. Training of scientists to improve their capabilities to communicate with nearby scientific communities, decision makers and the general public.
- **SHE**: cordinate and improve collaboration of the European laboratories involved in the synthesis of **s**uper-**h**eavy **e**lements by exploiting the research capabilities, by sharing equipment and extending collaborations, define common research goals and new R&D.
- **TNET**: the **t**heory **net**work will coordinate the nuclear-structure and reaction-theory work within **EURONS**. Bring together theorists and experimentalists. Training and visits in the development of modern computational techniques, use of European computational infrastructures.

5. Present status of EURONS and next steps

All information described above and in particular the submitted proposal is available in a transparent way from a web based document server at GSI (http://www.gsi.de), which is operational for all activity coordinators. The project will start on 1. Jan. 2005. As one of the first actions, a meeting of the **EURONS** PCC will be held early in 2005. A consortium agreement, in spirit in accordance with the standard suggested way, is presently in preparation.

6. Final remarks and conclusions

As was discussed in the final session of the present meeting, an I3 of this magnitude, reaching coherence and integration at this remarkably high level, requires not only large enthusiasm and good spirit, but is a big collaborative effort in itself. Streamlined project controlling measures, largely using electronic data management, will be used in order to maximize resources for innovation.

It is to note that a substantial part of the requested EU contribution is for financing manpower, mainly at the post-doctoral level. Thus **EURONS** will stimulate increased employment possibilities in Europe, and possibly provides long-term perspectives for the best talents.

EURONS represents a quantum leap of European integration in nuclear physics. It will significantly contribute to develop the next generation of European researchers. In accordance with the long-range plan for the field, presently elaborated by NuPECC, it will significantly contribute to the preparation of the future landscape of competitive facilities for the 2010 horizon.

HADRON STRUCTURE: THE PHYSICS PROGRAM OF HAPNET*

P.J. MULDERS

Department of Physics and Astronomy, Faculty of Sciences, Vrije Universiteit, De Boelelaan 1081, 1081 HV Amsterdam, the Netherlands, E-mail: mulders@few.vu.nl

We discuss a number of important scientific issues in hadron physics which are addressed in the HAPNET collaboration (Hadronic Physics Network in Experiment and Theory), the successor of the successful HAPHEEP (Hadronic Physics with High Energy Electromagnetic Probes) and ESOP (Electron Scattering off Confined Partons) networks.

1. Hadronic Physics Network in Experiment and Theory

This network has been submitted in the 6th framework as successor of the successful ESOP network in the 5th and HAPHEEP in the 4th framework. The proposal involved 14 teams from 11 countries and asked for 14 Ph.D. students (Early Stage Researchers, 36 months each) and 12 Postdocs (Experienced Researchers, 24 months each), providing them with training while working on a number of projects together with senior researchers as well as through participation in dedicated schools. The network passed all threshold, but could not yet be financed.

The physics addressed by the network focusses on Quantum chromodynamics (QCD). Quantum chromodynamics is the theory describing the strong interactions. Our present understanding of the elementary structure of matter is described in the Standard Model of Particle Physics. The theory has proven consistent with a tremendous number of measurements with increasing accuracy. The frontiers in our knowledge of the field are the search for the physics beyond the standard model, for which experiments will be carried out at the highest possible energies, and the investigation

*invited talk at the 10. conference on problems in theoretical nuclear physics, cortona (italy), 6-9 oct. 2004

of the interactions within the standard model, including the determination of masses, mixing and coupling constants and understanding the strong interactions that govern the structure of hadrons. For the latter dedicated experiments are needed in which among others the choice of beams and targets, the inclusion of polarization and specific detector options are important. In theoretical physics considerable efforts are needed to formulate a self-consistent theory for all interactions, as well as to understand the details and consequences of the basic interactions.

The structure of protons and neutrons is the result of the strong interactions. A part of the standard model is the theory of Quantum Chromodynamics (QCD). It has convincingly been shown that it describes the strong interactions between the coloured quarks and gluons, which are the elementary building blocks of the protons and neutrons that, in turn, are the constituents of the atomic nuclei and hence the basic building blocks of at least the visible mass in the universe. Quarks and gluons, however, do not appear as free particles, but stay confined into hadrons, divided into baryons (like proton and neutron) and mesons (like pions, that are being exchanged in nuclei) at distance scales larger than about 1 fm = 10-15 m. Understanding the phenomenon of confinement in hadrons is one of the basic quests of the 21th century.

Study of structure of hadrons requires intensive interaction between experiment and theory. The quarks and gluons interact with each other via the strong force through the interchange of gluons. Although the microscopic underlying theory is known, the quark and gluon structure of hadrons is far from understood. This shows both in theoretical and experimental work in the field. Theoretical calculations are complex because the interaction is strong, prohibiting the use of perturbation theory. Experiments cannot be performed with quarks and gluons, as they do not appear as free particles, but they can only be performed with hadrons. It has turned out that progress requires besides development of novel theoretical methods and experimental techniques, a close collaboration of theorists and experimentalists.

Probing structure of hadrons requires dedicated experiments. Within theory a variety of non-perturbative methods has been employed to study various aspects of hadron structure, the understanding of the possible kinds of hadrons, their spectra, their charge distributions, their quark and gluon content. We mention the use of large-scale lattice gauge calculations, the use of more or less sophisticated quark models and the construction of effective theories. To test the accuracy and range of validity of models and calculations, accurate data are needed for very specific asymmetries involving additional degrees of freedom such as spin which requires polarization experiments and the latest techniques in particle detection.

Interpretation of results requires dedicated theoretical work. On the experimental side, the key to the study of hadron structure in a variety of scattering processes has been the search for the right identifiers, e.g. the known underlying electron-quark interaction in deep inelastic scattering pins down the initial state, the production of specific particles identifies the quark flavours and polarization is used to select specific spin states. In order to interpret the results one must be able to describe the measured results in terms of quark and gluon properties, which requires a detailed formalism to make sure that all theoretical refinements are under control.

Hadron physics is an essential ingredient in nuclear physics, particle physics and astrophysics. In nuclear physics the essential degrees of freedoms are hadrons (nucleons and pions), but their substructure has become increasingly important to explain precision experiments. In particle physics, the emphasis is on the particles in the standard model (quarks and leptons), but the confinement of the quarks in hadrons requires understanding the structure of hadrons. In the same way as the understanding of nuclear physics is essential to understand astrophysical processes such as stellar evolution and nucleosynthesis in the big bang, standard model physics, including e.g. the phase structure of QCD, will likely turn out to be important to understand high-energy and high-density astrophysical phenomena.

2. Objectives of HAPNET

First main objective: Building and strengthening interactions between experiment and theory. The first objective is the transfer of knowledge and training of skills to young researchers in the field, in both experimental and theoretical research. With the increasing complexity of experimental and theoretical research, there is a trend of decreasing interaction. The network identifies milestones in terms of planning, producing and interpreting data. It engages experienced experimentalists and theorists to work jointly towards reaching those milestones. The early-stage researchers hired by the network will be trained at considering both the experimental and theoretical aspects of their research, thus securing the long-range health of the field. The network focuses on hadron physics, but includes groups and institutes working on nuclear physics, particle physics and astrophysics covering a broad spectrum of topics and research methods. At the present time it is important that young persons learn about the multidisciplinary implications of their field of research. It is at these boundaries that many new discoveries are made.

Second main objective: Understanding of the QCD structure of hadrons.

The second objective is improving our understanding of the structure of hadrons within the field theoretical framework of quantum chromodynamics on the basis of novel experiments at present and future facilities. This means looking for a variety of specific observables and ways to measure them. Specific in this context means that it must become clear which properties of quarks and gluons or which correlations between them are addressed. This is essential to enable comparison of data with predictions or calculations in models of hadrons or in lattice calculations. On the theoretical side, the range of validity and the implications of specific predictions or calculations must be critically investigated. On the experimental side, participation in the preparation of experiments, development of detectors, as well as performing relevant experiments is an integral part of this objective.

3. Physics issues

In the field of hadron physics a number of far-reaching developments have taken place. Experimental techniques have been or potentially can be improved considerably, in particular when it comes to polarization of beams and targets in scattering experiments, detection of particles, increasing luminosity and using advanced methods in the data analysis. The possibilities to do advanced lattice computations opens new ways to perform ab initio calculations in QCD. Furthermore, new perturbative and nonperturbative approaches in quantum field theory are incorporated into model calculations. Most importantly, however, is the realization that the most successful attempts to understand the quark and gluon structure of hadrons within QCD, involve combined effort of theorists and experimentalists.

Specific physics issues, which are proposed for study within the network, are:

(1) The spectrum of QCD and the global spatial structure of hadrons: confinement, exotic hadrons, the role of gluons, elastic and transition form factors

(2) Decoding the quark and gluon structure of hadrons: parton distribution functions (PDF's) and fragmentation functions (FF's).

(3) Spatial and angular momentum structure of hadrons at quark-gluon level: generalized parton distributions (GPD's).

Each of the physics issues covers many aspects of hadron physics, which is important from the perspective of research training and transfer of knowledge. The activities of the network are well focused by selecting a number of coherent tasks. These tasks aim at investigating the above topics to the extent possible at existing facilities and, equally important, also examining the necessary upgrading of these facilities for carrying out improved measurements, as well as looking at possible experiments at new facilities. Along this line of activities the young researchers, in particular early-stage researchers, can profit greatly from the expertise of more experienced scientists, some of whom are also working on strengthening the infrastructure in Europe via the recently approved Integrated Infrastructure Initiative network 'Study of Strongly Interacting Matter (Hadronic Physics, I3HP). This entails a massive research effort combining all of the European hadron physics community. It is one of the aims of the proposed network to match these efforts with a dedicated training programme.

4. Goals and breakthroughs

Issue 1: The spectrum of QCD and the global spatial structure of hadrons: confinement, exotic hadrons, the role of gluons, elastic and transition form factors.

The spectrum of QCD, i.e. masses and lifetimes of hadrons, remains full of surprises as shown recently by the discovery [1] of an exotic baryon, manifestly consisting of five valence quarks, the pentaquark state $\Theta^+(1540)$. A number of basic properties of hadrons, determined from form factor measurements in exclusive processes, such as electromagnetic, flavour and weak charges of hadrons and the spatial distributions of these charges, represent a challenge for models [2] and lattice calculations [3]. For each of these hadron properties the collective response of the confined quarks and gluons can be considered. The essential role played by confinement requires nonperturbative approaches to QCD such as building models, formulating and solving effective field theories [4] or performing lattice gauge calculations. The network aims at bringing together groups that are experts in these fields. In particular the collaboration of experimental groups with theory groups involved both in phenomenological models of QCD and in lattice gauge calculations is expected to lead to breakthroughs. Lattice gauge theories already now provide information towards modelling that is complementary

Physics issue 1
Local operators (coordinate space densities) $O(x)$, evaluated between momentum states, can be expanded in formfactors $G(t)$, schematically

$$\langle P'|O(x)|P\rangle = e^{i\,q\cdot x}\,[G_1(t) - i\,q_\mu\,G_2^\mu(t)]\,,$$

where $t = q^2 = (P - P')^2$ is the momentum transfer squared. The form factors can be considered as the Fourier transform of the coordinate space densities. The form factors at $t = 0$ constitute forward matrix elements (corresponding to static properties),

$$G_1(0) = \langle P|O(x)|P\rangle,$$
$$G_2^\mu(0) = \langle P|x^\mu\,O(x)|P\rangle.$$

Examples of such local matrix elements are the charge, the axial charge, the magnetic moment, the mass, the angular momentum or the spin of the nucleon.

to direct experimental data. The reliability of this information can be estimated from the quality of lattice gauge theory results for those quantities that are directly measurable.

Issue 2: Decoding the quark and gluon structure of hadrons: parton distribution and fragmentation functions.

Parton distribution and fragmentation functions constitute a link between experiment and theory. Theoretically, their structure in terms of quark and gluon field operators within QCD is known. They can be extracted from particular combinations of unpolarized or polarized cross sections in inclusive or semi-inclusive high-energy scattering processes such as leptoproduction or electron-positron annihilation. At the highest energies, three quark distribution functions and two gluon distribution functions are needed to characterize the quark and gluon structure of the nucleon, including spin degrees of freedom. Concerning the gluon structure, a major breakthrough is expected through the measurement of their contribution to the nucleon spin by the COMPASS collaboration.

In processes in which two or more hadrons play a role, such as in hadron-hadron collisions, the transverse momentum of partons becomes important and manifests itself in azimuthal spin asymmetries. The dominant fragmentation function for a pion, moreover, has an unusual (odd) time-reversal behaviour that is accessible experimentally by measuring single-spin asymmetries. The goal within the network is to establish if indeed the mea-

Physics issue 2
Parton distribution and fragmentation functions involve forward matrix elements of nonlocal operators (correlators),

$$\langle P|O\left(x-\frac{y}{2},x+\frac{y}{2}\right)|P\rangle = \langle P|O\left(-\frac{y}{2},+\frac{y}{2}\right)|P\rangle.$$

In particular squares of the form

$$O\left(x-\frac{y}{2},x+\frac{y}{2}\right) = \Phi^\dagger\left(x-\frac{y}{2}\right)\Phi\left(x+\frac{y}{2}\right)$$

in which Φ stands for some quark or gluon field operator, are useful. The Fourier transform of those nonlocal matrix elements are the relevant objects appearing in cross sections at high energies. They constitute momentum space densities of Φ-ons,

$$\int dy\, e^{i\,p\cdot y}\,\langle P|\Phi^\dagger\left(-\frac{y}{2}\right)\Phi\left(+\frac{y}{2}\right)|P\rangle = |\langle P-p|\Phi(0)|P\rangle|^2 = f_\Phi(p).$$

sured single spin asymmetries [5] can be described via universal transverse-momentum dependent distribution and fragmentation functions. This requires new measurements and coherent efforts of theorists and experimentalists in the analysis phase. An expected breakthrough is establishing the scale dependence of single-spin asymmetries, which requires the understanding of field theoretical issues such as the colour gauge link structure in the description [6]. Clarification of various theoretical issues is needed for the interpretation of experimental studies of single spin azimuthal asymmetries and to augment existing model estimates [7].

Issue 3: Spatial and angular momentum structure of hadrons at quark-gluon level: generalized parton distributions.

Generalized parton distributions constitute a further link between experiment and theory [8]. In contrast to usual parton distribution and fragmentation functions, they are relevant in exclusive scattering processes [9,10]. One of their salient features is that they contain simultaneous information about the momentum distribution of quarks and gluons along a reference direction and about their spatial distribution in the directions perpendicular to it. In this way, a fully three-dimensional picture of parton dynamics can be unraveled [11]. By the same virtue, generalized parton distributions provide the only known access to the orbital angular momentum of quarks and gluons, which needs to be added to the intrinsic angular momentum in order to understand the full spin decomposition of the nucleon.

Physics issue 3

Off-forward matrix elements of nonlocal operators (correlators and densities) can be measured in hard exclusive processes. Schematically one has

$$\int dy\; e^{i\,p\cdot y}\,\langle P'\,|\,\Phi^\dagger\left(x-\frac{y}{2}\right)\,\Phi\left(x+\frac{y}{2}\right)\,|\,P\rangle$$
$$= e^{i\,\Delta\cdot x}\,\left[f_1(t,p) - i\,\Delta_\mu\,f_2^\mu(t,p)\right],$$

where $t = \Delta^2 = (P' - P)^2$. Via sum rules the connection with form factors is established,

$$\int dp\; f_1(t,p) = G_1(t),$$
$$\int dp\; f_2^\mu(t,p) = G_2^\mu(t),$$

while in the forward limit distribution functions are found,

$$f_1(0,p) = \int dy\; e^{i\,p\cdot y}\,\langle P\,|\,\Phi^\dagger\left(x-\frac{y}{2}\right)\,\Phi\left(x+\frac{y}{2}\right)\,|\,P\rangle,$$
$$f_2^\mu(0,p) = \int dy\; e^{i\,p\cdot y}\,\langle P\,|\,x^\mu\,\Phi^\dagger\left(x-\frac{y}{2}\right)\,\Phi\left(x+\frac{y}{2}\right)\,|\,P\rangle.$$

A multi-step procedure is required to obtain the physics information residing in these quantities. It involves taking high-quality experimental data [12], subsequent theoretical and phenomenological analysis are in order to relate measured cross sections to the generalized parton distributions, and finally to confront information about these functions with the nonperturbative dynamics of quarks and gluons in QCD. The proposed network aims to contribute to all steps in this chain. An expected breakthrough is to achieve a better understanding of the interplay between the longitudinal and transverse variables in generalized parton distributions, with information both from lattice QCD calculations [13] and from experimental data to be obtained in the years ahead.

5. Scientific originality of the HAPNET proposal

Hadron physics is unique and addresses fundamental and longstanding issues such as the description of relativistic bound states in which almost massless quarks build massive hadrons, such as the nucleons that constitute almost all of the visible mass in the universe. Other issues are the understanding of colour confinement and spontaneous symmetry breaking. The network brings together groups that are directly or indirectly involved in a variety of experiments at several facilities in Europe, not only the large scale facilities such as DESY (Hamburg) and CERN (Geneva) but also smaller-scale facilities such as Frascati, Bonn, GSI (Darmstadt) and Mainz as well as facilities outside Europe such as TJNAF (Newport News, Virginia) in the USA. Furthermore, it involves a sizeable fraction of leading theorists working in hadron physics. Together they ensure the diversity in methods, experiments and phenomenology needed in order to come to a fundamental understanding of the field theoretical working of QCD at the level of hadrons.

Issue 1. The spectrum of QCD and the global spatial structure of hadrons: confinement, exotic hadrons, the role of gluons, elastic and transition form factors The experimental and theoretical investigation of the spectrum of QCD and the spatial structure of hadrons continue to provide a most fertile ground for understanding QCD in the confinement region. Precise new data concerning the form factors of the nucleon and mesons (e.g. the neutron G_E^n), deviation from the dipole form, the search for exotic states, the detailed investigation of the strangeness form factors of the nucleon, its polarizabilities and N-Δ transition densities will be obtained in the next few years [14]. The interpretation of these data either through improved QCD inspired models or through lattice gauge calculations will provide a reliable source of quantitative information useful for experimental activities. Several new ideas and techniques in the field of lattice gauge calculations will be implemented such as the use of chirally improved fermions (domain wall and overlap fermions) to study unquenching effects on various quantities such as (transition) form factors using light enough pions. The implementation of chiral fermions together with increase in computer speed allows lattice calculations using pion masses in the range of 200-300 MeV, while until recently only pion masses down to about 500 MeV could be used. The use of a realistic pion mass is crucial for understanding pionic contributions from first principles. Since such calculations are just beginning, considerable progress is expected in the nearby future.

Issue 2. Decoding the quark and gluon structure of hadrons: parton distribution and fragmentation functions Of the quark distributions, the unpolarized quark distributions are well established and reasonably accurately known for the various quark and antiquark flavours. The longitudinal spin distributions are also known, but much work on the flavour-spin decomposition remains to be done. This is part of the programme of the HERMES collaboration working at DESY. For the transverse spin distributions first indirect data are emerging. Using models for hadron structure and lattice calculations, and knowing its scale dependence from perturbative QCD calculations, a clear theoretical picture has emerged as well, but experimental confirmation is eagerly awaited. It is foreseen that the first measurements will be performed during the lifetime of the network with active involvement of participants (HERMES at DESY and COMPASS at CERN). A difficulty with the transverse spin distribution is its chiral-odd nature, which prohibits measurements in inclusive deep inelastic leptoproduction, the preferred experiment for measurements of quark distributions. Access to the function requires semi-inclusive (1- or 2-particle inclusive) processes [15]. This requires knowledge of fragmentation functions for quarks or gluons into hadrons. For the most abundantly produced particle, the pion with spin 0, there is no leading (collinear) chiral-odd fragmentation function. There is, however, a fragmentation function, the Collins function, that involves transverse momenta of the quarks. This function appears in semi-inclusive leptoproduction and other processes [16,17], its most striking experimental signature being the appearance in single spin asymmetries. Its universality is presently being studied in models [18] and by using field theoretical methods.

Issue 3. The angular momentum structure of quarks and gluons in hadrons: generalized parton distributions. The theoretical formalism relating generalized parton distributions to experimental observables is well established, and several general properties of these functions in QCD are known. An outstanding problem is a better understanding of their dependence on the two longitudinal momentum (scaling) variables and on the momentum transfer (which is related to the transverse degrees of freedom) and in particular on the interplay between these variables. Whereas so far this question has been addressed at the level of models and of constraints from known elastic form factors, it will in the course of the network become possible to include in this investigation both first principle calculations from lattice QCD and experimental results on relevant kinematical distributions in exclusive processes. Further outstanding theoretical issues to be addressed in the network are how to achieve an adequate accuracy in relating cross sections with generalized parton distributions (controlling in particular corrections of higher twist and of higher orders in the strong coupling [19]), and the identification of a set of observables that will allow one to disentangle generalized parton distributions with different spin and flavour structure. Such a separation is certainly required to achieve the long-term goal of evaluating the orbital angular momentum of quarks and antiquarks in the nucleon. These theoretical efforts will be matched by experimental work in preparing and performing measurements of exclusive cross sections and event distributions at a sufficient level of detail. The first dedicated experiments to perform such measurements will be carried out during the lifetime of this network at TJNAF and DESY.

6. Research Method

A variety of both theoretical and experimental research methods are covered in the network [20]. They are selected because of the role they can play in addressing the physics issues mentioned and at the same time ensure training and knowledge transfer to young researchers. Novel methods and techniques are incorporated in the network: expertise in the area of lattice gauge calculations [21], teams that look at more formal aspects of QCD [22,23,24,25] as well as teams that try to build models for hadrons [26]. It also has been ensured that expertise on developing experimental tools, performing measurements and analysing data is available. It is to be noted that most of these methods have general applicability to many disciplines, both in fundamental and especially in applied research and applications, providing valuable training for the future generation of scientists.

Method 1. Developing experimental tools.

Part of the measurements foreseen can be performed at existing facilities such as the HERMES experiment at DESY, the COMPASS experiment at CERN and the experimental facilities of TJNAF and Mainz. Usually modest modifications, such as the installation of a transversely polarized target or the upgrade of the CLAS detector at TJNAF, are sufficient to extend the measurements to address theoretical questions arising from earlier experiments. The experiments trying to access the generalized parton distributions, however, require the development of sizable additional detection equipment. In particular, the construction of instruments for the detection of recoil products emerging from deep-inelastic scattering events are needed to identify the exclusive processes which give access to the generalized parton distributions. The facilities mentioned above are engaged into the construction or design of such additional equipment. Moreover, at a longer time scale entirely new facilities will be needed to increase the precision of some of the existing data to a level where a distinction between competing theoretical models can be made. For the development of such new facilities, simulation and prototype studies will be performed. Some of these developments are also part of the networking activity 'Transversity' within the approved I3HP project. The emphasis in the network will be placed on the training aspects of these projects and on the necessary exchanges between the teams.

Method 2. Performing measurements. Experiments studying the structure of hadrons require the availability of intense polarized lepton beams having energies up to 200 GeV. Such beams are available at the lepton scattering facilities of DESY, CERN and TJNAF. Because of their complementary nature, the network intends to be involved in experiments at each of these facilities. The measurements themselves represent a rather large effort as the time needed to collect a significant data set varies from a few months (at TJNAF) to a couple of years (DESY). During data taking the available detectors are constantly monitored to enable a quasi-permanent check of the quality of the measurements. The detection instruments and tools to monitor the quality of the data coming in are largely available at the mentioned facilities. The challenge of measurements of this kind is to obtain internally consistent data that represent a robust data set involving beam and target polarization levels well in excess of 50%. The employment of cutting edge technology in the experimental arrangements (ultra high vacuum, state of the art electronics, cryogenics, high power laser, RF superconductivity etc) is a most important component of the training aspect

Method 3. Data analysis. The analysis of the data collected consists of several steps. Initially the quality of the data is verified by checking whether all sub-detectors were operating at their nominal settings during the measurements. Subsequently, the data are used to reconstruct particle tracks of which the charge, energy and momentum are determined. At this stage also the identity of the produced hadrons is evaluated. Having thus converted the data into events containing several well-identified tracks, the physics information can be extracted from the data. This involves the calculation of particle spectra (relevant for searches) cross sections, (transition) form factors, single- and double-spin asymmetries [27], first and second moments etc. Finally, a considerable effort is needed to determine the margins of uncertainty that must be associated with each of the aforementioned observables. The intensive use of computers for data reduction and simulation of the experiments fits nice with the broad training it provides to the young scientists.

Method 4. Formalism and development of theoretical tools. It is important to have systematic expansions of measurable quantities such as cross sections or asymmetries. In high-energy scattering processes in which a hard scale, such as the momentum transfer in leptoproduction, can be identified, one can write down an expansion in inverse powers of this hard scale. Each of the terms in the expansion contains matrix elements of specific quark and gluon field operators (characterized by their twist). Within each term contributions may be distinguished by looking at the order of the (running) strong coupling constant $\alpha_s(Q^2)$. At high energies, these contributions can be calculated in perturbative QCD. With increasing refinement of measurements, e.g. the production of specific particles or measurements of azimuthal asymmetries one can distinguish again different contributions. For these contributions one needs to include transverse momenta within the theoretical framework. From the various contributions one can often single out specific ones that rely on symmetries such as parity or time-reversal. For instance, in measurements where the time-reversal symmetry does not lead to constraints, single spin asymmetries are allowed.

Method 5. Modelling and simulation. With the exception of very few cases, the description of high-energy scattering processes contains one or even more components that are not calculable by means of perturbative QCD. These components are nonperturbative matrix elements that can be investigated successfully by models containing the crucial symmetries of QCD. For instance, the (almost exact) chiral symmetry of QCD for the light quarks plays an important role in the construction of realistic approaches. Based on this symmetry a systematic tool (Chiral Perturbation Theory) has been developed for calculations in the nonperturbative sector of QCD. Another powerful method is studying the predictions of QCD in the limit of a large number of colours. Such predictions are to a large extent model-independent, and have not only proven to be successful in describing the main features of, e.g., parton distribution functions but also of the hadron spectrum. Calculations and simulations of observables based on these methods are important for the theoretical understanding, but are also extremely useful in order to guide dedicated experiments.

Method 6. Lattice gauge calculations. Lattice Gauge Theory (LGT) allows computation of a large number of masses, hadronic matrix elements, and coupling constants, as well as certain properties of the QCD vacuum which are of central importance for phenomenological models of hadrons. The key element is the analytical continuation of QCD to imaginary times. It results in an exponential suppression of all resonance states for a given set of quantum numbers when a state is propagated in the Euclidean time direction. Thus the exact (within the accuracy of any given practical simulation) hadron wave functions are isolated, for which one can then calculate matrix elements of interest. While the range of applicability of this very powerful approach is continuously extended, models are indispensable for the description of many processes. Calculations in LGT are computationally very demanding and as such they have always been at the forefront of computer developments or they have triggered new developments.

7. Summary

In spite of the fact that the proposal has not been selected for funding, the physics issues will be very important ones in the next few years in the field of hadron physics.

References

1. CLAS Collaboration: S. Stepanyan et al., Observation of an exotic S=+1 baryon in exclusive photoproduction from the deuteron, hep-ex/0307018, submitted to Phys. Rev. Lett. 91 (2003) 252001
2. K. Goeke, M.V. Polyakov and M. Vanderhaeghen, Hard exclusive reactions and the structure of hadrons, Prog. Part. Nucl. Phys. 47 (2001) 401 [arXiv:hep-ph/0106012].
3. G. Bali, The $D_{sJ}(2317)$: What can the lattice say?, Phys. Rev. D 68 (2003) 071501
4. U. van Kolck, J. A. Niskanen and G. A. Miller, Charge symmetry violation in pn $\rightarrow d\pi^0$ as a test of chiral effective field theory, Phys. Lett. B 493 (2000) 65 [nucl-th/0006042]
5. A.Airapetian et al, Measurement of Single-spin Azimuthal Asymmetries in Semi-inclusive Electroproduction of Pions and Kaons on a Longitudinally Polarized Deuterium Target, Phys. Lett. B 562 (2003) 182 [hep-ex/0212039]
6. D. Boer, P.J. Mulders and F. Pijlman, Universality of T-odd effects in single spin and azimuthal asymmetries, Nucl. Phys. B 667 (2003) 201-241 [hep-ph/0303034]
7. Stanley J. Brodsky, Paul Hoyer, Nils Marchal, Stephane Peigne and Francesco Sannino, Structure functions are not parton probabilities, Phys. Rev. D 65 (2002) 114025 [hep-ph/0104291]
8. M. Garcon, An introduction to Generalized Parton Distributions, Eur. Phys. J. A 18 (2003) 389
9. P. Guichon et al., Pion production in deeply virtual Compton scattering, Phys. Rev. D 68 (2003) 034018
10. I.V. Anikin, D. Binosi, R. Medrano, S. Noguera, V. Vento, Single Spin Asymmetry Parameter from Deeply Virtual Compton Scattering of Hadrons up to Twist Three Accuracy. 1. Pion Case, Eur. Phys. J. A 14 (2002) 95-103
11. J.P. Ralston and B. Pire, Femtophotography of protons to nuclei with deeply virtual Compton Scattering, Phys. Rev. D 66 (2002) 111501.

12. CLAS Collaboration, Deeply Virtual Compton Scattering in Polarized Electron beam asymmetry measurements, Phys. Rev. Lett. 87 (2001) 182002.
13. The QCDSF Collaboration (M. Gockeler et al.), Generalized parton distributions from lattice QCD, Phys. Rev. Lett. 92 (2004) 042002 [hep-ph/0304249]
14. N. F. Spaveris et al. (OOPS collaboration), Measurement of the R(LT) response function for π^0 electroproduction at $Q^2 = 0.070$ (GeV/c)^2 in the N $\rightarrow \Delta$ transition, Phys. Rev. C 67 (2003) 058201.
15. D. Boer, R. Jakob, and M. Radici, Interference fragmentation functions in electron-positron annihilation, Phys. Rev. D 67 (2003) 094003
16. A.V. Efremov, K. Goeke and P. Schweitzer, Sivers vs. Collins effect in azimuthal single spin asymmetries in pion production in SIDIS, Phys. Lett. B 568 (2003) 63-72 [hep-ph/0303062]
17. M. Anselmino, U. D'Alesio, and F. Murgia, Transverse single spin asymmetries in Drell-Yan processes, Phys. Rev. D 67 (2003) 074010
18. A. Metz, Gluon-Exchange in spin-dependent fragmentation, Phys. Lett. B 549 (2002) 139-145 [hep-ph/0209054]
19. N. Kivel and L. Mankiewicz, NLO corrections to the twist 3 amplitude in DVCS on a nucleon in the Wandzura-Wilczek approximation: quark case, Nucl. Phys. B 672 (2003) 357-371.
20. G. van der Steenhoven, Concluding remarks on the QCD-N'02 Workshop, Nucl. Phys . A 711 (2002) 363 [hep-ex/0206071]
21. C. Alexandrou, Ph. de Forcrand and A. Tsapalis, Probing hadron wave functions in lattice QCD, Phys. Rev. D 66 (2002) 094503[hep-lat/0206026].
22. A.I. Karanikas, C.N. Ktorides, Polyakov's spin factor and new algorithms for efficient perturbative computations in QCD, Phys. Lett. B 500 (2001) 75-86.
23. L. Del Debbio, H. Panagopoulos and E. Vicari, Confining strings in representations with common n-ality, JHEP 0309 (2003) 034 [hep-lat/0308012].
24. V.M. Braun, G.P. Korchemsky, D. Mller, The uses of conformal symmetry in QCD, Prog. Part. Nucl. Phys. 51 (2003) 311 [hep-ph/0306057]
25. K. Golec-Biernat and A.M. Stasto, On solutions of the Balitsky-Kovchegov equation with impact parameter, Nucl. Phys. B 668 (2003) 345-363 [hep-ph/0306279].
26. Sergio Scopetta, Vicente Vento, Generalized Parton Distributions in Constituent Quark Models, Eur. Phys. J. A 16 (2003) 527-535
27. A. Airapetian et al, Evidence for Quark-Hadron Duality in the Proton Spin asymmetry A_1, Phys. Rev. Lett. 90 (2003) 092002 [hep-ex/0209018]

AUTHOR INDEX